INTRODUCTION TO
COMBINATORICS
SECOND EDITION

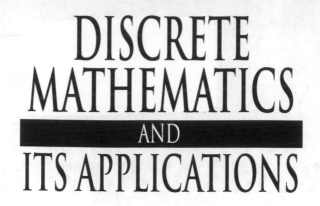

DISCRETE MATHEMATICS AND ITS APPLICATIONS

Toufik Mansour and Matthias Schork, Commutation Relations, Normal Ordering, and Stirling Numbers

Alasdair McAndrew, Introduction to Cryptography with Open-Source Software

Elliott Mendelson, Introduction to Mathematical Logic, Fifth Edition

Alfred J. Menezes, Paul C. van Oorschot, and Scott A. Vanstone, Handbook of Applied Cryptography

Stig F. Mjølsnes, A Multidisciplinary Introduction to Information Security

Jason J. Molitierno, Applications of Combinatorial Matrix Theory to Laplacian Matrices of Graphs

Richard A. Mollin, Advanced Number Theory with Applications

Richard A. Mollin, Algebraic Number Theory, Second Edition

Richard A. Mollin, Codes: The Guide to Secrecy from Ancient to Modern Times

Richard A. Mollin, Fundamental Number Theory with Applications, Second Edition

Richard A. Mollin, An Introduction to Cryptography, Second Edition

Richard A. Mollin, Quadratics

Richard A. Mollin, RSA and Public-Key Cryptography

Carlos J. Moreno and Samuel S. Wagstaff, Jr., Sums of Squares of Integers

Gary L. Mullen and Daniel Panario, Handbook of Finite Fields

Goutam Paul and Subhamoy Maitra, RC4 Stream Cipher and Its Variants

Dingyi Pei, Authentication Codes and Combinatorial Designs

Kenneth H. Rosen, Handbook of Discrete and Combinatorial Mathematics

Yongtang Shi, Matthias Dehmer, Xueliang Li, and Ivan Gutman, Graph Polynomials

Douglas R. Shier and K.T. Wallenius, Applied Mathematical Modeling: A Multidisciplinary Approach

Alexander Stanoyevitch, Introduction to Cryptography with Mathematical Foundations and Computer Implementations

Jörn Steuding, Diophantine Analysis

Douglas R. Stinson, Cryptography: Theory and Practice, Third Edition

Roberto Tamassia, Handbook of Graph Drawing and Visualization

Roberto Togneri and Christopher J. deSilva, Fundamentals of Information Theory and Coding Design

W. D. Wallis, Introduction to Combinatorial Designs, Second Edition

W. D. Wallis and J. C. George, Introduction to Combinatorics, Second Edition

Jiacun Wang, Handbook of Finite State Based Models and Applications

Lawrence C. Washington, Elliptic Curves: Number Theory and Cryptography, Second Edition

DISCRETE MATHEMATICS AND ITS APPLICATIONS

INTRODUCTION TO
COMBINATORICS
SECOND EDITION

WALTER D. WALLIS

SOUTHERN ILLINOIS UNIVERSITY

CARBONDALE, USA

JOHN C. GEORGE

GORDON COLLEGE, BARNESVILLE

GEORGIA, USA

CRC Press
Taylor & Francis Group
Boca Raton London New York

CRC Press is an imprint of the
Taylor & Francis Group, an **informa** business

A CHAPMAN & HALL BOOK

CRC Press
Taylor & Francis Group
6000 Broken Sound Parkway NW, Suite 300
Boca Raton, FL 33487-2742

First issued in paperback 2022

© 2017 by Taylor & Francis Group, LLC
CRC Press is an imprint of Taylor & Francis Group, an Informa business

No claim to original U.S. Government works

ISBN 13: 978-1-03-247699-5 (pbk)
ISBN 13: 978-1-4987-7760-5 (hbk)

DOI: 10.1201/9781315366890

Library of Congress Cataloging-in-Publication Data

Names: Wallis, W. D.. author. | George, J. C. (John Clay), 1959- author.
Title: Introduction to combinatorics / Walter D. Wallis and John C. George.
Description: Second edition. | Boca Raton : Taylor & Francis, 2017.
Identifiers: LCCN 2016030318 | ISBN 9781498777605
Subjects: LCSH: Combinatorial analysis–Textbooks.
Classification: LCC QA164 .W35 2017 | DDC 511/.6–dc23
LC record available at https://lccn.loc.gov/2016030318

Visit the Taylor & Francis Web site at
http://www.taylorandfrancis.com

and the CRC Press Web site at
http://www.crcpress.com

For Our Families

Contents

List of Figures

Preface

This book is a text for introductory courses on combinatorics, usually offered around the junior year of college. The audience for such a course consists primarily of mathematics majors, although the course is often taken by computer science students and occasionally by electrical engineering students.

The preparation of the students for these courses is very mixed, and one cannot assume a strong background. In some cases this will be the first course where the students see several real proofs, while others will have a good background in linear algebra, will have completed the calculus stream, and will have started abstract algebra. For this reason we have included two chapters on background material—one on sets, induction and proof techniques, and one on vectors and matrices. We have made them appendices so that students can get straight to the meat of the course; the better-prepared students can ignore the appendices entirely, while others can refer to them as needed.

Some textbooks contain biographical material on the mathematicians who invented the concepts that are studied, but students are often irritated by the interruption. Others omit this; the reader misses out on some interesting information, and a different set of students is unhappy. In an attempt to get around this problem, we have included some brief sketches, but have made them into a separate section of the book– a third appendix. We refer readers to the biographies when relevant to the topic being studied.

Not all books discuss the increasing role played by technology (computer algebra systems, among many other examples); we have added a discussion of *Maple*™ and *Mathematica*™ where appropriate and commented on other technological tools (e.g., spreadsheets and calculators). These are not emphasized, so that students or instructors who prefer not to have them do not need access to any of these tools. Similarly we have used more modern examples for some of the traditional problems and exercises (e.g., digital music tracks in MP3 players, rather than books on a shelf).

What Is Combinatorics Anyway?

Broadly speaking, combinatorics is the branch of mathematics that deals with different ways of selecting objects from a set or arranging objects. It tries to answer two major kinds of questions, namely, *counting* questions (how many ways can a selection or arrangement be chosen with a particular set of properties?) and *structural* questions (does there exist a selection or arrangement of objects with a particular set of properties?).

The Book: Problems and Exercises

Exercises are collected at the end of the chapters. They are divided into three parts. Exercise sets A and B are fairly straightforward, and the only difference is that answers and partial or complete solutions are provided for the A exercises. Many textbooks provide answers for every second exercise or for what appears to be a random selection, but we have found that the "two sets" model works well in the classroom. There is also a set of "Problems." These contain some more difficult or more sophisticated questions, and also a number of exercises where the student is asked to provide a formal proof when a result has been treated informally in the text. We have provided solutions for some Problems, and hints for some others. There are also Exercises, but not Problems, in the Appendices on background material.

The Second Edition

We have introduced two new chapters, one on probability and the other on posets. We have also attempted to correct errors from the first edition, and would like to thank several of our readers for pointing these out. In addition, there are numerous new illustrations, exercises, and problems.

The Book: Outline of the Chapters

We start by briefly discussing several examples of typical combinatorial problems to give the reader a better idea of what the subject covers. Chapter 1 also contains some basic information on sets, proof techniques, enumeration, and graph theory, topics that will reappear frequently in the book. The next few chapters explore enumerative ideas, including the pigeonhole principle and inclusion/exclusion, and also probability.

In Chapters 6 through 8 we explore enumerative functions and the relations between them. There are chapters on generating functions and recurrences, on some important families of functions, or numbers (Catalan, Bell, and Stirling numbers), and the theorems of Pólya and Redfield. These chapters also contain brief introductions to computer algebra and group theory.

The next six chapters study structures of particular interest in combinatorics: posets, graphs, codes, Latin squares, and experimental designs. The last chapter contains further discussion of the interaction between linear algebra and combinatorics.

We conclude with the appendices mentioned earlier, solutions and hints, and references.

Acknowledgments

This book has benefited significantly from the comments of Ken Rosen, and those of anonymous reviewers. We would also like to thank Robert Ross,

David Grubbs, and Bob Stern of Taylor & Francis Group for their assistance.

Every edition of any textbook will, unavoidably, contain errors. Each of us wishes to state categorically that all such errors are the fault of the other author.

Chapter 1

Introduction

Broadly speaking, combinatorics is the branch of mathematics that deals with different ways of selecting objects from a set or arranging objects. It tries to answer two major kinds of questions, namely the *existence* question (Does there exist a selection or arrangement of objects with a particular set of properties?) and the *enumerative* question (How many ways can a selection or arrangement be chosen with a particular set of properties?). But you may be surprised by the depth of problems that arise in combinatorics.

The main point to remember is that it really doesn't matter what sort of objects are being discussed. For example, we shall often assume that we are talking about sets of numbers, and sometimes use their arithmetical or algebraic properties. But these methods are used to prove results that we then apply to all sorts of objects.

In the first section we shall show you a few examples of combinatorial problems. In Section 1.2 we briefly summarize some ideas of and notations of set theory (if you need to review this material, see Appendix A). The remainder of the chapter introduces some basic combinatorial ideas.

1.1   Some Combinatorial Examples

Some of these problems have a recreational flavor, puzzles and so on, because they will be more familiar; but all these ideas have very serious applications. We address many of them in more detail in subsequent chapters.

Passwords

We start with an enumerative problem. Enumeration (the theory of counting) is an important part of combinatorics.

Most computer systems require passwords, in order to protect your information and your privacy. For example, the social networking program Youface™ requires everybody to have an eight-character password made up of letters and digits. The passwords are case-sensitive, but the letter O is not allowed (to avoid confusion with zero), so there are 60 symbols available. How many different passwords could you choose?

There are 60 choices for the first character. For each of those there are 60 possible second characters, for 3600 possible 2-character starts, and so on. In all, there are 60^8, or about 168 trillion, passwords. This calculation is an example of the multiplication principle, which we'll discuss further in Section 1.3, later in this chapter.

Suppose a hacker wants to break into your Youface account. She has a program that can generate and test a thousand passwords per second. Is she dangerous?

Well, if she tests every possibility, it will take her over 5,000 years. So no, your password is pretty secure.

However, you need to be careful about passwords. Dr. John Walters, a computer science professor that we shall meet again in this volume, always uses the login *jwalters*. Having a poor memory, he always starts his password with his initials, *jw*, and ends with the last two digits of the year. The hacker found this out, so in 2009 she worked out that his password had the form *jw****09*. There are $60^4 = 12,360,000$ passwords of this form. That sounds like a lot, but in 3.6 hours she could run every possible password. Even if she is very unlucky, and her computer does not find the actual password until very late in the run, she could still hack into his account in an afternoon.

In order to protect people like Dr. Walters, Youface introduced some further requirements. Your password cannot begin with the same two symbols as your username, and the last two digits cannot be the last two digits of the year. In 2010, when he rebuilt his account, Dr. Walters could not choose any password of the form *jw******* or *******10*.

How many possible passwords remained? We start with 60^8 possibilities, and subtract the number of passwords that were banned. There were 60^8 passwords originally. There are 60^6 passwords of type *jw*******, and 60^6 of type *******10*. Subtracting these numbers leaves $60^8 - 2 \times 60^6$. However, we have taken away some passwords twice: those of form *jw****10*. There are 60^4 passwords like this, so we have "oversubtracted" 60^4 from the total. So the final result is $60^8 - 2 \times 60^6 + 60^4$.

This method of counting—subtract all objects in one class, do the same with those in another class, then add back those that were common to both classes—is the first case of the *Principle of Inclusion and Exclusion*. We shall deal with this further in Chapter 5.

The Pancake Problem

Another good example of a combinatorial problem is this: Into how many regions can the plane be divided by n straight lines, given that no lines are parallel and at most two lines intersect at any point? These conditions ensure the maximum number of regions. This is sometimes called the *Pancake Problem* because we may draw a large enough circle (the "pancake") surrounding all points of intersection of the lines, so that the problem may be described as: *What is the maximum number of pieces remaining after cutting a pancake with n cuts (none parallel and no three concurrent)?*

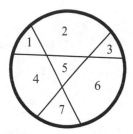

FIGURE 1.1: Pancake with three slices.

The first thing we do is define a notation and try a few examples. This is almost always a good start to any kind of problem. We let P_n be the maximum number of pieces. Then it is easy to see that $P_0 = 1$ (with no cuts, the pancake is still one piece) and $P_1 = 2$. When we notice that $P_2 = 4$, we see the progression $1, 2, 4$ and wonder whether we might have powers of two; perhaps, we think, $P_n = 2^n$? This would make sense if each cut passes through every region, so that each region is cut into two. However, this does not happen, and we find $P_3 = 7$, as we can see from Figure 1.1. A few more tries will give us $P_4 = 11$ and $P_5 = 16$. Clearly, another approach is needed.

How many new regions do we get by adding a new cut? If we have $n - 1$ cuts in the pancake, and add another cut, the rules require that the new cut must intersect each of the $n - 1$ old cuts. If we start the cut on the edge of the pancake in the middle of a region, we cut along the pancake and cut off a new region when we reach the first of the $n - 1$ old cuts. Then as we continue cutting in a straight line, we intersect each of the other old cuts, cutting off a new region when we do. Then at the end, we reach the edge of the pancake, and we cut off a final piece. This means that we have added n pieces; one for each of the original $n - 1$ cuts, and one when we reach the edge of the pancake. This tells us that $P_n = P_{n-1} + n$. This is a formula for P_n, but not a nice one because we cannot easily calculate, for example, P_{100} without a great deal of effort. We will work a little harder and see whether we can find a more efficient formula.

We start by repeating the formula for larger values of n. We get $P_1 = P_0 + 1$, and $P_2 = P_1 + 2 = P_0 + 1 + 2$, then $P_3 = P_0 + 1 + 2 + 3$, $P_4 = P_0 + 1 + 2 + 3 + 4$, and so on. The pattern becomes clear, and we write $P_n = P_0 + (1 + 2 + \cdots + n)$. Now we nearly have our answer. The reader may be familiar with the formula for $1 + 2 + \cdots + n$ or of the first n terms of any arithmetic progression; the story goes that the eminent mathematician Carl Friedrich Gauss (see Appendix C) found the sum of the first one hundred integers almost immediately by realizing how the formula worked (for more details, see [16], p. 509). The "two-rows proof" is as follows:

$$
\begin{array}{cccccc}
1 & 2 & 3 & \cdots & n-1 & n \\
n & n-1 & n-2 & \cdots & 2 & 1
\end{array}
$$

Each of these rows clearly sums to the same value, S. If we add the columns, we find that each column adds to $n+1$, and there are n columns; thus the sum of all elements in both rows is $2S = n(n+1)$. Thus $1+2+\cdots+n = n(n+1)/2$.

This gives us, finally, our formula for P_n, the maximum number of pieces after n cuts.

$$
P_n = 1 + \frac{n(n+1)}{2}
$$

Sudoku

Over the last few years we have seen a great surge of interest in the *Sudoku puzzle*. For those few readers who have been vacationing on another planet, the puzzle consists of a 9×9 array, partitioned into 3×3 subarrays (*subsquares*), in which some cells contain entries chosen from $\{1, 2, 3, 4, 5, 6, 7, 8, 9\}$. The object is to complete the array so that every row, every column and every subarray contains each symbol precisely once.

Sudoku was apparently invented by an American architect, Howard Garns (see Appendix C), in 1979, but did not achieve great popularity until 1986, when it started appearing in Japanese publications. Worldwide circulation dates to about 2004. A nice history can be found in [137].

We'll use the phrase *Sudoku square* for a completed 9×9 square that could be the solution to a Sudoku puzzle. In a well-posed puzzle, there is one and only one Sudoku square that completes the original partial array.

Figure 1.2 shows a typical Sudoku puzzle in the top left-hand position. It needs a 1 in column 3. There are already 1s in the top left and bottom left subsquares, so there is only one possible space in the column; the top right array shows the 1 in place.

Now look at the top right subsquare. It needs a 2, 4, 5, and 7. The only possible place for a 4 in this subsquare is the middle top. At first, it looks as though the 5 can go in any of the other three vacant cells. However, the second row is not available (the 2 must go there), and the 7 must go in the last column (there is already a 7 in the seventh column, and the 4 filled up the eight column). So the 5 only has one possible home. The completion of the top right subsquare is shown in the lower left copy.

Continuing in this way, we obtain the Sudoku square shown at the lower right of the figure.

The methods used in solving Sudoku are combinatorial in nature. However, not many of the millions who play Sudoku realize that a Sudoku square is an example of a combinatorial object, called a *Latin square*, that is widely used in statistical and geometric applications. We shall study Latin squares in Chapter 13.

9			6	2	8			1
	1						8	3
		2			1	6	9	
7		8			4		3	
5		9				4		6
	6		3			7		8
	5	4	7			8		
1	9						5	
3			2	5	9			4

9			6	2	8			1
	1						8	3
		2			1	6	9	
7		8			4		3	
5		9				4		6
	6	1	3			7		8
	5	4	7			8		
1	9						5	
3			2	5	9			4

9			6	2	8	**5**	**4**	1
	1					**2**	8	3
		2			1	6	9	**7**
7		8			4		3	
5		9				4		6
	6	1	3			7		8
	5	4	7			8		
1	9						5	
3			2	5	9			4

9	7	3	6	2	8	5	4	1
6	1	5	9	4	7	2	8	3
8	4	2	5	3	1	6	9	7
7	2	8	1	6	4	9	3	5
5	3	9	8	7	2	4	1	6
4	6	1	3	9	5	7	2	8
2	5	4	7	1	3	8	6	9
1	9	7	4	8	6	3	5	2
3	8	6	2	5	9	1	7	4

FIGURE 1.2: A Sudoku puzzle.

The Tower of Hanoi

Virtually every student in combinatorics and computer science runs into the Tower of Hanoi problem. Because it is such a basic example of combinatorial reasoning and so quickly yields to straightforward methods of attack, and because it may be used to illustrate such ideas as recurrence relations and recursive programming and has ties to such topics as Gray codes, it finds its way into textbooks in programming, algorithms, and discrete and combinatorial mathematics. The puzzle itself predates the modern study of these topics; it was first described in 1883 by Édouard Lucas (see Appendix C), who also publicized the Fibonacci numbers (as discussed in Chapter 6). Lucas introduced the puzzle under the name M. Claus. One year later, Henri de Parville published the following fanciful legend concerning the puzzle [26]:

> In the great temple at Benares, beneath the dome which marks the center of the world, rests a brass plate in which are fixed three diamond needles, each a cubit high and as thick as the body of a bee. On one of these needles, at the creation, God placed sixty-four discs of pure gold, the largest disk resting on the brass plate, and the others getting smaller and smaller up to the top one. This is the Tower of Bramah. Day and night unceasingly the priests transfer the discs from one diamond needle to another according to the fixed and immutable laws of Bramah, which require that the priest on duty must not move more than one disc at a time

and that he must place this disc on a needle so that there is no smaller disc below it. When the sixty-four discs shall have been transferred from the needle on which at the creation God placed them to one of the other needles, tower, temple, and Brahmins alike will crumble into dust, and with a thunderclap the world will vanish.

(translation by W. W. Rouse Ball [92]). The legend, oddly, places the towers in a temple to the god Brahma in Benares (modern Varanasi, India), which renders the reference to Hanoi (in Vietnam, over 1000 miles away) puzzling.

In essence, the puzzle is this; we have three "needles" or "spindles" and a set of n differently sized disks with holes in the middle to allow the disks to be placed over the needles. The disks are arranged on the leftmost needle in order of diminishing size (smallest on the top). We are to move the entire column to a different needle (say, the rightmost) according to the rules:

- At each play, precisely one disk is moved.
- No disk may be placed over a spindle that already has a smaller disk.

An illustration for the case $n = 3$ is shown below. Commercial versions of the puzzle typically include from five to eight disks.

The problem is twofold; first, what is the most efficient way of moving the tower? Second, how many plays does this most efficient way take for a tower of n disks? One might add a third (or more properly, *first*) question: Can it be done at all? A careful analysis will answer all three questions.

First of all, observe that the problem is trivial in the cases of $n = 1$ (simply move the disk from the starting needle to the ending needle) and $n = 2$ (move the smallest disk to the middle needle; then the larger disk to the ending needle; then place the smallest over the largest). Either of these may form the basis case for a proof by mathematical induction. (If you are unfamiliar with induction methods, or need to revise this material, see Section 1.2, below.)

Suppose that it is known that we can solve the puzzle for n disks, and we have $n+1$ disks before us. We may use the solution for n disks to move the top n disks to the middle needle, ignoring for the moment the largest disk. Now we may freely move the largest disk from the leftmost needle to the rightmost needle. At this point, we use the solution for n disks to move the n disks

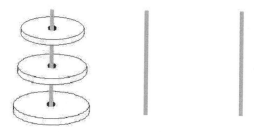

FIGURE 1.3: The Tower of Hanoi puzzle.

from the middle needle to the rightmost needle. This will cause no problems since the disk on the rightmost needle is the largest, so that any disk may be placed on it without violating a rule.

It is at least intuitively clear that this technique gives us the most efficient way to move a stack of n disks. Clearly the basis case uses the fewest possible moves (one move for one disk, three moves for two disks). Also, it seems clear that there is no shorter way to move $n + 1$ disks than by moving the stack of n disks first. Without going through a rigorous proof, it is at least very plausible that our approach uses the fewest moves possible.

Now the hard part: How many plays does this method take for n disks? We denote this (still unknown) quantity by H_n. Clearly, we have $H_1 = 1$ and $H_2 = 3$. For any $n > 2$, we can compute $H_n = 2 \cdot H_{n-1} + 1$. This follows because we use H_{n-1} plays to move the $n - 1$ disks to the middle needle, one play to move the largest disk to the rightmost needle, and another H_{n-1} plays to move the $n - 1$ disks from the middle to the rightmost needle.

This kind of formula for H_n, like the formula for P_n of the previous example, is called a *recurrence* because the expression we are solving for "recurs" in the equation. This does not give us a simple formula for H_n, but it gives us a way to calculate it. We will run through a few values.

n	1	2	3	4	5
H_n	1	3	7	15	31

At this point, we might be tempted to conjecture, based on the numbers shown, that H_n is always one less than a power of two, or more precisely, $H_n = 2^n - 1$. How might we prove this? Again, proof by induction is the easiest approach with which we are familiar (another method will be presented in Chapter 6). The basis is $n = 1$; suppose that $H_n = 2^n - 1$ and use the recurrence to find H_{n+1}, which our guess says should be $2^{n+1} - 1$. The recurrence tells us that $H_{n+1} = 2 \cdot H_n + 1$. The induction hypothesis allows us to replace H_n by $2^n - 1$, yielding $H_{n+1} = 2 \cdot (2^n - 1) + 1$. We simplify to obtain $H_{n+1} = 2^{n+1} - 1$, as required.

The Seven Wargamers

A group of friends (Adam, Beth, . . .) meet frequently to play the wargame *Vampire Cat Wars*, a game for four individual players (no partnerships) in which the players are vampires who have to beat off attacks by the others. They decide to conduct a tournament over several sessions. In each session four of the players will take part in one game. They would like it if each competitor played against each of the others equally—in this case let's call the competition *equitable*. Various questions come up: for example, what is the minimum number of games in such a tournament?

When they first decided to organize a tournament, there were seven players, initials A, B, C, D, E, F, G. At first, they thought maybe everyone could play everyone else once. If Adam were going to play every other competitor exactly once, he would play in two matchups; writing the initials of each player to represent a game, the schedule could include ABCD and AEFG. But whom will Beth play? One matchup is already given, ABCD. She cannot meet A, C or D again, so the only possibility is BEFG. But a schedule cannot include both AEFG and BEFG: pairs like EF would meet twice.

So they decided to try a schedule where everyone plays everyone else twice. Everybody will have four games (each person must meet six opponents twice each—12 appearances—and each game accounts for three). We may as well assume Adam plays in ABCD. He must play against Beth again, and the two matches AB... could have a total of four, three or two players in common. If we try four players—that is, use the matchup ABCD again—Adam's other two games must be AEFG twice. But Beth has been scheduled against A, C and D twice, so her other two matchups must each be BEFG, and EF (as well as other combinations) has occurred too many times. Similarly, if there are two games with three players in common, say ABCD and ABCE, we strike problems: Adam plays twice more, and must meet both F and G twice. The only possibility is ADFG, AEFG. But the same argument shows that we must have BDFG and BEFG. Now F and G have played each other too often.

Okay, this argument shows that the two AB... matchups have no other member in common, and the same must apply to all other pairs. So no two games can have three players in common. Try to start with ABCD and ABEF. Adam's other two matchups have the form A...G, in order for him to meet Greg twice. They cannot use B again, and ACDG would have three in common with ABCD, so they try ACEG and ADFG (or ACFG and ADEG, which would be equivalent, just exchange two names). The same argument, avoiding three in common with the first two games, leads to either BCEG or BCFG, and BCEG clashes with ACEG, so they chose BCFG and BDEG. The final matchup is CDEF, and the schedule is

ABCD, ABEF, ACEG, ADFG, BCFG, BDEG, CDEF.

The players are very happy—and then Helga joins their group. What now? And what if other players join up?

Let's write v for the number of players, and say each pair plays together λ times. Then each player has $\lambda(v-1)$ instances of an opponent. Each matchup accounts for three of these. So each player takes part in $\lambda(v-1)/3$ games. You cannot schedule partial games, so $\lambda(v-1)$ must be a multiple of 3. So, for $v = 8$, λ must be 3 or 6 or .... For the 8-player version, the smallest possible case is $\lambda = 3$, in which case there will be 14 games and each player is in three of them.

The wargamers didn't feel up to calculating a schedule for 14 games, so they started playing the 7-player version again. Each week they scheduled one game. Each week the other three players came to watch, and played a game with Helga. And, at the end of seven weeks, they realized they had played an equitable 8-player tournament

ABCD, ABEF, ACEG, ADFG, BCFG, BDEG, CDEF,
EFGH, CDGH, BDFH, BCEH, ADEH, ACFH, ABGH.

The problem here is, given a set with v elements (the players), to choose a collection of subsets of equal size (4 in the example, more generally written k) such that every pair of elements occurs in the same number of subsets (our λ). This is called a *balanced incomplete block design*, and we'll study them in Chapter 14. The word *balanced* refers to fact that pairs occur together equally often, *incomplete* to the fact that k is smaller than v (or else there would be no problem), and the equal-size subsets are called *blocks*. It is called a *design* because the main application is in the design and statistical analysis of experiments. The numbers v, k and λ are the *parameters* of the design (together with b, the number of blocks, and r, the number of blocks in which a given object occurs; but b and r can be calculated from the other parameters). It is common to refer to these designs by specifying their parameters, for example, one says "a (v, b, r, k, λ)-design"; the wargamers' first example is a $(7, 7, 4, 4, 2)$-design.

We have already seen that some sets of parameters are impossible, because r and b have to be integers. But there are other cases, for example $v = 15, k = 5, \lambda = 2$ (for which $r = 7$ and $b = 21$, both whole numbers) for which there is no design; and there is an infinitude of parameter-sets for which we do not know whether a design exists.

The second example—the 8-player schedule—is an $(8, 14, 7, 4, 3)$-design in which the blocks can be partitioned into sets, where every object belongs to exactly one member of each set. This special case is called a *resolvable* balanced incomplete block design. These designs are of special interest for statisticians and also arise in finite models of Euclidean geometry.

The Hat Game

Here is the outline of a new television game show. There is a team of three contestants. Just before the show begins, the players are taken into separate rooms. Each is blindfolded and a hat—either red (R) or black (B)—is put onto his or her head. The blindfolds are removed, but the player cannot see which color hat they have been given. The players are then taken into the studio and line up behind three desks; they can see each others' hats, but still not their own. No communication between them is allowed.

The M.C. says, "Players, do you think your hat is red or black? Please write down your answer. If you don't know, you can pass—don't write anything."

Then the players' "votes" are revealed. If any of them has made the wrong guess, the team loses. If they all pass, they lose. If at least one player gets the right answer, and there are no wrong answers, they win three million dollars.

As you would expect, the players meet beforehand to discuss their options and choose a strategy. The organizers say they decide on the hat colors by tossing a coin for each player, so that the allocations are independent and each player has a 50-50 chance of red or black. We'll assume they are telling the truth, and we'll also assume the players don't try to concoct any type of cheating scheme. What is the team's best plan?

One obvious strategy would be for the team to nominate one of their number who will always say "red" while the other two players always pass. This would give them a 50% chance of winning the prize. Can they do better? Most people, and in fact most mathematicians, initially think not. Since each person's hat color is independent of the other players' colors and no communication is allowed, it seems impossible for them to learn anything just by looking at one another; all the players can do, it seems, is guess.

But in three out of four cases, we expect two of the players will have hats of the same color and the third player's hat will be the opposite color. The group can win every time this happens if the players follow the following strategy: Once the game starts, look at the other two players' hats. If you see two hats of different colors, pass. If they are the same color, guess that your hat is the opposite color. Every time the hat colors are distributed two and one, one player will guess correctly and the others will pass, and the team will win the game. When all the hats are the same color, all three players will guess incorrectly and the group will lose; but on average this will happen in only a quarter of cases.

There are eight ways the hats can be distributed: RRR, RRB, RBR, RBB, BRR, BRB, BBR and BBB (from left to right). Suppose the game was played eight times, and suppose it just so happened that every possible combination occurred once. In the first and last cases, all three players make the wrong guess; in the other six there is just one correct guess. So, of the twelve guesses, half are right and half are wrong, as you would expect.

How about more players? With four team members, the easiest strategy is to tell one player to stay silent throughout. The other three don't even look at that player! They play as they would if member number four was not there: each looks at the remaining two players and proceeds as if it was a team of size three. They still have a 75% chance of winning.

The Hat Game problem was introduced by Todd Ebert in his dissertation [28], and is in fact related to error-correcting codes. It has been discussed in the press [90], and a recent survey paper is [17]. We'll examine the problem further, and explain the connection to error-correcting codes in Chapter 12.

The Bridges of Königsberg

Leonhard Euler (see Appendix C), a pioneering Swiss mathematician and physicist, is seen by most as the leading mathematician of the eighteenth century. Euler (pronounced OY-ler) founded the area of combinatorics called *graph theory* (see Section 1.4, Chapter 10 and Chapter 11) and invented the concepts of Latin squares and orthogonality (see Section 13.1). He also made equally significant contributions to analytic geometry, trigonometry, geometry, calculus and number theory.

In 1735, Euler spoke to the St. Petersburg Academy on a problem in recreational mathematics. The Prussian city of Königsberg was set on both sides of the Pregel River. It included a large island called the Kneiphof, and the river branched to the east of it. So there were four main land masses—let's call them the four *parts* of the city, and label them A, B, C, D—connected to each other by seven bridges, as shown in the rough map in Figure 1.4:

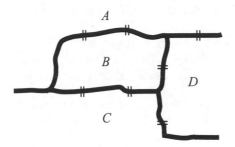

FIGURE 1.4: Königsberg and its bridges.

The problem was to find a walk through the city that would cross each bridge once and only once, and visit all four parts. The only way to go between the different pieces of land was to cross the bridges.

So we have a combinatorial problem. We have a set with seven elements—the seven bridges—and we need to find out whether they can be arranged in a certain way, specifically so that the corresponding bridges could be traversed in the given order.

The solution is very easy. Look for example at the north bank of the river, part A. Whenever you walk over a bridge that leads to A, your next bridge must take you back out of A. So walking into A and out again uses up two bridges (we say "uses them up," because you cannot cross a bridge twice). There are three bridges; the walk must take you onto A and off again once, and there is one bridge left over. The only possibility is that the walk must start or finish in A (but not both).

The same argument can be applied to C and D (with three bridges each), and to B (five bridges, so you walk through B twice). Each part of town is either the start or finish. But a walk can have only one start, and only one finish, and there are four parts of town! This is clearly impossible, so there is no such walk.

Of course, Euler proved that no walk is possible. But he did much more. As we said, he essentially invented graph theory and showed how it could be used to represent any set of islands and bridges, or indeed any set of towns and roads joining them, and concocted an algorithm for traversability problems. We shall look at his methods in Section 11.1.

Six Degrees of Separation

We have all experienced the following phenomenon: meeting a perfect stranger, and subsequently finding out that we have a common acquaintance. While this is very rare, a little further investigation may well show that the new person is a "friend of a friend" or a "friend of a friend of a friend." Guglielmo Marconi, the pioneer of radio, thought that these phenomena were becoming far more common because of the growth of communication. In his 1909 speech accepting the Nobel Prize, he suggested that the average two people could be connected through a chain of at most five acquaintances (that is, six links).

This idea has been explored by many people since. In particular, Harvard psychologist Stanley Milgram [75, 117] studied the problem experimentally. The following description of his experiment is based on [136].

1. Milgram typically chose individuals in Omaha, Nebraska and Wichita, Kansas to be the starting points and Boston, Massachusetts to be the end point of a chain of correspondence.
2. Information packets were sent to randomly selected individuals in Omaha or Wichita. They included letters, which detailed the study's purpose, and basic information about a target contact person in Boston, a roster on which they could write their own name, and business reply cards that were pre-addressed to Harvard.
3. Upon receiving the invitation to participate, the recipient was asked whether he or she personally knew the contact person in Boston. If so, the person was to forward the letter directly to that person. (Knowing someone "personally" was defined as knowing them on a first-name basis.)
4. If the person did not personally know the target, they were to think of a friend or relative they knew personally who they thought more likely to know the target. They then signed the roster and forwarded the packet to the friend or relative. A postcard was also sent to Harvard.
5. When and if the package eventually reached the contact person in Boston, the researchers could examine the roster to count the number of times it had been forwarded from person to person. Additionally, for packages that never reached the destination, the incoming postcards helped identify the break point in the chain.

This experiment has a number of obvious flaws. Many people refused to participate. In one case, 232 of the 296 letters never reached the destination. However, completed chains seemed to run to six or fewer links. Subsequent experiments have shown that this number is about right, and the phrase "six degrees of separation" has become fashionable.

We can look at the world as a huge communication network. Smaller communication networks include social networks like *Facebook* and *MySpace*, the internal communications (or *intranet*) of a large company, the physical network of wires connecting components of a computer system, and so on. The average length of communication links could be very small (for example, if all pairs are directly linked) or very large (some computer networks consist of a ring in which each component is linked only to the pair of components adjacent in the ring). Networks with a small average length are called *small world networks* (as in the phrase, "it's a small world"). The analysis of these networks—for example, finding out whether a particular type of network exhibits "small world" properties—is basically a combinatorial problem, and again graph theory comes into play.

1.2   Sets, Relations, and Proof Techniques

A *set* is a collection of objects with a well-defined rule, called the *membership law*, for determining whether a given object belongs to the set. The individual objects in the set are called its *elements* or *members* and are said to belong to the set. If S is a set and s is one of its elements, we denote this fact by writing

$$s \in S.$$

We say a set S is a *subset* of a set T if every member of T is also a member of T.

While nearly all readers will be familiar with basic set theory, you may wish to review the standard set ideas and notations in Appendix A. In particular, you need to know about the standard operations on sets: union, intersection, complementation; you need to know de Morgan's Laws; and you should be familiar with the standard number sets, the set of integers, denoted $\mathbb{Z}$, the *rational numbers* $\mathbb{Q}$, the *real numbers* $\mathbb{R}$ and the *complex numbers* $\mathbb{C}$.

If n is a positive integer, the *residue classes modulo n* are the n sets $[0], [1], \ldots, [n-1]$, where $[i]$ is the set of all integers that leave remainder i on division by n. Members of the same residue class are called *congruent modulo n* and we write $y \equiv x \pmod{n}$. When no confusion arises we simply write x instead of $[x]$.

If you have done a little abstract algebra, you will know that the set $\mathbb{Z}_n$ of residue classes modulo n is an example of a finite *group*. We discuss groups further in Chapter 8.

We often illustrate sets and operations on sets by diagrams. A set R is represented by a circle, and the elements of R correspond to points inside the circle. If we need to show a universal set (for example, if complements are involved) then this universal set is shown as a rectangle enclosing all the other sets. These illustrations are called *Venn diagrams*, because they were popularized by John Venn (see Appendix C) in 1880 [118, 119], although similar diagrams were studied earlier by Leibniz and Euler, and related ideas were studied by Lull as far back as the thirteenth century (see, for example, [5]). It has recently become fashionable to refer to "Euler diagrams," but Venn was primarily responsible for their modern usage.

For example, the Venn diagrams or $R \cup S$, $R \cap S$, $\overline{R}$, $R \backslash S$ and $R \cap S \cap T$ are shown in Figure 1.5.

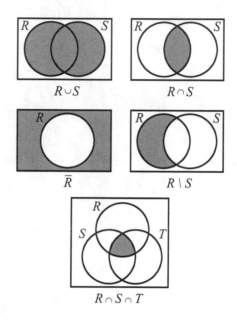

FIGURE 1.5: Some Venn diagrams

We define the *Cartesian product* (or *cross product*) $S \times T$ of sets S and T to be the set of all ordered pairs (s, t) where $s \in S$ and $t \in T$:

$$S \times T = \{(s, t) : s \in S, t \in T\}.$$

There is no requirement that S and T be disjoint; in fact, it is often useful to consider $S \times S$.

A (*binary*) *relation* ρ from a set S to a set T is a rule that stipulates, given any element s of S and any element t of T, whether s bears a certain relationship to t (written $s \rho t$) or not (written $s \not\rho t$). Alternatively, we can define a binary relation ρ from the set S to the set T as a set ρ of ordered pairs (s, t), where s belongs to S and t belongs to T, with the notation that $s \rho t$ when $(s, t) \in \rho$ and $s \not\rho t$ otherwise. This means that, formally, a binary relation from S to T can be defined as a subset of the Cartesian product $S \times T$.

A relation α on a set A is called an *equivalence relation* on A if and only if it is reflexive, symmetric, and transitive.

The obvious equivalence relation is equality, on any set. In sets other than number sets, equal objects are often called "equal in all respects." More generally, an equivalence relation can be considered as a statement that two objects are "equal in some (specified) respects." One example, on the integers, is the relation α, where $a\,\alpha\,b$ is true if and only if $a = \pm b$—a and b have the same absolute value. Another, is congruence, on the set of all plane triangles.

Equivalence relations give us an alternative way to discuss multiple elements. You can view a multiset as an object based on a set of distinct elements on which an equivalence relation has been defined, so that the elements of a multiset are the equivalence classes of the underlying set.

One of the most important proof techniques for working with finite sets or with sets of integers is the method of *mathematical induction*, the application of the *principle of mathematical induction*:

Suppose the proposition $P(n)$ satisfies

  (i) $P(1)$ is true; and
  (ii) for every positive integer n, whenever $P(n)$ is true, then $P(n + 1)$ is true.

Then $P(n)$ is true for all positive integers n.

If you are not familiar with induction, or need to review the topic, we treat it in detail in Appendix A.

1.3 Two Principles of Enumeration

In this section we introduce two easy rules for enumerating the numbers of elements of certain sets or certain types of arrangements.

The addition principle states that if we are building an arrangement of objects starting with a set of mutually exclusive beginning states, then the number of arrangements is the sum of the number of arrangements starting from each beginning state. This rather complicated-sounding sentence is far more trivial than it might appear. For instance, suppose that a young woman wants to go on a job interview, and has only one clean dress skirt and one clean pair of dress slacks. With the skirt, she can create three presentable outfits; with the pants, five. It follows that she has $3 + 5 = 8$ ways to dress appropriately.

The multiplication principle states that if we are building an arrangement of objects in stages, and the number of choices at each stage does not depend on the choice made at any earlier stage, then the number of arrangements is the product of the number of choices at each stage. Again, this principle is simpler than it appears; if a restaurant, for example, offers two kinds of salad, nine entrees, and seven desserts, the number of meals consisting of one salad and one entree and one dessert is $2 \times 9 \times 7 = 126$.

Although these principles seem painfully obvious, each may be proved by results from set theory; for instance, the multiplication principle follows from the fact that the cardinality of a Cartesian product of sets is the product of the cardinalities of the underlying sets. However, we prefer to emphasize the uses rather than the proofs of these principles.

Example 1.1: We wish to have lunch at a fast-food place, and there are three within walking distance. One place offers eight soft drinks, seven sandwiches, and two choices of a "side dish" (French fries or onion rings); another offers only six drinks, but 10 sandwiches and two sides; and the last (a coffee shop) offers 16 drinks, five sandwiches, and only one side. How many different lunches are possible?

The addition principle says that (because we can eat lunch at only one place) the answer is the sum of the number of ways to eat lunch at each place. The multiplication principle says that each place has a number of options equal to the product of the options at each stage: drink, sandwich, and side. So the first restaurant offers us $8 \times 7 \times 2 = 112$ possibilities; the second, $6 \times 10 \times 2 = 120$; and the third, $16 \times 5 \times 1 = 80$. Altogether, then, there are $112 + 120 + 80 = 312$ possibilities for lunch.   □

The next few examples will introduce some mathematical concepts that may already be familiar to the reader. These concepts arise frequently in enumeration problems. The reader should already be familiar with $n!$ to denote the product $1 \cdot 2 \cdots \cdots n$. By convention, we assume $0! = 1! = 1$.

Example 1.2: A set has n distinct elements. How many subsets does it have?

We line the elements of the set up in some order; beneath each, we will place a mark if the element is in our subset, and no mark if the element is not. We will make n decisions as we go through the list of elements, and each decision has two alternatives, mark or no mark. Since no decision will affect any other decision, we see that there are 2^n subsets of a set of n elements. □

Example 1.3: Suppose we have five tunes stored on our portable digital music player, and we wish to listen to each of them one after another. How many ways are there to arrange the playlist?

We have five tunes, so there are five possibilities for the first tune. Once we have chosen this tune, then *regardless of which tune we choose to play first* there are four tunes left (because we don't wish to repeat a tune). After we choose the second tune, there are three possibilities left, and so on. The result is that we may choose a playlist in $5! = 5 \times 4 \times 3 \times 2 \times 1$ ways, by the multiplication principle. In more general terms, we find that there are $n!$

ways to order n distinct objects; so we can arrange seven books on a shelf in 7! ways, place 12 people into a line in 12! ways, and so forth. □

Example 1.4: A license plate consists of three letters of the alphabet and three digits from 0 to 9, with the letters preceding the digits. How many license plates are possible? How many are possible if we do not use any letter twice? How many are possible if we may use letters twice but not digits? How many if we do not use either twice?

We imagine six boxes in a row as shown.

For the first box, we may choose any of the 26 letters, so we have 26 choices. For the first question, where we may repeat letters and digits, we have 26 possible choices for each of the next two boxes as well. The last three boxes each have 10 possibilities. It follows from the multiplication principle that there are $26^3 \cdot 10^3$ possible license plates. If we may not repeat a letter, then we must work a little harder; the second box has only 25 possible letters, and the third has only 24. We then get $26 \cdot 25 \cdot 24 \cdot 10^3$ possible plates. In the same way, there are $26^3 \cdot 10 \cdot 9 \cdot 8$ plates where letters may be repeated but digits may not; and $26 \cdot 25 \cdot 24 \cdot 10 \cdot 9 \cdot 8$ possible plates where neither may be repeated. □

Example 1.5: How many n-digit numbers in base k are there if we do not allow a leading digit of 0? How many numbers without leading zeros in base k are there with at most n digits?

We recall that a number in base k has possible digits $0, 1, \ldots k - 1$, so that each of the n digits may be assigned any of the k symbols, except for the first. We apply the multiplication principle to see that there are $(k - 1) \cdot k^{n-1}$ possible numbers. To find how many numbers of at most n digits in base k there are, we may simply permit leading zeros; so the number 1 (in any base) might be considered a "five-digit number" 00001. We find that there are k^n such numbers. □

Example 1.6: A large corporation gives employees ID codes consisting of two or three letters and two digits. How many codes are possible if we may use any of the 26 upper-case letters and any of the digits from 0 to 9, and letters precede digits?

The multiplication principle tells us that there are $26^2 \cdot 10^2$ possibilities for the two-letter codes, and $26^3 \cdot 10^2$ for the three-letter codes. The addition principle tells us that there are

$$26^2 \cdot 10^2 + 26^3 \cdot 10^2 = 67,600 + 1,757,600 = 1,825,200$$

codes. □

1.4   Graphs

A great deal of mathematics has been done concerning collections of subsets of two elements of a larger set. In that case the members of the universal set are often represented graphically, as points in a diagram, and the set $\{x, y\}$ is represented by a line drawn joining x and y. Provided the universal set is not empty, such a structure is called a *graph*. The sets are called *edges* (or *lines*), and the elements of the universal set are *vertices* (or *points*). The universal set is the *vertex-set* of the graph. For any graph G, we write $V(G)$ and $E(G)$ for the sets of vertices and of edges of G.

The edge $\{x, y\}$ is simply written xy, when no confusion can arise; x and y are called its *endpoints*. When x and y are endpoints of some edge, we say they are *adjacent* and write $x \sim y$ for short; the vertices adjacent to x are called its *neighbors*. The set of all neighbors of x is its (open) *neighborhood*, $N(x)$. If x and y are *not* adjacent we write $x \nsim y$.

Two vertices either constitute a set or not, so a graph can never contain two edges with the same pair of vertices. However, there are some applications where two edges joining the same vertices might make sense. For this reason we sometimes talk about *networks* or *multigraphs* in which there can be several edges joining the same pair of vertices; those edges are called *multiple edges*.

Another generalization is to allow *loops*, edges of the form xx. There is no very good term for a graph-type structure in which loops are allowed, and we will usually call one of these a *looped graph* or *looped network* or *network with loops* although strictly speaking it is not a graph or network at all. Moreover, when no confusion arises, the word *graph* can be used for any generalized graph.

Any binary relation can be represented by a diagram. If ρ is a binary relation on the set S, the elements of S are shown as vertices, and if $x\rho y$ is true, then an edge is shown from x to y, with its direction indicated by an arrow. Provided the set S is finite, all information about any binary relation on S can be shown in this way. Such a diagram is called a *directed graph* or *digraph*, and the edge together with its arrow is called an *arc*. If ρ is symmetric, the arrows may be dropped and the result is a graph (possibly with loops).

Several families of graphs have been studied. Given a set S of v vertices, the graph formed by joining each pair of vertices in S is called the *complete graph* on S and denoted K_S. K_v denotes any complete graph with v vertices. As you would expect, we often call K_3 a *triangle*. The *complete bipartite graph* on V_1 and V_2 has two disjoint sets of vertices, V_1 and V_2; two vertices are adjacent if and only if they lie in different sets. We write $K_{m,n}$ to mean a complete bipartite graph with m vertices in one set and n in the other. $K_{1,n}$ in particular is called an *n-star*. Figure 1.6 shows copies of K_6 and $K_{3,4}$.

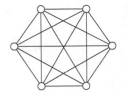 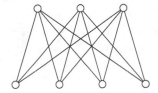

FIGURE 1.6: K_6 and $K_{3,4}$.

Suppose H is a graph all of whose vertices and edges are vertices and edges of some graph G—that is, $V(H) \subseteq V(G)$ and $E(H) \subseteq E(G)$. Then H is a *subgraph* of G; we write $H \le G$. Every graph G has itself as a subgraph; if H is a subgraph of G but $H \ne G$, H is a *proper* subgraph of G, and we write $H < G$. In particular, if S is some set of vertices of G then $\langle S \rangle_G$ is the subgraph consisting of all edges of G with both endpoints in S. If G is a complete graph whose vertex-set contains S then the subscript "G" is dropped, and $\langle S \rangle$ is the complete subgraph based on S. Any subgraph of a complete bipartite graph is itself called *bipartite*.

Instead of saying Figure 1.6 shows two graphs, we could say it is a single graph that consists of two separate subgraphs, with no edges joining one part to the other. We call such a graph *disconnected*; a graph that is all in one piece is called *connected*. The separate connected parts of a disconnected graph are called its *components*.

The graph G is trivially a subgraph of the complete graph $K_{V(G)}$. The set of all edges of $K_{V(G)}$ that are *not* edges of G will form a graph with $V(G)$ as its vertex set; this new graph is called the *complement* of G, and written $\overline{G}$. More generally, if G is a subgraph of H, then the graph formed by deleting all edges of G from H is called the *complement of G in H*, denoted $H - G$. The complement $\overline{K}_S$ of the complete graph K_S on vertex set S is called a *null graph*; we also write $\overline{K}_v$ as a general notation for a null graph with v vertices. Figure 1.7 shows a graph and its complement.

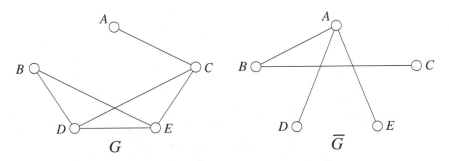

FIGURE 1.7: A graph and its complement.

We define the *degree* or *valency* $d(x)$ of the vertex x to be the number of edges that have x as an endpoint. If $d(x) = 0$, then x is called an *isolated* vertex. A graph is called *regular* if all its vertices have the same degree. If the common degree is r, it is called *r-regular*. In particular, a 3-regular graph is called *cubic*.

THEOREM 1.1 (Sum of Degrees)

In any graph or multigraph, the sum of the degrees of the vertices equals twice the number of edges (and consequently the sum of the degrees is an even integer).

Proof. Suppose the graph or multigraph has e edges; label the edges, say $y_1, y_2, \ldots, y_e$. Consider a list in which each edge appears twice, once for each endpoint. For example, if y_1 has endpoints x_4 and x_7, you might make entries $y_1 : x_4$ and $y_1 : x_7$. Vertex x will appear in precisely $d(x)$ entries, so the total number of entries equals the sum of the degrees of the vertices. On the other hand, each edge appears twice, so the total number of entries equals twice the number of edges. $\square$

COROLLARY 1.1.1

In any graph or multigraph, the number of vertices of odd degree is even. In particular, a regular graph of odd degree has an even number of vertices.

Suppose you encounter two graphs with exactly the same structure but different vertex-sets. For example, look at Figure 1.7 again; the two subgraphs with vertex-sets $\{B, D, E\}$ and $\{C, D, E\}$ are both triangles. There is no real difference between them until you need to consider the meaning of the vertices. Formally, we say two graphs G and H are *isomorphic* if there is a one-to-one correspondence between their vertex-sets $V(G)$ and $V(H)$ such that two vertices of G are adjacent if and only if the corresponding vertices of H are adjacent. For example, any two complete graphs with v vertices are isomorphic.

Not all graphs with the same number of vertices are isomorphic; for example, the graph G of Figure 1.7 and its complement are quite different, but both have five vertices.

1.5   Systems of Distinct Representatives

An important topic when discussing collections of subsets of size greater than 2 is the existence of *systems of distinct representatives*. If $\mathcal{D} = \{B_1, B_2, \ldots, B_k\}$ are any k sets, we define a system of distinct representatives, or

SDR for $\mathcal{D}$ to be a way of selecting a member x_i from each set B_i such that $x_1, x_2, \ldots$ are all different.

As an example, consider the sets

$$\{124, 124, 134, 235, 246, 1256\}.$$

One system of distinct representatives for them is

$$1, 2, 3, 5, 4, 6$$

(where the representatives are listed in the same order as the sets). There are several others. On the other hand, the sets

$$\{124, 124, 134, 23, 24, 1256\}$$

have no SDR.

If the collection of sets $\mathcal{D}$ is to have an SDR, it is clearly necessary that $\cup_{B \in \mathcal{D}} B$ have at least as many elements as there are sets in $\mathcal{D}$. The example above shows that this is not sufficient.

THEOREM 1.2 (Philip Hall's Theorem) [52]
A collection $\mathcal{D}$ of sets has a system of distinct representatives if and only if it never occurs that some n sets contain between them fewer than n elements.

Proof. We proceed by induction on the number of sets. If $\mathcal{D}$ consists of one set, the result is obvious. Assume the theorem to be true for all collections of fewer than b sets. Suppose $\mathcal{D}$ has k sets $B_1, B_2, \ldots, B_k$ and between them they have v elements $\{1, 2, \ldots, v\}$; and suppose $\mathcal{D}$ satisfies the hypothesis that the union of any n sets has size at least n, for $1 \leq n \leq k$. By induction, any n of the sets will have an SDR, provided that $n < k$. We distinguish two cases.

(i) Suppose no n sets contain between them fewer than $n + 1$ elements, for $n < k$. Select any element $x_1 \subset B_1$, and write $B_i^* = B_i \backslash \{x_i\}$, for $i \in \{1, 2, \ldots, k\}$. Then the union of any n of the B_i^* has at least n elements, for $n = 1, 2, \ldots, k-1$. By the induction hypothesis, there is an SDR $x_2, x_3, \ldots, x_k$ for $B_2, B_3, \ldots, B_k$, so $x_1, x_2, x_3, \ldots, x_k$ is an SDR for the original design $\mathcal{D}$.

(ii) Suppose there is a collection of n of the sets whose union has precisely n elements, for some n less than k. Without loss of generality, take these sets as $B_1, B_2, \ldots, B_n$. For $i > n$, write B_i^* to mean B_i with all members of $B_1 \cup B_2 \cup \ldots \cup B_n$ deleted. It is easy to see that the design with sets $B_{n+1}^*, B_{n+2}^*, \ldots, B_k^*$ satisfies the conditions of the theorem (if $B_{n+i_1}^*, B_{n+i_2}^*, \ldots, B_{n+i_t}^*$ were t sets whose union has fewer than t elements, then $B_1, B_2, \ldots, B_n, B_{n+i_1}^*, B_{n+i_2}^*, \ldots, B_{n+i_t}^*$ would be $n + t$ sets of $\mathcal{D}$ whose union has fewer than $n+t$ elements, which is impossible). From the induction hypothesis both sets have SDRs, and clearly they are disjoint, so together they comprise an SDR for $\mathcal{D}$. $\square$

Exercises 1A

1. In each case, represent the set in a Venn diagram.

    (i) $R \cup S \cup T$ (ii) $\overline{R} \cup S \cup T$

2. Show that the Hat Game team has at least a 75% chance of winning, for any number of players greater than three.

3. Find a formula for the sum of the first n odd positive integers,

$$1 + 3 + \ldots + (2n - 3) + (2n - 1)$$

    (i) using Gauss's formula for the sum of the first n positive integers;
    (ii) using induction.

4. The seven wargamers decide to play a new game, where players compete in groups of three. Show that they can find an equitable schedule of seven games.

5. Prove by induction that

$$\sum_{i=1}^{n} i(i+1) = \tfrac{1}{3}n(n+1)(n+2).$$

6. Define the relation τ on the integers by $m \tau n$ whenever $m + 2n$ is divisible by 3. Determine whether τ is an equivalence relation, and if not which of the properties (reflexive, transitive, symmetric) it lacks.

7. A phone store has four models of Samsung phones, three kinds of HTC phones, and two models of Apple phones. In how many ways can we purchase a phone from this store?

8. A computer store has four brands of desktops, three sizes of monitor, and six models of printer. In how many ways can we take home a system consisting of one tower, one monitor, and one printer?

9. In order to gain a high-paying career in the exciting field of basketweaving, we wish to take a class in Underwater Basketweaving and a class in Mountaintop Basketweaving. The local junior college offers three sections of each course. The nearby university offers six sections of Underwater and four of Mountaintop. A somewhat reputable online school has ten sections of Underwater and two of Mountaintop. Assuming that we will take both courses from the same school and there are no other conflicts, how many ways can we do this?

10. We have recorded eight movies and we want to watch three of them tonight. How many ordered sequences of three of the eight movies are possible?

11. A state has license plates consisting of two letters and four digits. How many license plates are possible?

12. Suppose a set X has $2n + 1$ elements. How many subsets of X have n or fewer elements?

13. What are the degrees of the vertices in these graphs?

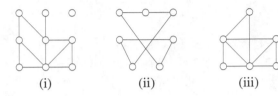

     (i)   (ii)   (iii)

14. How many edges does the star $K_{1,n}$ have? What are the degrees of its vertices?

15. A graph has 48 edges. There are no isolated vertices. What are the minimum and maximum numbers of vertices the graph can have?

16. Eight people attend a meeting. At the beginning various people shake hands. In total there are eighteen handshakes.
    (i) Show that there is at least one person who shook hands at least five times.
    (ii) Is it possible there were exactly three people who shook hands an odd number of times?

17. Do there exist SDRs for the following sets?
    (i) $12, 145, 12, 123$
    (ii) $123, 145, 12, 13, 23$.

Exercises 1B

1. The seven wargamers have become more ambitious. They wish to play a tournament as follows: each week two groups of three (with no one in common) play; the seventh member acts as host. Find a schedule for them to do this in seven weeks.

2. In each case, represent the set in a Venn diagram.

    (i) $R \cup (S \cap T)$ (ii) $\overline{(R \cap S)} \cup T$

3. Use induction to prove that, when $r \neq 1$, $a + ar + ar^2 \ldots + ar^{n-1} = a(1 - r^n)/(1 - r)$.

4. A ball is dropped from a height of 10 feet. Each time it hits the ground, it bounces back to half the previous height. How far will it have traveled when it hits the floor for the tenth time?

5. A furniture store has in stock four square dining tables, six expandable rectangular dining tables, and two circular dining tables. In how many ways can we purchase a dining table from this store?

6. A furniture store has three kinds of dining tables, six different sets of dining room chairs, and three kinds of sideboard. In how many ways can we purchase a dining room set consisting of a table, a set of chairs, and a sideboard?

7. To fulfill graduation requirements, we need a history course and an English course. At the local junior college, there are three courses that satisfy the history requirement and two that satisfy the English requirement. At the nearby university, there are four history and four English courses that satisfy requirements. An online institution offers two of each. Assuming that the credits transfer and we only wish to enroll in one institution, in how many ways can we complete the graduation requirements for these two courses?

8. Define the relation ϕ on the integers by $m \phi n$ whenever $m + 3n$ is divisible by 4. Determine whether ϕ is an equivalence relation, and if not which of the properties (reflexive, transitive, symmetric) it lacks.

9. There are nine new videos we want to see on YouTube, but we only have time to watch four of them right now. How many ways are there to watch four of the nine videos in some particular order?

10. A state has a license plate that consists of two letters, three digits, and one letter. How many license plates are possible?

11. What are the degrees of the vertices in these graphs?

      (i)   (ii)   (iii)

12. The *n-wheel* W_n has $n+1$ vertices $\{x_0, x_1, \ldots, x_n\}$; x_0 is joined to every other vertex and the other edges are

$$x_1 x_2, x_2 x_3, \ldots, x_{n-1} x_n, x_n x_1.$$

How many edges does W_n have? What are the degrees of its vertices?

13. A graph has 9 vertices and each vertex has degree at least 4. What are the minimum and maximum numbers of edges the graph can have?

14. Draw a graph with nine edges and:
    (i) five vertices;
    (ii) six vertices.

15. Do there exist SDRs for the following sets?
    (i) $123, 124, 134, 1234, 234$
    (ii) $123, 145, 167, 13, 23, 24.$
    (iii) $123, 145, 167, 13, 23, 12.$

Problems

Problem 1: A relation ρ on a set S is said to be *circular* if and only if for every $x, y, z \in S$, $x \rho y$ and $y \rho z$ together imply that $z \rho x$. Show that a reflexive circular relation is an equivalence relation.

Problem 2: Find the error in the following "proof."
Theorem. *A symmetric and transitive relation must be reflexive.*
Proof. Let ρ be a symmetric and transitive relation on a set A. If $a \rho b$, then $b \rho a$ because a is symmetric. But $a \rho b$ and $b \rho a$ together imply $a \rho a$, since ρ is transitive. Therefore ρ is reflexive.

Problem 3: Show by induction that $2^n \geq n^2$ for integer $n \geq 4$.

Problem 4: Prove by induction: if n people stand in line at a counter, and if the person at the front is a woman and the person at the back is a man, then somewhere in the line there is a man standing directly behind a woman.

Problem 5: Assume that the sum of the angles of a triangle is π radians. Prove by induction that the sum of the angles of a convex polygon with n sides is $(n-2)\pi$ radians when $n \geq 3$.

Problem 6: Prove that no graph has all its vertices of different degrees.

Problem 7: Five married couples attend a dinner party. During the evening some of the people shake hands; no one shakes hands with his or her spouse.

At the end of the party, the hostess asks each person (other than herself, of course), "with how many people did you shake hands?" It turns out that no two of them shook hands with the same number.

(i) With how many people did the hostess shake hands?

(ii) With how many people did her husband shake hands?

(iii) Generalize this to the case of n couples.

Problem 8: We have five student organizations $O_1, O_2, \ldots O_5$, each of which wishes to send a student ambassador to the National Congress of Useless Student Organizations. National Congress policy decrees that no student may represent two organizations, so we require an SDR of the memberships of the organizations. Of all students in these organizations, only five (A, B, C, D, and E) are interested in being ambassadors. Now, O_1 contains A and C; O_2 contains A, B, C, and D; O_3 contains A and C; O_4 contains B, D, and E; and O_5 contains A, C, and D.

(i) Find an SDR from among the interested students.

(ii) Find a student and organization such that if the student resigns from the organization, no SDR is possible.

(iii) Find a different student and organization such that if the student resigns then no SDR is possible.

(iv) Is it possible that if one of these five students were to join one other organization, part (ii) would have no solution? Prove your answer.

Chapter 2

Fundamentals of Enumeration

In this chapter we explore some applications of the two principles that were introduced in Section 1.3.

2.1 Permutations and Combinations

In practice, two particular applications of the Multiplication Principle arise many times. Because of this, they have their own notations.

Example 2.1: An antiques dealer has twelve different pieces of iridescent glass and wishes to display five pieces in a window display. In how many ways can she arrange five of the twelve pieces in some order in the window?

Since there are twelve pieces to choose from, she has 12 ways to put a piece in the first spot. There are 11 ways to put a piece in the second spot (because after the first piece has been placed, only 11 remain); 10 for the third; and so forth. The Multiplication Principle tells us that the total number of ways is $12 \times 11 \times 10 \times 9 \times 8 = 95,040$. □

This argument may be generalized nicely. We assume that the reader is aware that a *permutation* is an ordered subset, and a *combination* is an unordered subset, of a set.

THEOREM 2.1 (Counting Permutations)
The number of ways to arrange k objects in order from a set of n objects is $n!/(n-k)!$, commonly denoted $P(n,k)$ or $[n]_k$.

Proof. There are n possible choices for the first object, $(n-1)$ for the second, and so forth, until we find that there are $n-k+1$ possible choices for the kth object. The Multiplication Principle tells us that there are $n(n-1)\ldots(n-k+1)$ arrangements altogether. When we divide $n!$ by $(n-k)!$, this is precisely the result. □

Example 2.2: How many ways are there for an awards committee to award three distinct scholarships among 12 students entered?

The formula gives us at once $12!/9! = 1,320$. □

We wish to ask what happens if we do not wish to order the k chosen objects. Thus, if we have three (identical) scholarships and 12 students, how many ways can we award the scholarships? The answer will be much smaller than the $1,320$ that we got for the previous example, because we do not distinguish among scholarship winners. Thus, if students A, B, and C are selected to win a scholarship, that one case corresponds to six cases in the last example; for we might have A first, then B and C last, or A, then C, then B, and so on. Because there are three winners, there are 3! ways to order them. We can employ the Multiplication Principle by asking how many ways there are to order the three students. Let the number of ways to choose 3 of 12 students be denoted $\binom{12}{3}$ or $C(12,3)$. Both notations are commonly used; we say "12 choose 3." Then there are $C(12,3)$ ways to choose the students, and 3! ways to order the chosen students; this says that $C(12,3) \times 3! = P(12,3)$. It follows that $C(12,3)$ is just $P(12,3)/3! = 220$. This example also may be generalized.

THEOREM 2.2 (Counting Combinations)
The number of ways to choose k objects from a set of n distinct objects is $\binom{n}{k} = n!/(k!(n-k)!)$, also commonly denoted $C(n,k)$.

Proof. By Theorem 2.1, there are $P(n,k)$ ways to choose k elements in a particular order, and there are $k!$ ways of ordering the given k elements. It follows that $\binom{n}{k} \times k! = P(n,k)$, and dividing by $k!$ completes the proof. □

Example 2.3: An automobile manufacturer produces eight models of sports car. In how many ways can we choose three models to be blown up during the exciting climax of our next action movie?

The solution is $\binom{8}{3} = 8!/(3! \cdot 5!) = 56$. □

Example 2.4: A school decides to offer a new bioinformatics program, and decides to appoint a committee consisting of two mathematicians, three biologists, and three computer scientists to plan the implementation. If there are six mathematicians, ten biologists, and five computer scientists available to serve on this committee, in how many ways can the committee be formed?

There are six mathematicians who could serve and we must choose two of them; this may be done in $C(6,2) = 15$ ways. We select the biologists in one of $C(10,3) = 120$ ways. The computer scientists are chosen in one of $C(5,3) = 10$ ways. By the Multiplication Principle, we have a total of $15 \times 120 \times 10 = 18,000$ possible committees. □

Example 2.5: For the new bioinformatics program of the previous example, we find that Dr. John Walters of computer science and Dr. Walter Johns of mathematics cannot serve together due to their tendency to become distracted and make dreadful puns during committee deliberations. How many committees are there that contain at most one of these two individuals?

We could calculate the answer by doing cases, but it is easier to subtract the possible committees that contain both these professors. The committees that contain both involve a choice of one of the five mathematicians in addition to Dr. Walter Johns, two of the four computer scientists in addition to Dr. John Walters, and three biologists. There are $C(5,1){\cdot}C(10,3){\cdot}C(4,2) = 3,600$ such committees. We subtract these from the $18,000$, so we have $18,000 - 3,600 = 14,400$ possible committees. You should observe that we get the same result from calculating that there are $C(5,2){\cdot}C(10,3){\cdot}C(4,3)$ committees that contain neither professor, $C(5,2)\cdot C(10,3)\cdot C(4,2)$ committees that contain John Walters but not Walter Johns, and $C(5,1){\cdot}C(10,3){\cdot}C(4,3)$ that contain Walter Johns but not John Walters, and then adding the results. □

2.2   Applications of $P(n,k)$ and $\binom{n}{k}$

Here we discuss some problems whose solutions involve counting permutations and combinations, as well as a generalization of the concepts. In particular, the student is probably aware that the symbol $C(n,k)$ is called a *binomial coefficient*. A "subscripting technique" proves the theorem that explains this name.

THEOREM 2.3 *(The Binomial Theorem)*

$$(x+y)^n = \sum_{k=0}^{n} \binom{n}{k} x^k y^{n-k}$$

Proof. We write the product as $(x_1 + y_1)(x_2 + y_2)\ldots(x_n + y_n)$. Then each term of the product will be of the form $a_1 a_2 \ldots a_{n-1} a_n$, where each a is either an x or a y. To find a term that will contribute to the coefficient of $x^k y^{n-k}$, we need for k of the as to be x and the other $n-k$ of them to be y. There are clearly $\binom{n}{k}$ ways to pick k of the as to be x. Removing the subscripts means that we can collect all such terms together to get $\binom{n}{k} x^k y^{n-k}$. This works for each value of k from 0 to n. □

COROLLARY 2.3.1

$$\binom{n}{k} = \binom{n}{n-k}$$

Proof. Clearly $(x+y)^n = (y+x)^n$. $\square$

COROLLARY 2.3.2

$$\sum_{k=0}^{n} \binom{n}{k} = 2^n$$

Proof. Set $x = y = 1$ in Theorem 2.3. $\square$

COROLLARY 2.3.3

$$\sum_{k=0}^{n} (-1)^k \binom{n}{k} = 0$$

Proof. Set $x = 1$ and $y = -1$ in Theorem 2.3. $\square$

There are many other identities and formulas related to binomial coefficients. Some of these are explored in the exercises and problems. We mention one last identity as it is important.

THEOREM 2.4 (Pascal's Identity)

$$\binom{n-1}{k-1} + \binom{n-1}{k} = \binom{n}{k}$$

Combinatorial identities often have two or more kinds of proofs. One kind of proof for a theorem such as this is the *algebraic* proof, where we simply write both sides of the equation, expand in factorials, and transform one side into the other by means of elementary (and reversible) algebraic operations. Another approach is to use a *combinatorial* proof, which involves showing that each side of the equation counts the same set, or counts sets between which a bijection exists. The combinatorial proof is often preferred because it provides greater insight into the meaning of the identity.

Proof. Consider a set of n distinct objects, one of which we distinguish with a label of some kind from the others. We count the number of ways to select k of these objects. Clearly, there are $\binom{n}{k}$ ways to do this. Now, if we wish to separate these subsets of size k according to whether or not they contain the distinguished object, how many of each kind are there? The sets containing the distinguished object may be formed by taking the object and choosing $k-1$ of the remaining objects of the $n-1$; so there are $\binom{n-1}{k-1}$ subsets of size k that contain our distinguished element. Similarly, if we do not include the distinguished object, there are $\binom{n-1}{k}$ ways of choosing k objects from the $n-1$ not-distinguished objects. Since every subset either does or does not contain the distinguished object, the identity follows. $\qquad\square$

This method of proving an identity is sometimes called the *distinguished element method*. We will see many other identities proved in this way.

Binomial coefficients are usually presented in a triangular array, called *Pascal's Triangle* (although it certainly predates Pascal; see [16] or [24], which specify earlier Chinese, Indian, and European sources). In the figure below, the entry in row n and column k is $\binom{n}{k}$.

$n\backslash k$	0	1	2	3	4	5	6
0	1						
1	1	1					
2	1	2	1				
3	1	3	3	1			
4	1	4	6	4	1		
5	1	5	10	10	5	1	
6	1	6	15	20	15	6	1

Example 2.6: Suppose we live at the lower-left corner of the diagram of Figure 2.1, and we need to walk to the upper-right corner along the streets represented by the vertical and horizontal lines. In how many ways can we make this walk? We assume that we never take shortcuts through parking lots or other open areas, and each block is walked either north or east (so we never retrace our steps or detour south or west).

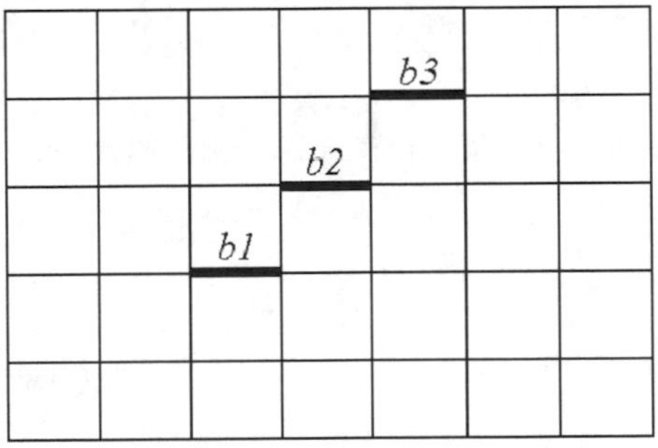

FIGURE 2.1: How many ways to walk?

We may describe our path by a string of 12 characters, each of which is either N or E to represent the direction we take. So if we start our string with "NNEE," this would indicate that we first walk two blocks north and then walk two blocks east. The string "EEEEEEENNNNN" indicates that we walk 7 blocks east along the southernmost street, then turn left and walk 5 blocks north to reach our goal. Thus, our path is a string of 12 characters, of which 7 must be "E" and 5 must be "N." How many such strings are there? Clearly $C(12, 5) = 792$. □

Example 2.7: In how many ways can we make this walk if we determine to walk along the block marked $b1$ in Figure 2.1? How many ways if the block marked $b1$ is blocked off and we must avoid it?

The first question is simple enough; to guarantee we walk along block $b1$, we must walk from our starting point (denoted $(0,0)$) to the westmost point of the block (at $(2,2)$); then walk from the end of that block (at $(3,2)$) to our destination at $(7,5)$. The Multiplication Principle gives us the number of ways to do this; it is $C(4,2) \cdot C(7,3) = 210$.

The second question requires us to take the 792 total ways to walk from one corner to the other and subtract the 120 ways that entail walking down block $b1$; this gives us 582 ways to walk without going down block $b1$. □

We will see other problems involving counting lattice paths like this elsewhere in the book. You should also see the exercises and problems at the end of the chapter that make use of the blocks labeled $b2$ and $b3$.

2.3   Permutations and Combinations of Multisets

Theorem 2.2 tells us how to select some k items from a set of n items. It is also possible that we wish to select more than one "copy" of some items.

Example 2.8: Consider the two questions: How many ways can we rearrange the letters of the word *KEPT*? How many distinct rearrangements are there of the word *KEEP*?

The first question is simple. Because there are four distinct letters, we may arrange them in 4! different ways, using Theorem 2.1. The second question, however, requires some effort. We begin by using subscripts, as we did in the proof of Theorem 2.3. The word KE_1E_2P now consists of four distinct symbols, and there are accordingly 4! distinct rearrangements. However, when we remove the subscripts, there are 2! ways of permuting the two Es. Thus, we have $4!/2! = 12$ rearrangements. This is a small enough number we can check by listing them explicitly. We find *EEKP, EEPK, EKEP, EKPE, EPEK, EPKE, KEEP, KEPE, KPEE, PEEK, PEKE,* and *PKEE*. Each of these turns into two permutations if we leave the subscripts on the *E*s. □

Example 2.9: In how many distinct ways can we rearrange the letters of *MADAM IM ADAM*?

There are 11 letters, of which four are M, 4 are A, two are D, and one is I. Thus, we have $11!/(4! \cdot 4! \cdot 2! \cdot 1!) = 34,650$. We divide by 4! for the Ms, 4! for the As, and 2! for the D. $\square$

Example 2.10: In almost every combinatorics textbook, the reader is shown the number of distinct ways to rearrange the letters of the word *MISSISSIPPI*. How many ways are there?

Since there is one *M*, four each of *I* and *S*, and two *P*s, we get $11!/(1!4!4!2!)$ or $34,650$. $\square$

THEOREM 2.5 (*Permutations of Multisets*)

The number of distinct permutations of n objects of r types, with k_1 objects of the first type, k_2 of the second, and so on, where two objects of the same type are considered identical, is $\dfrac{n!}{k_1!k_2! \dots k_r!}$.

Proof. As in the preceding example, we apply subscripts to all objects of a given type, so that we have n distinct objects. This gives us $n!$ ways to arrange the distinct objects. Now, we divide by $k_1!$ to enumerate the ways in which the objects of the first type might be rearranged, and by $k_2!$ to rearrange objects of the second type, and so on. The result follows. $\square$

This theorem may be described as a theorem about permutations of multisets; the reader will recall that in Chapter 1, we described a multiset as being able to contain more than one "copy" of a given element. Thus, PEEK contains one P, one K, and two copies of E. Theorem 2.5 counts the number of permutations of a multiset.

The binomial theorem may be generalized to powers of more involved sums. We define $n!/(k_1!k_2! \dots k_r!)$ where $k_1 + k_1 + \dots k_r = n$ to be the *multinomial coefficient* $\binom{n}{k_1 k_2 \dots k_r}$.

THEOREM 2.6 (*The Multinomial Theorem*)

$$(x_1 + x_2 + \dots + x_r)^n = \sum_{k_1 + k_1 + \dots + k_r = n} \binom{n}{k_1 \ k_2 \ \dots \ k_r} x_1^{k_1} x_2^{k_2} \dots x_r^{k_r}$$

where the sum is over all ways to write n as a sum of non-negative integers $k_1 + k_2 + \dots + k_r$.

The proof is similar to that of Theorem 2.3 and is omitted. $\square$

Here we look at a problem whose solution involves $\binom{n}{k}$. Ordinarily, $\binom{n}{k}$ counts the number of ways of selecting k of n *distinct* elements. We wish to divide up n *identical* objects into k distinct classes.

Example 2.11: If we are given a dozen identical chocolate eggs and wish to arrange them into four distinct Easter baskets, in how many ways can we do this? In how many ways can we do this if each basket is to receive at least one egg?

We are not concerned with fairness; we might give all twelve eggs to one basket, for instance, and do not need to put three eggs in each. Because the baskets are distinct, a solution in which all 12 eggs go into the first basket is different from one in which all 12 go into the second, third, or fourth baskets. We set this problem up by lining up the twelve eggs, and inserting three dividers between eggs to indicate which eggs go into which basket. Thus a diagram like this indicates three eggs in each basket:

$$OOO|OOO|OOO|OOO$$

If all twelve eggs were to go into the first basket, the diagram would be this:

$$OOOOOOOOOOOO|||$$

Each vertical divider separates the eggs that go into one basket from the eggs that go into the next. A little thought will reveal that each diagram corresponds to a distinct way to arrange the eggs, and each arrangement of eggs is represented by exactly one diagram. Thus we have a one-to-one correspondence between diagrams and ways to arrange the eggs, and we may count the diagrams to solve the problem.

How many diagrams are there? With 12 Os and three $|$s, we have 15 symbols altogether; and three of these symbols must be $|$. There are $\binom{15}{3} = 455$ ways to arrange the eggs. Now, if each basket must contain at least one egg, we merely place one of our twelve eggs into each basket. This leaves us 8 eggs and four distinct baskets (each with an egg in it), and we repeat the problem by taking 8 eggs and three dividers, to get $\binom{11}{3} = 165$ ways.   □

We use the reasoning of the foregoing examples to establish the next theorem.

THEOREM 2.7 (Sorting Identical Objects into Distinct Classes)
We can sort n identical objects into k distinct classes in $\binom{n+k-1}{k-1}$ ways.

Proof. We create a string of length $n+k-1$ that will consist of n objects and $k-1$ dividers. The ith divider will separate the objects that go into class i from the objects that go into class $i+1$. Because there are $C(n+k-1, k-1)$ such strings, there are $C(n+k-1, k-1)$ ways to separate the objects.   □

Example 2.12: We are organizing a party for small children and need to buy 18 party hats; the store has 5 different designs. In how many ways can we select 18 hats, assuming any hats with the same design are identical and the store has an unlimited number of hats of each design?

At first glance, this problem does not appear to be solvable using the theorem. However, with a little thought we can see that it may be stretched to apply. Imagine that we have 18 identical hat boxes, and we wish to place the hat boxes into one of five categories depending on the design of hat that it will hold. Then by Theorem 2.7 there are $C(18+5-1,5-1) = C(22,4) = 7,315$ possibilities. □

This example illustrates an idea that arises often enough in different contexts that it deserves its own theorem. Just as Theorem 2.5 dealt with the permutations of a multiset, this deals with combinations of a multiset.

THEOREM 2.8 *(Combinations of Multisets)*
The number of ways of selecting n (not necessarily distinct) items from a set of k distinct items where there is no limit on the number of repetitions is $C(n + k - 1, k - 1)$.

Proof. We imagine n boxes. Each box is to hold one item, and each is to be classified into one of k classes, depending on which kind of item the box is to hold. Again, we use Theorem 2.7 to show that there are $C(n+k-1, k-1)$ ways in which to classify the boxes. We fill each box with a copy of the object of that type, and after discarding the boxes, we have our desired combination. There is clearly a bijection between the ways to classify the boxes and the ways to select the items from the multiset. □

Another common counting problem is that of *cyclic permutations*. Suppose we wish to seat n people at a round table with identical seats. At first glance, this laudably Arthurian arrangement might be accomplished by lining up our people in one of the $n!$ possible ways, and simply seating them in order. However, this actually overcounts the number of ways to seat people. Suppose we moved each person one or more places to the left. This would be the same seating arrangement, because the same people would be seated next to each other. Since there are n ways to choose the seat where a selected person sits, and any of the n seating arrangements we obtain by rotating people left or right are equivalent, we see that there are actually $n!/n = (n-1)!$ distinct cyclic seating arrangements.

Example 2.13: Suppose that we are to seat 10 children at a circular table. First of all, in how many distinct ways can this be done? In how many if two of the children are siblings and always sit next to one another? In how many, if two of the children are *mischievous* siblings who on no account may sit next to one another if sanity is to be preserved?

The first question is a simple matter of applying the formula; there are $10!/10 = 9! = 362,880$ distinct seatings. The second question takes some more thought. Suppose we "glue" the siblings c_0 and c_1 together into a single unit. We may do this in two distinct ways; we may glue them so that c_0

is on c_1's left, or on c_1's right. Having done this, we have the two siblings as one unit, so there are 9 "units" to be seated in a circle. We arrive at the figure $2 \cdot 8! = 80,640$ by the Multiplication Principle. Finally, to count the arrangements in which the children are kept separate, we subtract from the total number of arrangements the arrangements in which the children are adjacent; this leaves $282,240$ arrangements. $\square$

There are very many works that discuss the uses of binomial coefficients, identities involving them, and interesting uses. Among others, the reader may benefit from [41] and [63].

2.4 Applications and Subtle Errors

We introduce a number of counting problems by use of a standard deck of playing cards. There are four *suits* (clubs, diamonds, hearts, and spades); each suit consists of thirteen *ranks*, the ace (or one), the deuce (or two), the trey (three), the four through ten, the jack, the queen, and the king. Many card games involve sets (*hands*) of varying numbers of cards; for instance, bridge consists of four hands of thirteen cards each and draw poker consists of a variable number of hands of five cards each. We will consider some subtleties of counting using cards and similar devices.

Example 2.14: For instance, one poker hand is known as *two pair*; this means that of the five cards, we have two cards of one rank (say, the eight of clubs and the eight of diamonds); two cards of a different rank (such as deuces of clubs and spades); and a fifth card of a different rank (say, jack of clubs). How many such hands are there?

An unsuspecting student might reason as follows: We have thirteen ways to select the rank of the first pair; that is $\binom{13}{1}$ or 13. Once we have chosen that rank, we choose the suit of the first card of that rank; there are four suits. Then we choose the suit of the second card of that rank, with three suits left. Then there are twelve ways to select the rank of the second pair, four choices for the suit of the first card, and three for the suit of the second card. Finally, there are 11 ranks left for the fifth card, and four possible suits for that card. Applying the Multiplication Principle, we appear to have $13 \times 4 \times 3 \times 12 \times 4 \times 3 \times 11 \times 4$ or $988,416$. This is *tragically wrong!*

To see why this does not work, let us see how the sample hand given above might be selected. We might choose an 8 of the thirteen ranks, then choose a club for the first suit, then choose a diamond for the second. Then we might choose a deuce from the 12 remaining ranks, and choose club for our first suit (from four possibilities) and spade for our second (from the three remaining).

Finally, of the 11 remaining ranks, we would choose the jack and choose clubs from the four possible suits.

Suppose instead we chose a deuce first, and chose a spade for the first suit and a club for the second. Then we chose an 8, and chose a diamond for the first suit and a club for the second. Then we again chose the jack and clubs for the fifth card. Notice that, even though we made distinctly different choices, we still obtain the same hand! This means that each possible hand is being counted more than once.

How might we choose this hand correctly? The problem arose because the poker hand does not care whether the eight or the deuce was dealt first; and it makes no difference for the eight whether we choose diamonds or clubs first. Thus, we must choose both pairs at once. Now consider that we want two ranks for our two pairs, and we wish to choose them from the thirteen ranks available; that is $\binom{13}{2}$ or 78. Now we choose two suits for one of the pairs; we can do this in $\binom{4}{2}$ or 6 ways. Another $\binom{4}{2}$ selects the two suits for the second pair. Now, as before, we have 11 ranks remaining and four suits for each rank. Now our total is $78 \times 6 \times 6 \times 11 \times 4$ or $123,552$. This is exactly one-eighth of the total we had before (and is correct). □

Example 2.15: A *flush* is a poker hand in which each of the five cards is of the same suit; a *straight flush* is a flush in which the cards are consecutive, as the 4, 5, 6, 7, and 8 of clubs. A *royal flush* is a straight flush consisting of the 10 through ace. How many of each of these hands are there?

Clearly, there are only four royal flushes; one for each suit. (The royal flush is the strongest hand in poker because it is the rarest.) How many straight flushes are possible if we do not count royal flushes? We count according to the lowest card. The ace may be the lowest or the highest card, so a straight flush may have as its lowest card ace, deuce, trey, and on through 9. Since a straight flush is entirely determined by its lowest card, and since there are 9 possible lowest cards in each of 4 suits, we have $4 \times 9 = 36$ straight flushes that are not royal flushes. Finally, to count flushes, we choose a suit in 4 ways; we choose five cards from that suit in $\binom{13}{5}$ ways; and we subtract the straight and royal flushes from this total. This gives us $5,108$ possible flushes that are not straight flushes or royal flushes. □

Example 2.16: A hand of bridge consists of 13 cards; each player has a partner. How many possible ways are there for a player and that player's partner to have two hands? Assume it makes no difference if the players exchange hands.

There are $\binom{52}{13}$ possible bridge hands for the player, and the player's partner can then have $\binom{39}{13}$ possible hands from the remaining cards. We count the configuration as the same if the two players switch hands, so the correct answer is half the product of these two numbers, $\frac{1}{2}\binom{52}{13}\binom{39}{13}$, or about 2.6×10^{21}. □

Closely tied to the concept of set enumeration is the concept of *probability*. We shall discuss probability in the next Chapter.

2.5   Algorithms

Because permutations have such a broad range of applications, we present two algorithms that may be used to generate permutations. Easy modifications of these algorithms may be used to produce particular kinds of permutations, or to produce combinations. The first algorithm will produce a list of all $n!$ permutations on n letters.

Permutation-Generating algorithm: If $n = 1$, then return the list consisting only of the letter a_1. If $n > 1$, then use the algorithm to generate the list of all permutations on $n - 1$ letters, using the symbols $a_1, \ldots a_{n-1}$. Make n copies of each permutation on this list. For each permutation, there are n places where the new symbol a_n may be placed (before the first symbol, between the first and the second, and so on until it is placed at the end). For the ith copy of this permutation, place a_n in the ith position, and repeat until all permutations have been treated this way.

This will become clearer with an example. Consider the case $n = 2$; we use the algorithm to generate all permutations consisting of one letter, and we make n copies of this permutation. This gives us two copies of the permutation a_1.

The first copy gets an a_2 in front; the second copy gets an a_2 behind it. This gives us the two permutations $a_2 a_1$ and $a_1 a_2$, which are all the permutations on two letters. Now we will work through the case of $n = 3$ in more detail.

The algorithm has given us both permutations on two letters, and we make three copies of each, which we display in the diagram below.

a_1	a_2		a_2	a_1
a_1	a_2		a_2	a_1
a_1	a_2		a_2	a_1

We now interleave the $\mathbf{a_3}$ (boldfaced for clarity) within each to generate all six permutations on three letters.

$\mathbf{a_3}$	a_1		a_2		$\mathbf{a_3}$	a_2		a_1	
	a_1	$\mathbf{a_3}$	a_2			a_2	$\mathbf{a_3}$	a_1	
	a_1		a_2	$\mathbf{a_3}$		a_2		a_1	$\mathbf{a_3}$

The result is the set of six permutations on three letters. This algorithm is fairly well known, and is discussed in, among other sources, [43] and [13]; our approach follows that of Brualdi[13].

Often, what is desired is not the set of all permutations on n letters, but just one, perhaps chosen at random. While this could easily be accomplished with the foregoing algorithm (merely generate all $n!$ permutations on n letters and select one at random), the time and space required for even moderate-sized values of n is prohibitive. Accordingly, we present an algorithm to select a permutation on n letters at random.

Random permutation algorithm: Create a list of all n letters in any order. Each letter is unmarked. Now, for each i between 1 and n, choose an unmarked letter at random. Now, place that letter in position i of the permutation and mark the letter in the list. Continue until all letters have been chosen. No letter may be used twice because at each step the selected letter is marked so that it cannot be chosen at a later step.

Exercises 2A

1. A program for generating random computer passwords gives one lower-case letter, one upper-case letter, two digits (0 through 9), and three letters that may be upper-case or lower-case. How many possible passwords can this program produce?

2. At one time, area codes were three-digit numbers with the second digit 0 or 1 and the first digit greater than 1. How many area codes were possible under this system? How many are possible if we don't allow 1s in both the second and third digit (so as not to conflict with 911, 411, etc.)?

3. A bookstore has a large box of very old books from an estate sale, and the owner wishes to display one book from the box in each of five window displays. If there are 24 books in the box, how many ways can this be done?

4. A number is called a *palindrome* if its digits read the same forwards as backwards.
   (i) Find the number of palindromes on seven digits.
   (ii) Find the number of seven-digit palindromes that have no digit appearing more than twice.

5. How many of the first million positive integers contain exactly one 3, one 5, and one 7?

6. A student decides that time is available to join no more than one social networking site. Of four sites recommended, the student has three acquaintances on the first, five on the second, six on the third, and four on the fourth. Once joined to any site, the student has for each acquaintance the option to befriend or not to befriend that acquaintance. How

many possible outcomes are there of the process of choosing a site and befriending or not befriending each acquaintance on the site?

7. A student decides that time is available to join exactly two social networking sites. Of four sites recommended, each has four of the student's acquaintances. Once joined to a site, the student has for each acquaintance the option to befriend or not to befriend that acquaintance. How many possible outcomes are there of the process of choosing two sites and befriending or not befriending each acquaintance on the site? Assume that the acquaintances on each site are different.

8. A student decides that time is available to join exactly two social networking sites. Of four sites recommended, two have four of the student's acquaintances, and two have five. Once joined to a site, the student has for each acquaintance the option to befriend or not to befriend that acquaintance. How many possible outcomes are there of the process of choosing two sites and befriending or not befriending each acquaintance on the site? Assume that the acquaintances on each site are different.

9. A student wishes to celebrate with a good friend. She has four friends from which to choose, and one of these friends is under the age of 21. The restaurants that she may choose from include three that do not serve anyone under 21 due to state liquor laws, and four with no such restriction. How many possible ways can the student celebrate assuming she takes only one friend out? How many ways if she can afford to take two of her friends out?

10. There are six tutors who are available to handle the 2:00 PM shift at the Tutoring Center. In how many ways can we select three to assign to that shift?

11. Compute $P(6,2)$, $P(7,2)$, and $P(8,2)$.

12. Find the row of Pascal's triangle corresponding to $n = 7$.

13. Use Pascal's triangle to expand the expression $(x^2 + x^{-1})^5$.

14. Show that $\sum_{k=1}^{n}(-1)^{k-1}\binom{n}{k} = 1$ when n is a positive integer.

15. Refer to Figure 2.1. Compute the number of ways to walk from the southwest corner to the northeast corner if we require that we pass along block $b2$. Use this result to count the number of ways if block $b2$ is closed off and cannot be entered.

16. In how many distinct ways can we arrange the letters in the phrase *SLEEPING STEEPLES*?

17. Use the multinomial theorem to expand $(3x + 5y + 7z)^3$.

18. We wish to distribute 10 identical memory cards among three teen-agers for use in their tablet computers. In how many ways can we do this? In how many ways if each of the teenagers is to get at least one memory card?

19. A popular brand of mobile phone cover comes in eight different colors. We wish to buy three covers. In how many ways can we do this if any two covers of the same color are considered identical?

20. Using the permutation-generating algorithm to generate all permutations of $\{1, 2, 3, 4, 5, 6\}$, which permutation precedes 356142? Which one follows it? Which follows 415326?

Exercises 2B

1. A program for generating computer passwords produces two lower-case letters, one upper-case letter, three digits, and two letters that may be either upper-case or lower-case, in that order. How many possible passwords can this program generate?

2. A game store has 10 new games; the manager wants to devote each of three distinct displays, one to each of three of the new games. How many ways can the manager select three games, one for each of the displays, in order?

3. At one time, seven-digit telephone numbers never had a zero or a one in the first three digits. How many telephone numbers (without area code) are possible in this system? How many are possible given that we cannot have any beginning with 555 or with the second and third digits being 11 (so as not to conflict, for instance, with 911 or 411)?

4. Find the number of 5-digit numbers with no repeated digits.

5. Palindromes were defined in Exercise 2A.4. Find the number of palindromes on 10 digits.

6. A *binary sequence* is a sequence of 0s and 1s. Find the number of binary sequences of length n that have an even number of elements 0.

7. There are eight tutors available to handle the 4:00 PM shift at the Tutoring Center. In how many ways can we choose three of them to assign to that shift?

8. Compute $P(5, 3)$, $P(6, 3)$, and $P(7, 3)$.

9. Find the row of Pascal's triangle corresponding to $n = 8$.

10. Use Pascal's triangle to expand the expression $(2x^2 + 1)^4$.

11. Refer to Figure 2.1. Compute the number of ways to walk from the southwest corner to the northeast corner if we require that we pass along block $b3$. Use this result to count the number of ways if block $b3$ is closed off and cannot be entered.

12. How many distinct rearrangements of the letters *SNIP TIN NIT PINS* are there?

13. Use the multinomial theorem to expand $(x + 2y + 3z)^3$.

14. There are seven identical memory cards that we wish to distribute among four teenagers for use in their MP3 music players. In how many ways can we do this? In how many ways if each of the teenagers is to get at least one memory card?

15. A popular brand of pocket calculator comes in five different colors. We wish to buy four calculators. In how many ways can we do this if any two of the same color are considered identical?

16. Using the permutation-generating algorithm to generate all permutations of $\{1, 2, 3, 4, 5, 6\}$, which permutation follows 2436135? Which one precedes it? Which precedes 641253?

Problems

Problem 1: Find the total number of integers with distinct digits.

Problem 2: Palindromes were defined in Exercise 2A.4. Prove that any palindrome of even length is divisible by 11.

Problem 3: Prove the *Vandermonde Convolution*:

$$\binom{n + m}{k} = \sum_{i=0}^{k} \binom{n}{i}\binom{m}{k - i}$$

Problem 4: Find a combinatorial proof of the identity:

$$k\binom{n}{k} = n\binom{n - 1}{k - 1}$$

Problem 5: Find an algebraic proof of the identity:

$$k\binom{n}{k} = n\binom{n-1}{k-1}$$

Problem 6: Find a four-letter and a five-letter word where the five-letter word has fewer distinct rearrangements than the four-letter word. (It is even possible to find two such words with similar meanings!)

Problem 7: Prove the identity:

$$\sum_{k \text{ odd}} \binom{n}{k} = \sum_{k \text{ even}} \binom{n}{k}$$

Problem 8: Show that the number of odd-order subsets of a set of n elements is 2^{n-1}. Use this result to find the number of even-order subsets.

Problem 9: Using Figure 2.1, find the number of ways to walk from the southwest to the northeast corner if we *must* pass through block $b1$ but *cannot* pass through $b2$.

Problem 10: How many distinct ways are there to rearrange the letters of the phrase:

(i) *BANDS BAN BANDIT BRAND BANDANAS?*

(ii) *THE SIXTH SHEIK'S SHEEP'S SICK?*

Problem 11: We have four Republicans and four Democrats who wish to sit at a round conference table to discuss a bill. In how many ways can they be seated? In how many ways can they be seated if we require that no two compatriots (Republicans or Democrats) sit next to each other? How many ways can we seat four Republicans, four Democrats, and an Independent if no members of the same party sit next to each other?

Problem 12: We have n pairs of twins. In how many ways can we choose a committee of $n-1$ people from this set of $2n$ people so that no pair of twins both serve on the committee?

Problem 13: *This problem is for those of you who have been wondering at the improbably small numbers in problems and exercises up to now. "What kind of person," you ask, "would only have five tunes on a digital music player that can hold 3000 tunes? Who on Earth only knows four people with whom they would go to a restaurant?" Very well, then, try this.*
A student wishes to take two of her 10 friends to a restaurant to celebrate; three of these friends are vegetarian, and two insist on eating meat. Of the 10 restaurants available, three are vegetarian only and two do not offer enough vegetarian entrees to interest her vegetarian friends. How many ways can

she take two of her friends out to celebrate without violating any of these restrictions?

Problem 14: Modify the algorithm that produces a random permutation on n letters to obtain an algorithm to produce a random permutation of k of the n letters, for any given $k \leq n$.

Problem 15: Produce an algorithm to choose a random subset of order k of a set of n letters.

Chapter 3

Probability

3.1 Introduction

In this chapter we explore some aspects of probability. We talk about chance in various ways: "there is a good chance of rain today," "they have no chance of winning," "there is about one chance in three," and so on. In some cases the meaning is very vague, but sometimes there is a precise numerical meaning. *Probability* is the term for the precise meaning.

3.2 Some Definitions and Easy Examples

If we have a set S, we may ask about the consequences of choosing an element of the set blindly in such a way that any element of the set has the same likelihood of being chosen as any other. When discussing such choices, we often call the process of choosing an *experiment*, and we usually call S the *sample space*. The elements of S are often called the *outcomes*. We define an *event* to be a subset E of the sample space. The *probability* of E is then just $|E| / |S|$.

One way to think about the probability that an event will happen. Suppose the same circumstances were to occur a great many times. In what fraction of cases would the event occur? This fraction is the *probability that the event occurs*. So probabilities will lie between 0 and 1; 0 represents impossibility, 1 represents absolute certainty. Of course, a probability can also be expressed as a percentage.

For example, consider the question: What is the chance it will rain tomorrow? We could ask, if the exact circumstances (current weather, time of year, worldwide wind patterns, etc.) were reproduced in a million cases, in what fraction would it rain the next day? And while we cannot actually make these circumstances occur, we can try to get a good estimate using weather records.

Many problems involve a standard deck of playing cards, which we defined in Section 2.4. Recall that such a deck contains 52 cards and is divided into four suits: diamonds and hearts are red, clubs and spades are black. (There are also decks containing jokers.) Each suit contains an ace, king, queen, jack, 10, 9, 8, 7, 6, 5, 4, 3 and 2, and these are called the 13 *denominations*. The cards are ordered as shown, with the ace highest; for some purposes the ace also counts as the low card (the "one"). When a problem says "a card is dealt from a standard deck" it is assumed that all possible cards are equally likely, so the probability of any given card being dealt is $\frac{1}{52}$. As there are four cards of each denomination, the answer to a question like "what is the probability of dealing a queen," or any other fixed denomination, is $\frac{4}{52}$, or $\frac{1}{13}$.

Example 3.1: Suppose we have a game in which each of two players draws a card from a standard deck of cards, and whoever draws the card of higher rank wins; in the case that both cards have the same rank, we draw again. If we draw first, and get (say) a jack, what is the probability that we win the game?

We will win the game if the other player draws any of the cards below a jack; we will tie or lose otherwise. In order to satisfy the assumption that all elements of the sample space (all cards) have the same likelihood of being drawn, we assume that the cards were well-shuffled. There are 40 cards below the jack if we count the ace as low, but that is not usual; it is more common to let the ace be the highest rank, so that there are nine ranks (deuce through 10) below the jack. Because there are four of each rank, that leaves 36 cards. Since after we have drawn our jack there are 51 cards remaining in the deck, the probability of a win is 36/51 or (reducing) 12/17. □

Example 3.2: In the above game, suppose we see that our opponent, having drawn first, draws a nine. What is the probability that we will not lose?

We consider the question in two ways. First, we may take the probability that we win and add it to the probability that we have a tie. In this case, we may win with any rank from among10, jack, queen, king or ace; there are 20 such cards. We may tie with any other nine, and there are three nines besides the one our opponent has drawn. This gives us 23 "non-losing" cards, so that our probability of not losing is 23/51 (where the 51 is, as with the previous example, the number of cards remaining after our opponent's nine is removed from the deck).

Another approach works as follows; there are 51 cards remaining in the deck, and we will lose if we draw any of deuce through eight (seven ranks); this gives us a probability of losing of 28/51. If we subtract these 28 losing cards from the 51 cards remaining, we have $(51 - 28)/51 = 23/51$ as our probability of not losing. □

The last example demonstrates a standard technique in probability, the technique of *complementary events*. Two events are complementary if they form a partition of the sample space S, or if one is the complement (in the set theory sense) of the other. To make the statement more compact, we introduce the probability function $P(E)$ that assigns to each event E its probability.

THEOREM 3.1 (Probabilities of Complements)
If E_1 and E_2 are complementary events, then $P(E_1) + P(E_2) = 1$.

Proof. If the sample space S contains n possible outcomes, we observe that because E_1 and E_2 are complementary it follows that $|E_1| + |E_2| = n$. Thus $P(E_1) + P(E_2) = (|E_1| + |E_2|)/n = 1$. □

Example 3.3: The Florida Lotto™ is a game in which players choose six different numbers between 1 and 53 inclusive. If the player gets the same six numbers as were randomly drawn by the lottery officials, he or she wins the jackpot. If the player only gets five of the six numbers, it is a win of a few thousand dollars; if only four numbers, less than a hundred dollars; and five dollars for three numbers. Find the probability of each outcome.

The probability of getting the jackpot is quite small; the sample space consists of all possible $C(53,6)$ ways to select six numbers from 1 to 53. This is $22,957,480$. Since only one of these will win the jackpot, the probability is therefore just over one in twenty-three million. The probability of getting five of the six is somewhat more encouraging, though still small; we choose five of the six winning numbers to go on a ticket, and then we choose one of the 47 non-winning numbers to go with them. This gives us $C(6,5) \cdot C(47,1)$ or 282 ways to get five of the six winning numbers. Accordingly, our probability is $282/22957480 = 0.000012$ or just over a thousandth of a percent. Four of six gives us more leeway (as well as smaller prize amounts); we have $C(6,4) \cdot C(47,2)$ possible tickets, or $16,215$ ways to get four of the six winning numbers. The probability here is $16215/22957480$ or 0.0007, about seven percent of one percent. Finally, to get three of the six may be done in $C(6,3) \cdot C(47,3) = 324,300$ ways, giving a probability of about 1.4%. □

There are also problems that involve ordinary dice, as used for example in games like Monopoly. These have six faces, with the numbers 1 through 6 on them. Dice can be biased, so that one face is more likely to show than another. If we roll an ordinary, unbiased die, what is the probability of rolling a 5? The six possibilities are equally likely, so the answer is $\frac{1}{6}$. If the die were biased, you might try rolling a few hundred times and keeping records.

There are also other dice, with 4, 8, 12 and 20 sides. Usually the die is a regular polyhedron, and each side contains one number; an n-sided die has each of $1, 2, \ldots, n$ once.

We say an event is *random* if you cannot predict its outcome precisely. This does *not* mean the chances of different outcomes are equal, although sometimes people use the word that way in everyday English. For example, if a die is painted black on five sides, white on one, then the chance of black is $\frac{5}{6}$

and the chance of white is $\frac{1}{6}$. This is random, although the two probabilities are not equal.

When we talk about the *outcomes* of a random phenomenon, we mean the distinct possible results; in other words, no more than one of them can occur, and one must occur. The set of all possible outcomes is called the *sample space*. Each different outcome will have a probability. These probabilities follow the following rules:

1.   One and only one of the outcomes will occur.

2.   Outcome X has a probability, $P(X)$, and $0 \leq P(X) \leq 1$.

3.   The sum of the $P(X)$, for all outcomes X, is 1.

3.3   Events and Probabilities

Much of the language of probability developed from the study of experiments. An *experiment* is defined to be any activity with well-defined, observable outcomes or results. For example, a coin flip has the outcomes "head" and "tail." The act of looking out the window to check on the weather has the possible results "it is sunny," "it is cloudy," "it is raining," "it is snowing," and so on. Both of these activities fit the definition of an "experiment."

The set of all possible outcomes of an experiment is called the *sample space*. Each subset of the sample space is called an *event*. For our purposes, an event is simply a set of outcomes. For example, when we talk about the weather we might say there are four possible outcomes: *sunny, cloudy, rain* or *snow*. So we could say the sample space is the set {*sunny, cloudy, rain, snow*}. The event "today is fine" consists of the outcomes *sunny* and *cloudy*. The event "It is not snowing" consists of *sunny, cloudy* and *rain*. The different events can occur at the same time (for example, on a sunny day), so we say the events are not *mutually exclusive*.

When we roll a die, we are primarily interested in seeing which number is uppermost, so the sample space is the set $S = \{1, 2, 3, 4, 5, 6\}$. The event "an odd number is rolled," for example, is the subset $\{1, 3, 5\}$, and this event has three outcomes. A six-element set has 2^6 subsets, so there are 2^6 events that describe the number that is rolled.

Remember, probability 0 means the particular event *cannot* occur—impossibility. And probability 1 means the event *must* occur—certainty.

The probability of an event equals the sum of the probabilities of its outcomes.

Example 3.4: Consider a die that is weighted so that 1 is rolled one time in 5 and all other rolls are equally likely. What is the probability of rolling an even number?

$P(1) = 0.2$, so the probabilities of the other five rolls must add to 0.8. They are equal, so $P(2) = P(3) = P(4) = P(5) = P(6) = 0.16$. So the probability of an even roll is $0.16 + 0.16 + 0.16 = 0.48$.

Example 3.5: Walter and John roll a die. If the result is a 1 or 6, John wins \$2 from Walter; otherwise Walter wins \$1 from John. What is the event "John wins"? What is the event "Walter wins"?

We defined the event in a die roll, above to be the set of results that produce the event. So the event "John wins" is $\{1, 6\}$, and the event "Walter wins" is $\{2, 3, 4, 5\}$.

The elementary enumeration methods from Chapter 2 will come in handy when computing probabilities. For example, if event 1 has x possible outcomes, event 2 has y possible outcomes, and the two events are not related, then the compound event "event 1 followed by event 2" has xy possible outcomes. For example, "roll a die, look at score, roll again" has $6 \times 6 = 36$ outcomes. If an act has n possible outcomes, then carrying it out k times in succession, independently, has n^k possible outcomes.

One convenient way to represent this sort of situation is a *tree diagram*.

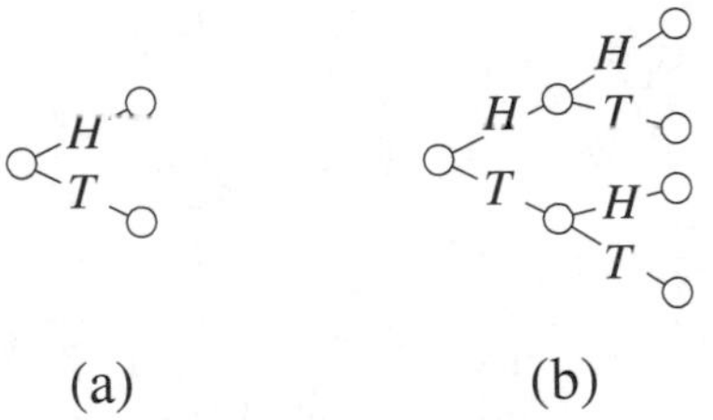

(a) (b)

FIGURE 3.1: Tree diagrams for coin tossing.

Consider the experiment where a coin is tossed; the possible outcomes are "heads" and "tails" (H and T). The possible outcomes can be shown in a diagram like the one in Figure 3.1(a). A special point (called a *vertex*) is drawn to represent the start of the experiment, and lines are drawn from it to further vertices representing the outcomes, which are called the *first generation*. If the experiment has two or more stages, the second stage is drawn onto the outcome vertex of the first stage, and all these form the *second generation*, and so on. The result is called a *tree diagram*. Figure 3.1(b) shows the tree diagram for an experiment where a coin is tossed twice.

Example 3.6: A coin is flipped three times; each time the result is written down. What is the sample space? Draw a tree diagram for the experiment.

The sample space is

$$\{HHH, HHT, HTH, HTT, THH, THT, TTH, TTT\}.$$

The tree diagram is shown in Figure 3.2(a).

The situation is more complicated where the outcome of the first event can influence the outcome of the second event, and so on. Again, tree diagrams are useful. We modify the previous example slightly.

Example 3.7: A coin is flipped up to three times; you stop flipping as soon as a head is obtained. What is the sample space? Draw a tree diagram for the experiment.

The sample space is $\{H, TH, TTH, TTT\}$. The tree diagram is shown in Figure 3.2(b).

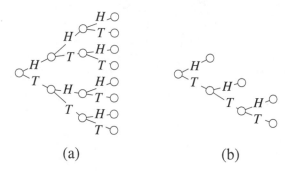

(a)   (b)

FIGURE 3.2:   Tree diagrams for Examples 3.6 and 3.7.

Example 3.8: An experiment consists of flipping three identical coins simultaneously and recording the results. What is the sample space?

The sample space is $\{HHH, HHT, HTT, TTT\}$. Notice that the order is not relevant here, so that for example, the events HHT, HTH and THH of Example 3.6 are all the same event in this case.

One common application of tree diagrams is the *family tree*, which records the descendants of an individual. The start is the person, the first generation represents his or her children, the second the children's children, and so on. (This is why the word "generation" is used when talking about tree diagrams.)

Since events are sets, we can use the language of set theory in describing them. We define the union and intersection of two events to be the events whose sets are the union and intersection of the sets corresponding to the two events. For example, in Example 3.5, the event "either John wins or the roll is odd" is the set $\{1,6\} \cup \{1,3,5\} = \{1,3,5,6\}$. (For events, just as for sets, "or" carries the understood meaning, "or both.") The complement of an event is defined to have associated with it the complement of the original set in the sample space; the usual interpretation of the complement $\overline{E}$ of the event E is "E doesn't occur." Venn diagrams can represent events, just as they can represent sets.

The language of events is different from the language of sets in a few cases. If S is the sample space, then the events S and $\emptyset$ are called "certain" and "impossible," respectively. If U and V have empty intersection, they are disjoint sets, but we call them *mutually exclusive* events (to be consistent with ordinary English usage).

Example 3.9: A die is thrown. Represent the following events in a Venn diagram:

A: An odd number is thrown;

B: A number less than 5 is thrown;

C: A number divisible by 3 is thrown.

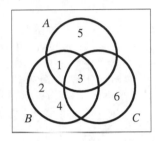

Example 3.10: A coin is tossed three times. Events A, B and C are defined as follows:

A: The number of heads is even;

B: The number of tails is even;

C: The first toss comes down heads.

Write down the outcomes in A, B, C, $A \cup C$, $B \cap C$ and $\overline{C}$. Are any of the sets mutually exclusive? Write a brief description, in words, of $B \cap C$.

$$
\begin{aligned}
A &= \{HHT, HTH, THH, TTT\}, \\
B &= \{HHH, HTT, THT, TTH\}, \\
C &= \{HHH, HHT, HTH, HTT\}, \\
A \cup C &= \{HHH, HHT, HTH, HTT, THH, TTT\}, \\
B \cap C &= \{HHH, HTT\}, \\
\overline{C} &= \{THH, THT, TTH, TTT\}.
\end{aligned}
$$

A and B are mutually exclusive, as are A and $B \cap C$. $B \cap C$ consists of all outcomes with a head first and the other two results equal.

Suppose all the different outcomes are equally likely. Say S is the sample space, and E is the set of all outcomes which we can describe as saying a certain event happens.

For example, rolling a die, say the event is "score 5 or more." Then

$$
\begin{aligned}
S &= \{1, 2, 3, 4, 5, 6\} \\
E &= \{5, 6\}
\end{aligned}
$$

and
$$P(\text{event happens}) = \frac{|E|}{|S|}.$$

So, in our example, the probability equals 2/6 or 1/3.

3.4   Three Interesting Examples

Probabilities can sometimes be surprising. We shall discuss three examples.

Birthday Probabilities

Suppose there are 30 people in a room. What is the probability that two of them share the same birthday (day and month)? (To keep the arithmetic simple, we shall ignore leap years. The probabilities are very similar.) Most people would say at first that the probability is small, but this is wrong.

First, list the birthdays of the 30 people in the order of their names (alphabetical order). If there is no restriction, each person has 365 possible birthdays. So there are 365^{30} possibilities.

Suppose no two of them have the same birthday. There are 365 choices for the first person's birthday, 364 for the second, 363 for the third, and so on. With no repeats the total number of possible lists of birthdays is

$$365 \times 364 \times \ldots \times 336.$$

Therefore the probability of the event "no two have the same birthday" is

$$\frac{365 \times 364 \times \ldots \times 336}{365^{30}}$$

which is approximately .294. So the probability of the "birthday coincidence" is about

$$1 - .294 = .706,$$

or approximately 70%.

The Game Show

This example was made popular by columnist Marilyn vos Savant.

Consider a TV game show: the contestant chooses between three doors marked doors 1, 2, 3. Behind one door is a new car; behind each of the other doors is something almost worthless (a week's supply of Kleenex tissues; two movie tickets; a pet goat); call it a booby prize. The contestant chooses a door, and she gets the prize behind it.

After the choice is announced but before the door is opened, the host opens one of the other two doors (not the one the contestant chose). Behind the door we see a goat. The host then asks, "Do you want to stay with your original choice? Or would you rather switch to the third door?"

Well, should the contestant stay or switch? Or doesn't it matter? Most people would say it doesn't matter.

To analyze the problem, we need to agree that the host always acts the same way; he always opens a door to show a goat, then offers the switch. And on any given night the chance that the car is behind any particular door is one in three. Moreover, we assume that when the contestant's first choice is the door with the car, there is an equal chance that the host will open either of the other two doors; he doesn't open the lower-numbered one more often, or anything like that.

Suppose the doors are numbered 1, 2, and 3. The contestant chooses door 1. We shall write $C1$ to mean "the car is behind door 1," $C2$ for "car behind door 2" and $C3$ for "car behind door 3." Similarly, $H2$ means "host opens door 2," and $H3$ means "host opens door 3". (He will not ever open door 1.)

Now suppose the game is played 300 times. We expect the car to be behind each door in 100 cases. If the car is behind door 1, then we expect the host will open door 2 in 50 cases and open door 3 in 50 cases. In the 100 cases where the car is behind door 2, he always opens door 3; in the 100 cases where the car is behind door 3, he always opens door 2. So we can represent the data by

	$C1$	$C2$	$C3$
$H2$	50	0	100
$H3$	50	100	0

Now suppose the host opens door 2. This tells us that tonight is one of the 150 cases represented in the row H2. Of those 150 cases, the car is behind door 1 in 50 cases and behind door 3 in 100 cases. So the odds in favor of switching are 2 to 1. The same reasoning applies when he opens door 3. So it is always best to switch.

This problem is based on the game " Let's Make a Deal," with host Monty Hall, so it is often called the *Monty Hall problem*. Mr Hall has pointed out that in the real world, the conditions 1 through 4 do not always apply.

Another Game Show

This example was also published by Marilyn vos Savant. We shall call it the "three marble problem."

In this game there are three contestants, Alice, Bruce, and Cathy. The host produces a bag containing three marbles, colored red, blue and yellow. The host draws one, then puts it back in the bag. Then the contestants each draw one at random: Alice first, then Bruce, then Cathy. If the marble drawn is the same color as one that has already been drawn, that contestant wins, and the contest is over. Otherwise the marble is replaced and the next candidate draws. The question is, who has the best chance of winning?

The easiest way to analyze this is to ask what would happen if the game is played N times. Obviously, Alice has probability $\frac{1}{3}$ of drawing the same color as the host, so she will expect to win in $\frac{1}{3}N$ games. In the remaining games, Bruce draws. There are now two colors that have been drawn, and if Bruce duplicates either, he wins. He expects to draw in $\frac{2}{3}N$ games, and the chance of him winning in those cases is $\frac{2}{3}$, so the expected number of wins is $\frac{2}{3} \times \frac{2}{3}N$. So his probability is $\frac{4}{9}$. If he loses, then all three colors have been drawn, so Cathy wins in all the remaining cases, and her probability is $1 - \frac{1}{3} - \frac{4}{9} = \frac{2}{9}$. Therefore, Bruce has the best chance of winning.

3.5 Probability Models

A *probability model* consists of a sample space together with the list of all probabilities of the different outcomes.

One example is a fair die, whose outcomes 1, 2, 3, 4, 5, 6 have probabilities

$$
\begin{array}{ll}
P(1) = \frac{1}{6} & P(2) = \frac{1}{6} \\
P(3) = \frac{1}{6} & P(4) = \frac{1}{6} \\
P(5) = \frac{1}{6} & P(6) = \frac{1}{6}
\end{array}.
$$

This is a perfectly good probability model.

Another is a biased die, whose outcomes 1, 2, 3, 4, 5, 6 have probabilities

$$
\begin{array}{ll}
P(1) = 0.15 & P(2) = 0.19 \\
P(3) = 0.18 & P(4) = 0.16 \\
P(5) = 0.14 & P(6) = 0.18
\end{array}.
$$

This is still a good probability model, because the probabilities still add to 1.

In some cases there is no numerical value associated with the outcomes. However, in many probability models, each outcome is a number, or has a numerical value. For example, this is true of rolling dice.

For example, suppose you are playing a game with three outcomes, call them A, B, C. If the outcome is A, you win \$10. If it is B, you win \$2. And if it is C, you lose \$15. The game is completely random, and observations show that A and C each occur 40% of the time, while B is the result of 20% of plays.

If the game were played 100 times, your best guess would be that A came up 40 times, B 20 and C 40. So someone who played 100 times might expect to win \$$(10 \times 40 + 2 \times 20)$, or \$440, and lose \$$15 \times 40 = \$600$. The net loss is \$160, or \$1.60 per play. You would say this is the *expected* cost of a play.

In general, suppose an occurrence has a numerical value associated with outcome, and a probability also. The *expected value* of the occurrence is found by multiplying each value by the associated probability, and adding.

This is also called the *mean value*, or *mean* of the occurrence. The mean is most commonly denoted m.

Example 3.11: You are playing a game where you draw cards from a standard deck. If you draw a 2, 3, ..., or 10, you score 1 point; a jack, queen or king is worth 3; and an ace is 5 points. How many points would you get for a typical draw?

The probability of a draw worth 1 point is 36/52, or 9/13, since 36 of the 52 cards are worth 1 point, Similarly, the probabilities of values 3 and 5 are 3/13 and 1/13 respectively. So the expected value is

$$\frac{9}{13} \times 1 + \frac{3}{13} \times 3 + \frac{1}{13} \times 5 = 23/13.$$

Similar calculations can be applied to any probability model where the outcomes have values associated with them.

3.6 Bernoulli Trials

An experiment with exactly two outcomes is called a *Bernoulli trial*. Independent repetitions of an experiment with exactly two possible outcomes are also called Bernoulli trials. These are named in honor of the seventeenth century Swiss mathematician Jacob Bernoulli. Sometimes the two outcomes are labeled "success" and "failure," but these are merely labels and should not be interpreted literally. For example, in Example 3.3 above, event A (the number of heads is even) could be called either a success or a failure.

Suppose the probability of a success is p and the probability of a failure is q. Obviously, $p + q = 1$. These numbers are often expressed as *odds*: the odds *in favor* are $p{:}q$, and the odds *against* are $q{:}p$. These odds can also be expressed as fractions: for example, the odds are p/q in favor. Often, the odds are simplified to be the ratio of integers. For example, if the probability of success is .6 and that of failure is .4, then the odds are .6/.4 which equals $3/2$, and we say "the odds are 3 to 2 in favor"; if the odds in favor are .4/.6, it is common to say "the odds are 3 to 2 against."

A *Bernoulli process* is a sequence of experiments, each of which is a Bernoulli trial with the same probability of success. The associated random variable is the number of successes. In a Bernoulli process consisting of n trials, with probabilities of success and failure p and q, the probability of k successes is

$$P(k) = \binom{n}{k} p^k q^{n-k} = \binom{n}{k} p^k (1 - p)^{n-k},$$

and because of the binomial coefficient, this set of probabilities is often called a *binomial distribution*.

Example 3.12: Suppose a fair coin is tossed four times. What is the probability that a head occurs exactly twice?

Assume that a head counts as a success. From the formula, with $p = q = .5$, $n = 4$ and $k = 2$, the probability is

$$P(k) = \binom{n}{k} p^k q^{n-k} = \binom{4}{2} 0.5^2 \times 0.5^2 = 6 \times \frac{1}{2^4} = \frac{3}{8}.$$

3.7 The Probabilities in Poker

In regular poker, a standard deck of cards is used. Each person receives a set of five cards (the *hand*), and the players bet against each other; the player with the most valuable hand wins. There are various methods—in some cases the players may discard some cards and replace them, in others more than five cards are dealt and the player chooses which excess cards to delete—but we will concentrate on the case where five cards are dealt to each player, what is called *straight draw poker*. In the other cases the end result depends on the final five-card hand. The number of possible final hands is

$$\binom{52}{5} = 2,598,960.$$

A hand consisting of five consecutive cards in the same suit is called a *straight flush*. There are 10 such hands in each suit (the ace can be either high or low) so there are 40 possible straight flushes. These hands are ranked by their highest cards; in the $\{5, 4, 3, 2, ace\}$ case the five is the high card. If the cards are consecutive in value but not in the same suit, the hand is a *straight*; if they are the same suit but not consecutive, it is a *flush*. Again, these hands are evaluated by their high cards. Given two hands of the same kind, the higher card is better.

A *pair* consists of two cards with the same numerical value; *three of a kind* and *four of a kind* are defined similarly. A hand containing three of a kind of one denomination and a pair of another is called a *full house*; two full houses are ranked according to the three of a kind. In the other cases, the hand with the higher significant card (the four of a kind denomination in four of a kind, and so on) is better. A hand containing two pairs of different denominations is called *two pairs* and is ranked according to the higher pair. A hand with no flush or straight and no two cards with the same number is simply called *high card*.

It is not hard to calculate the number of hands of the different kinds, and the probability of receiving such a hand in straight draw poker:

Poker hand	Number	Prob
Straight flush	40	0.0015%
Four of a kind	624	0.024%
Full house	3,744	0.14%
Flush	5,108	0.20%
Straight	10,200	0.39%
Three of a kind	54,912	2.11%
Two pairs	123,552	4.75%
One pair	1,098,240	42.26%
High card	1,302,540	50.12%

In poker, the hands on the above list always beat the hands lower than them. This same ranking is used in all variations of poker.

3.8   The Wild Card Poker Paradox

Another form of poker involves *wild cards*. The simplest model is where one additional card, a joker, is added to the deck. A player who receives the joker can say it is any card she wishes. A card whose meaning can be decided by the player is called a *wild card*. There are other models—for example, all the twos are wild cards, or there are two or more jokers. But we will restrict ourselves to the one-joker case.

With a 53-card deck, there are 2,869,685 possible hands. We assume that, if a hand has a joker, the joker is assigned the value that makes the hand the strongest. For example, given three kings, a four and a joker, the joker would be a king, and the hand would be a four of a kind. The suit only needs to be specified when a flush or straight flush is involved. There is one new type of hand possible, a five of a kind, and it outranks all the other hands. The numbers of hands of the various types, and their probabilities, are easily calculated:

Poker hand	Number	Prob
Five of a kind	13	0.0045%
Straight flush	184	0.0064%
Four of a kind	3,120	0.11%
Full house	6,552	0.23%
Flush	7,804	0.27%
Straight	20,532	0.72%
Three of a kind	137,280	4.78%
Two pairs	123,552	4.31%
One pair	1,268,088	44.19%
High card	1,302,540	45.39%

Note the anomaly: The probability of three of a kind is significantly higher that the probability of two pairs. This is not surprising. Consider a hand such as 8, 6, 6, 3, joker. The player could make the joker a six, so that the hand becomes three of a kind, or an eight, making it two pairs. (It could also be made a three, but that would be an inferior two pair hand.) The three of a kind is the better hand, so it is the obvious choice. In all the hands consisting of one pair, a joker, and two non-matching cards, the result will always be three of a kind. So the stronger hand, the three of a kind, is more common.

At first sight, one may think that the solution to this is to decide that two pairs beats three of a kind in wild card poker. But if this is done, the player would convert the hands in question to two pairs. Then the number of hands evaluated as two pairs would be 205,920 (probability 7.18%), while the number of three of a kind hands would be only 54,912 (probability 1.91%).

This unsolvable problem, sometimes called the *wild card poker paradox*, was studied in 1996 by Emert and Umbach [30], and also by Gadbois [38], who showed that the problem is more marked if two jokers are used.

Exercises 3A

1. If the probability that it will rain today is 0.74, what is the probability that it will not rain?

2. A box contains eight marbles: five red, two blue and one white. One is selected at random. What is the probability that:

   (i) The marble is blue;

   (ii) The marble is not blue?

3. Two dice are rolled and the numbers showing are added. What is the probability that the sum is greater than nine?

4. An experimenter tosses four coins—two dimes and two nickels—and records the number of heads. For example, two heads on the nickels and one on the dimes is recorded $(1, 2)$.
   (i) Write down all members of the sample space.
   (ii) Write down all members of the following events:
       A: There are more heads on the dimes than on the nickels.
       B: There are exactly two heads in total.
       C: The number of heads is even.
   (iii) What are the probabilities of A, B and C?

5. A coin is tossed. If the first toss is a head, the experiment stops. Otherwise, the coin is tossed a second time.
   (i) Draw a tree diagram for this experiment.
   (ii) List the members of the sample space.
   (iii) Assume the coin is fair, so that there is a probability $\frac{1}{2}$ of tossing a head. What is the probability of the event, "the coin is tossed twice"?

6. A fair coin is tossed four times. What is the probability that it lands heads exactly once?

7. We roll two fair six-sided dice and wish to roll doubles, that is, to have the same number appear on each die. What is the probability of this in a given roll?

8. The word "SLEEP" is printed on five cards, one letter on each card. The cards are shuffled and laid out in random order. What is the probability that the rearrangement will contain the word "EEL" in order from left to right with no letters intervening (for example, PEELS or SPEEL)? What is the probability that "LEP" occurs in the same way?

9. Suppose a four-sided die and a six-sided die are rolled together, and the total is calculated. What are the possible values of the total? Assuming the dice are fair, what are the probabilities of the different totals?

10. A sample space contains four outcomes: s_1, s_2, s_3, s_4. In each case, do the probabilities shown form a probability model? If not, why not?
    (i) $P(s_1) = .2$, $P(s_2) = .5$, $P(s_3) = .2$, $P(s_4) = .1$.
    (ii) $P(s_1) = .2$, $P(s_2) = .5$, $P(s_3) = .6$, $P(s_4) = .2$.
    (iii) $P(s_1) = .1$, $P(s_2) = .2$, $P(s_3) = .5$, $P(s_4) = .2$.
    (iv) $P(s_1) = .6$, $P(s_2) = .2$, $P(s_3) = .5$, $P(s_4) = -.3$.
    (v) $P(s_1) = .1$, $P(s_2) = .3$, $P(s_3) = .3$, $P(s_4) = 1.2$.

11. A regular deck of cards is shuffled and one card is dealt.
    (i) What is the probability that the card is red?

   (ii) What is the probability that the card is an ace?

  (iii) What is the probability that the card is a red ace?

  (iv) What is the probability that the card is either red or an ace?

12. Electronic components are being inspected. Initially one is selected from a batch and tested. If it fails, the batch is rejected. Otherwise a second is selected and tested. If the second fails, the batch is rejected; otherwise a third is selected. If all three pass the test, the batch is accepted.

   (i) Draw a tree diagram for this experiment.

  (ii) How many outcomes are there in the sample space?

  (iii) What is the event, "fewer than three components are tested"?

Exercises 3B

1. Exactly one of three contestants will win a game show. The probability that Joseph wins is 0.25 and the probability that Alice wins is 0.4. What is the probability that Barbara wins?

2. A student is taking a four-question true/false test. If the student chooses answers at random, what is the probability of getting all questions correct?

3. Two fair dice are rolled. What are the probabilities of the following events?

   (i) The total is 7;

  (ii) The total is even;

  (iii) Both numbers shown are even.

4. The word "TOAST" is printed on five cards, one letter on each card. The cards are shuffled and laid out in a line, in random order. What is the probability that the rearrangement will contain the word "TOT" in order from left to right with no letters intervening (for example, TOTAS or STOTA)? What is the probability that "ATO" occurs in the same way?

5. An examination has two questions. Of 100 students 75 answer Question 1 without errors and 70 answer Question 2 without errors; 65 answer both questions without errors.

   (i) Represent these data in a Venn diagram.

  (ii) A student's answer book is chosen at random. What is the probability that:

    (i) Question 1 contains an error;

    (ii) Exactly one question contains an error;

    (iii) At least one question contains an error?

6. A bag contains two red, two green and three blue balls. In an experiment, one ball is drawn from the bag and its color is noted, and then a second ball is drawn and its color noted.

    (i) Draw a tree diagram for this experiment.

    (ii) How many outcomes are there in the sample space?

    (iii) What are the outcomes in the event, "two balls of the same color are selected"?

7. A die is rolled twice, and the results are recorded as an ordered pair.

    (i) How many outcomes are there in the sample space?

    (ii) Consider the events:

        E: The sum of the throws is 4;

        F: Both throws are even;

        G: The first throw was 3.

        (a) List the members of events E, F and G.

        (b) Are any two of the events E, F, G mutually exclusive?

    (iii) Draw a tree diagram for this experiment.

    (iv) Represent the outcomes of this experiment in a Venn diagram. Show the events E, F and G in the diagram.

    (v) Assuming the die is unbiased, what are the probabilities of E, F and G?

8. A card is drawn from a standard deck of cards, and then replaced. Then another card is drawn.

    (i) What is the probability that both are red?

    (ii) What is the probability that neither is red?

9. An experiment consists of studying families with three children. B represents "boy," G represents "girl," and for example BGG will represent a family where the oldest child is a boy and the two younger children are both girls.

    (i) What is the sample space for this experiment?

    (ii) We define the following events:

        E : The oldest child is a boy;

        F : There are exactly two boys.

        (a) What are the members of E and F?

        (b) Describe in words the event $E \cap F$.

    (iii) Draw a Venn diagram for this experiment, and show the events E and F on it.

10. At a local supermarket, the number of people in checkout lines varies. The probability model for the number of people in a randomly chosen line is

Number in line	0	1	2	3	4	5
Probability	0.08	0.15	0.20	0.22	0.15	0.20

What is the mean number of people in a line?

Problems

Problem 1: In Example 3.1 what is the probability of a roll of 4 or less with this die?

Problem 2: An experiment consists of flipping a quarter and noting the result, then flipping two pennies and noting the number of heads. What is the sample space? Draw a tree diagram.

Problem 3: Represent the following events in Venn diagrams:

A: A regular, unbiased die is thrown and the result is an even number;

B: John wins (in the game of Example 3.3);

C: Walter wins in that game.

Problem 4: In the three marble problem, suppose four different marbles are used instead of three. Which player has the best chance of winning? Assuming that each player draws a marble only once, what is the probability that there is no winner?

Problem 5: Suppose three regular, six-sided dice are rolled, and the two highest numbers shown are added. What is the probability that this sum is 9? What is the expected value of the sum?

Problem 6: Suppose we roll two fair six-sided dice and wish to roll doubles, as in Exercise 3A.6.

(i) What is the probability of *not* getting doubles in one roll? In two consecutive rolls? In three consecutive rolls?

(ii) What is the probability of getting doubles two or more times in six attempts?

Problem 7: Verify the calculations of the numbers of hands and probabilities for regular poker, given in Section 3.7.

Problem 8: Show that if event E_1 has expected value v_1, and E_2 has v_2, then the random event consisting of adding the values of E_1 and E_2 has expected value $v_1 + v_2$. Use this result to find the expected value obtained by rolling n fair k-sided dice when each die has faces numbered 1 through k, and adding the numbers showing on the upper faces.

Problem 9: Consider all the subsets with four elements of the set $\{0, 1, \ldots, 9\}$. The *sum* of such a set is the sum of the four elements. Without calculating all the sums or listing all the subsets, show that there must be at least 10 subsets with the same sum.

Chapter 4

The Pigeonhole Principle and Ramsey's Theorem

In this chapter we introduce a basic combinatorial principle called the *pigeon-hole principle*. It is used in proving the existence of objects that are difficult to construct. We then go on to the generalization of the principle.

4.1 The Pigeonhole Principle

Our basic idea can be expressed as follows: If more than n pigeons are roosting in n pigeonholes, then at least one pigeonhole contains more than one pigeon. We present this principle in a more formal version, which sounds more mathematical and less ornithological than the above.

THEOREM 4.1 (The Pigeonhole Principle)
(i) *If an s-set is partitioned into n disjoint subsets, where $s > n$, then at least one of the subsets contains more than one element.*
(ii) *Suppose $a_1, a_2, \ldots, a_n$ are real numbers such that $\sum_{I=1}^{n} a_i \geq b$. Then for at least one value of $i, i = 1, 2, \ldots, n$, $a_i \geq b/n$.*

Proof. (i) Suppose the statement is false. Let

$$S = S_1 \cup S_2 \cup \ldots \cup S_n \tag{4.1}$$

be a partition of S, where $|S| = s > n$, $S_i \cap S_j = \emptyset$ whenever $i \neq j$, $S_i \neq \emptyset$ and $|S_i| \leq 1$, for $i = 1, 2, \ldots, n$. But then $|S_i| = 1$ for $i = 1, 2, \ldots, n$, and by (4.1) $s = |S| = |S_1| + |S_2| + \ldots + |S_1| = 1 + 1 + \ldots + 1 = n$. Since $s > n$, this is a contradiction.
(ii) Again suppose the statement is false, so that we have $a_i < b/n$ for every $i, i = 1, 2, \ldots, n$. Then

$$\sum_{i=1}^{n} a_i < b/n + b/n + \ldots + b/n = b.$$

But $\sum_{i=1}^{n} a_i > b$ and again we have a contradiction. $\qquad \square$

We can also see that form (ii) of the principle can be used to deduce (i). Let $a_i = |S_i|$. Then by (4.1) we have

$$|S| = \sum_{i=1}^{n} a_i > n + 1;$$

by (ii) this implies that for at least one value of i, we have

$$a_i \geq \frac{n+1}{n} > 1.$$

4.2   Applications of the Pigeonhole Principle

Most applications of the pigeonhole principle involve the appropriate partitioning of a set into subsets. Sometimes we must also use some special feature of the problem and sometimes we need repeated applications of the pigeonhole principle.

Example 4.1: Show that in a set of eight people there are at least two whose birthdays fall on the same day of the week this year.

We partition the set of people into seven subsets, the first subset being those whose birthdays fall on a Monday this year, the second those whose birthdays fall on a Tuesday, and so on. Since there are more people in our set than there are days in the week, Theorem 3.1 tells us that at least one subset contains at least two people; that is, at least two people have birthdays on the same day of the week. □

Example 4.2: [45] Given N objects to be distributed between B boxes, what is a necessary and sufficient condition on N and B that in every distribution at least two boxes must contain the same number of objects?

Suppose no two boxes contain the same number of objects; what is the smallest number of objects we would need? If the first box contained no objects, the second box one object, and in general the ith box $(i-1)$ objects, then we have $N = 0+1+\ldots+(i-1)+\ldots+(B-1) = \frac{1}{2}B(B-1)$. Similarly, for any larger value of N, we can arrange the objects so that no two boxes contain the same number of objects. Hence, a necessary and sufficient condition for at least two boxes to contain the same number of objects is

$$N < B(B - 1).$$ □

Example 4.3: [62] Choose an integer n. Show that there exists an integer multiple of n that contains only the digits 0 and 1 (in the usual decimal notation).

Consider the integers $m_i = \frac{10^i - 1}{9}$, for $1 \leq i \leq n$. For every i, m_i is represented in the decimal scale by a string of i ones. We look at their residues modulo n: if some m_i is divisible by n, the problem is solved; if not, then the n integers $m_1, m_2, \ldots, m_n$ must belong to the $(n-1)$ residue classes congruent to $1, 2, \ldots, n - 1 (\mathrm{mod}\ n)$, so for some i and j, $1 \leq i < j \leq n$,

$$m_i \equiv m_j (\mathrm{mod}\ n).$$

Now $m_{ij} - m_i$ is divisible by n, and it is represented in the decimal scale by $(j - i)$ ones followed by i zeros. □

Example 4.4: [77] Let $A = \{a_1, a_2, \ldots, a_7\}$ be a set of seven distinct positive integers none of which exceeds 24. Show that the sums of the non-empty subsets of A cannot all be distinct.

Since A is a 7-set, it has $\binom{7}{1} + \binom{7}{2} + \binom{7}{3} + \binom{7}{4} = 98$ non-empty subsets with at most four elements. The lower bound on the sums of the non-empty subsets is obviously 1; the upper bound on the sums of the k-sets, for $k \leq 4$, is $21 + 22 + 23 + 24 = 90$. Since we have 98 sums that can take at most 90 values, at least two sums must be equal. □

Example 4.5: Suppose 70 students are studying 11 different subjects, and any subject is studied by at most 15 students. Show that there are at least three subjects that are studied by at least five students each.

As 70 students take 11 subjects, there must be at least one subject with at least seven students; call it subject A. At most 15 students take A, so at least 55 students study the remaining 10 subjects. Hence at least one of these subjects has at least six students; call it subject B. At most 15 students take B, so at least 40 students study the remaining nine subjects. Hence at least one of these subjects has at least five students; call it subject C. Then A, B, C are the required subjects. □

Example 4.6: [101] Let $a_1, a_2, \ldots, a_k$, $k > mn$, be a sequence of distinct real numbers. Show that if the number of terms of every decreasing subsequence is at most m then there exists an increasing sequence of more than n terms.

To each term a_i of the sequence we assign an ordered pair of integers (m, n) in the following way: starting with a_i, we choose the longest possible decreasing subsequence and we let m_i be its number of terms; similarly n is the largest number of terms of an increasing subsequence beginning with a_i. We claim that for distinct terms a_i and a_j, $i < j$, the corresponding pairs (m_i, n_i) and (m_j, n_j) are distinct. Suppose $a_i < a_j$. We know we have an increasing subsequence of n_j terms beginning with a_j. Since $a_i < a_j$, we can adjoin a_i at the beginning of this subsequence, so that we now have an increasing subsequence of $n+1$ terms, beginning with a_i. Hence $n_i \geq n_j + l$. Similarly, if $a_i > a_j$, then $m_i \geq m_j + 1$.

So we have $k > mn$ distinct ordered pairs (m_i, n_i) and since we assumed that the number of terms of every decreasing subsequence is at most m we have $1 \leq m_i \leq m$ for all i. If we now have $1 \leq n_i \leq n$ for all i, then we can have at most mn distinct ordered pairs. Since we have more than mn pairs that are distinct, we must have $n_i > n$ for at least one i. But this means we have at least one increasing subsequence of more than n terms.

In fact, this result is the best possible. For in the sequence $m, m-1, \ldots, 1, 2m, 2m-1, \ldots, m+1, 3m, 3m-1, \ldots, 2m+1, \ldots, nm, nm-1, \ldots, (n-1)m+1$, we have mn terms; the longest decreasing subsequence has m terms and the longest increasing subsequence has n terms.   □

Example 4.7: [76] A student solves at least one problem every day. Since she is anxious not to overwork she makes it a rule never to solve more than 730 problems per year. Show that she solves exactly n problems in some set of consecutive whole days for any given positive integer n.

Consider first the case where $n = 1$; this gives us a clue to the solution. Suppose that the student solves exactly two problems per day, throughout any year with 365 days. She thus solves 730 problems in the whole year and on no one day has she solved exactly one problem. But during a leap year, solving exactly two problems per day leads her to exceed her quota: there are at least two days of a leap year on each of which she must solve exactly one problem. Since leap years include all those with numbers divisible by 4 but not by 25 (as well as those with numbers divisible by 400), any eight consecutive years include at least one leap year and contain at least $(8 \times 365) + 1 = 2921$ days. We start from some fixed date and let S_i be the number of problems solved in the first i days from the starting date. Consider the sets of integers

$$S = \{S_i : 1, \leq i \leq N = 2921n\}$$

and

$$T = \{n + S_i : 1, \leq i \leq N = 2921n\}.$$

Notice that $S_i \neq S_j$ when $i \neq j$, since the student solves at least one problem per day. Similarly $n + S_i \neq n + S_j$ when $i \neq j$. Now

$$|S| + |T| = 2N = 5842n;$$

all the elements of S and T are positive integers, the largest of which is $n + S$, and

$$n + S_n \leq n + (8n \times 730) = 5841n.$$

Hence, there exists at least one positive integer j belonging to both S and T. So for some h and k,

$$j = S_h = n + S_k,$$

which means that on the days $k+1, k+2, \ldots, h$, the student solved exactly n problems.   □

4.3 Ramsey's Theorem—The Graphical Case

In discussing partitions of the edge-set of a graph it is common to refer to *coloring* the edges. A set of labels called *colors* is used. The set of all edges that receive a given color is called a *color class*.

Suppose the edges of a graph G are colored in k colors. We say a subgraph H of G is *monochromatic* if all its edges receive the same color; when no confusion can arise, the k subgraphs formed by all the edges of one color are called *the monochromatic subgraphs* for that coloring. A k-coloring of G is called *proper* with respect to H if G contains no monochromatic subgraph isomorphic to H in that coloring. If no subgraph is specified, "proper" will mean proper with respect to *triangles* (graphs isomorphic to K_3). We begin with an example for the case $k = 2$.

Example 4.8: Suppose G is a complete graph and its vertices represent people at a party. An edge xy is colored red if x and y are acquaintances, and blue if they are strangers. An old puzzle asks: given any six people at a party, prove that they contain either a set of three mutual acquaintances or a set of three mutual strangers.

To observe that the result is *not* true for fewer than six people, consider the complete graph K_5. It is easy to see that K_5 has a proper 2-coloring: in the example shown in Figure 4.1, color all the heavy edges red and all the other edges blue.

FIGURE 4.1: K_5 in two colors.

On the other hand, there is no proper 2-coloring of K_6. To see this, select a vertex x in any 2-coloring of K_6. There are five edges touching x, so there must be at least three of them that receive the same color, say red. Suppose xa, xb and xc are red edges. Now consider the triangle abc. If ab is red, then xab is a red triangle. Similarly, if ac or bc is red, there will be a red triangle. But if none is red, then all are blue, and abc is a blue triangle.

This proves that any 2-coloring of K_v must contain a monochromatic triangle whenever $v \geq 6$: if $v > 6$, simply delete all but six vertices. The resulting 2-colored K_6 must contain a monochromatic triangle, and that triangle will also be a monochromatic triangle in K_v. $\square$

LEMMA 4.2

There exists a number $R(p, q)$ such that any coloring of $K_{R(p,q)}$ in two colors c_1 and c_2 must contain either a K_p with all its edges in color c_1 or a K_q with all its edges in c_2.

Proof. We proceed by induction on $p + q$. The lemma is clearly true when $p + q = 2$, since the only possible case is $p = q = 1$ and obviously $R(1, 1) = 1$. Suppose it is true whenever $p + q < N$, for some integer N. Consider any two positive integers P and Q that add to N. Then $P + Q - 1 < N$, so both $R(P - 1, Q)$ and $R(P, Q - 1)$ exist.

Consider any coloring of the edges of K_v in two colors c_1 and c_2, where $v \geq R(P - 1, Q) + R(P, Q - 1)$, and select any vertex x of K_v. Then x must either lie on $R(P - 1, Q)$ edges of color c_1 or on $R(P, Q - 1)$ edges of color c_2. In the former case, consider the $K_{R(P-1,Q)}$ whose vertices are the vertices joined to x by edges of color c_1. Either this graph contains a K_{p-1} with all edges of color c_1, in which case this K_{p-1} together with x forms a K_p with all edges in c_1, or it contains a K_Q with all edges in c_2. In the latter case, the K_v again contains one of the required monochromatic complete graphs. So $R(P, Q)$ exists, and in fact $R(P, Q) \leq R(P, Q - 1) + R(P - 1, Q)$.   □

(In the example of the six people at the party, we proved that $R(3, 3) = 6$.)

The same argument can be used when there are more than two colors. The general result is the graphical version of a result called *Ramsey's Theorem*.

THEOREM 4.3 (Ramsey's Theorem, Graphical Case)

Suppose $p_1, p_2, \ldots, p_n$ are positive integers. Then there exists an integer

$$R(p_1, p_2, \ldots, p_n)$$

such that if $v \geq R(p_1, p_2, \ldots, p_n)$, then any coloring of the edges of K_v with n colors will contain a subgraph K_{p_i} with all its edges of color i, for at least one i.

Proof. We proceed by induction on n. From Lemma 4.2, the result is true for $n = 2$. Assume it is true for $n = t - 1$. Then $R(p_2, p_3, \ldots, p_t)$ exists. Write $p_0 = R(p_2, p_3, \ldots, p_t)$. Select $v \geq R(p_1, p_0)$ and color K_v in colors $1, 2, \ldots, t$. Then recolor all edges of color $2, 3, \ldots,$ or t in the new color 0. Either there is a K_{p_1} in color 1 or a K_{p_0} in color 0. In the latter case, revert to the original colors. Since $p_0 = R(p_2, p_3, \ldots, p_t)$, the induction hypothesis ensures that the K_{p_0} (and therefore the original graph) must contain a monochromatic K_{p_i} in color i, for some $i, 2 \leq i \leq t$. So the theorem is true for $n = t$.   □

In fact, we have shown that

$$R(p_1, p_2, \ldots, p_n) \leq R(p_1, R(p_2, p_3, \ldots, p_n)).$$

The numbers $R(p_1, p_2, \ldots, p_n)$ are called *Ramsey numbers*. In the particular case where all n constants p_i are equal, to p say, we shall use the notation $R_n(p)$.

In discussing individual small Ramsey numbers, it is often useful to consider the *monochromatic subgraphs* of a given coloring of a complete graph. We illustrate this with an example.

Example 4.9: Evaluate $R(3, 4)$.

Suppose K_v has been colored in red and blue so that neither a red K_3 nor a blue K_4 exists. Select any vertex x. Define R_x to be the set of vertices connected to x by red edges—that is, R_x is the neighborhood of x in the red monochromatic subgraph, and similarly define B_x in the blue monochromatic subgraph.

If $|R_x| \geq 4$, then either $\langle R_x \rangle$ contains a red edge yz, whence xyz is a red triangle, or else all of its edges are blue, and there is a blue K_4. So $|R_x| \leq 3$ for all x.

Next suppose $|B_x| \geq 6$. Then $\langle B_x \rangle$ is a complete graph on six or more vertices, so it contains a monochromatic triangle. If this triangle is red, it is a red triangle in K_9. If it is blue, then it and x form a blue K_4 in K_9.

It follows that every vertex x has $|R_x| \leq 3$ and $|B_x| \leq 5$, so $v \leq 9$. But $v = 9$ is impossible. If $v = 9$, then $|R_x| = 3$ for every x, and the red monochromatic subgraph has nine vertices each of (odd) degree 3, in contradiction of Corollary 1.1.1.

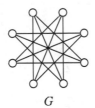 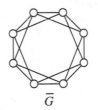

$$G \qquad\qquad \overline{G}$$

FIGURE 4.2: Decomposition of K_8 proving $R(3, 4) \geq 9$.

On the other hand, K_8 can be colored with no red K_3 or blue K_4. The graph G of Figure 4.2 has no triangle, and can be taken as the red monochromatic subgraph, while its complement $\overline{G}$ is the blue graph. (The construction of this graph will be discussed in Section 4.5, below.) So $R(3, 4) = 9$. □

4.4   Ramsey Multiplicity

We know that any 2-coloring of K_6 must contain a monochromatic triangle. However, it is not possible to find a coloring with exactly one triangle. (This is easily checked by exhaustion—see the Exercises—and follows from Theorem 4.4 below.) More generally, one can ask: what is the minimum number of monochromatic triangles in a k-coloring of K_v?

Such questions are called *Ramsey multiplicity problems*. The k-Ramsey multiplicity $N_{k,v}(n)$ of H in K_v is defined to be the minimum number of monochromatic subgraphs isomorphic to K_n in any k-coloring of K_v. Clearly $N_{k,v}(n) = 0$ if and only if $v < R_k(n)$.

The 2-Ramsey multiplicity of K_3 was investigated by Goodman [47], who proved the following theorem. Our proof follows that given by Schwenk [100].

THEOREM 4.4 (Goodman's Theorem)

$$N_{2,n}(3) = \binom{n}{3} - \left\lfloor \frac{n}{2} \left\lfloor \left(\frac{n-1}{2} \right)^2 \right\rfloor \right\rfloor.$$

Proof. Suppose K_v is colored in two colors, red and blue. Write R for the number of red triangles, B for the number of blue triangles, and P for the number of *partial* triangles—triangles with at least one edge of each color. There are $\binom{n}{3}$ triangles in K_v, so

$$R + B + P = \binom{v}{3}.$$

Since $N_{2,v}(3)$ equals the minimum possible value of $R + B$,

$$N_{2,v}(3) = \binom{v}{3} - \max(P).$$

Suppose the vertices of K_v are $x_1, x_2, \ldots, x_v$, and x_i is incident with r_i red edges. Then it is adjacent to $v - 1 - r_i$ blue edges. Therefore the K_v contains $r_i(v - 1 - r_i)$ paths of length 2 in which one edge is red and the other blue. Let us call these *mixed* paths. The total number of mixed paths in the K_v is

$$\sum_{i=1}^{v} r_i(v - 1 - r_i).$$

The triangle xyz can be considered as the union of the three paths xyz, yzx and zxy. Moreover, the paths corresponding to different triangles will all be different. If the triangle is monochromatic, no path is mixed, but a partial triangle gives rise to two mixed paths. So there are $2P$ mixed paths in the K_v, and

$$P = \frac{1}{2} \sum_{i=1}^{v} r_i(v - 1 - r_i).$$

If v is odd, the maximum value of $r_i(v-1-r_i)$ is $(v-1)^2/4$, attained when $r_i = (v-1)/2$. If v is even, the maximum of $v(v-2)/4$ is given by $r_i = v/2$ or $(v-2)/2$. In either case, the maximum is

$$\left\lfloor \left(\frac{v-1}{2}\right)^2 \right\rfloor,$$

so

$$P \quad \leq \quad \tfrac{1}{2} \sum_{L=1}^{v} \left\lfloor \left(\frac{v-1}{2}\right)^2 \right\rfloor$$

$$\leq \quad \tfrac{v}{2} \left\lfloor \left(\frac{v-1}{2}\right) 1^2 \right\rfloor,$$

and since P is an integer,

$$P \leq \left\lfloor \frac{v}{2} \left\lfloor \left(\frac{v-1}{2}\right)^2 \right\rfloor \right\rfloor.$$

So

$$N_{2,v}(3) \geq \binom{v}{3} - \left\lfloor \frac{v}{2} \left\lfloor \left(\frac{v-1}{2}\right)^2 \right\rfloor \right\rfloor.$$

It remains to show that equality can be attained.

If v is even, say $v = 2t$, then partition the vertices of K_v into the two sets $\{x_1, x_2, \ldots, x_t\}$ and $\{x_{t+1}, x_{t+2}, \ldots, x_{2t}\}$ of size t, and color an edge red if it has one endpoint in each set, blue if it joins two members of the same set. (The red edges form a copy of $K_{t,t}$.) Each r_i equals $v/2$. If $v = 2t + 1$, carry out the same construction for $2t$ vertices, except color edge $x_i x_{t+i}$ blue for $1 \leq i \leq \lfloor \frac{t}{2} \rfloor$. Then add a final vertex x_{2t+1}. The edges $x_i x_{2t+1}$ and $x_{i+t} x_{2t+1}$ are red when $1 \leq i \leq \lfloor \frac{t}{2} \rfloor$ and blue otherwise. In both cases it is easy to check that the number of triangles equals the required minimum. □

Substituting into the formula gives $N_{2,v}(3) = 0$ when $v \leq 5$, $N_{2,6}(3) = 2$, $N_{2,7}(3) = 4$, and so on.

The 3-Ramsey multiplicity of K_3 has not been fully investigated. We know that $R_3(3) = 17$, so the number $N_{3,17}(3)$ is of special interest. It has been shown that $N_{3,17}(3) = 5$; the argument involves discussion of many special cases.

The following theorem provides a recursive bound on $N_{k,v}(3)$.

THEOREM 4.5 (A Ramsey Multiplicity Bound)

$$N_{k,v+1}(3) \leq \left\lfloor \frac{v-1}{k} \right\rfloor + \left\lfloor \left(1 + \frac{3}{v}\right) N_{k,v}(3) \right\rfloor.$$

Proof. Suppose F is a k-coloring of K_v that contains $N_{k,v}(3)$ monochromatic triangles. Select a vertex x of F that lies in the minimum number of monochromatic triangles. Since there are v vertices, one can assume that x lies on at most $\lfloor \frac{3}{v} N_{k,v}(3) \rfloor$ monochromatic triangles. Since x has degree $v-1$ in K_v, there will be a color—say R—such that x lies on at most $\lfloor \frac{v-1}{k} \rfloor$ edges of color R.

Construct a k-coloring of K_{v+1} from F by adjoining a new vertex y. If z is any vertex other than x, then yz receives the same color as xz, and xy receives color R. Then xy lies in $\lfloor \frac{v-1}{k} \rfloor$ or fewer monochromatic triangles, all in color R. The original K_v contained $N_{k,v}(3)$ monochromatic triangles, so this is the number not containing x. Finally, the number of monochromatic triangles with y as a vertex but not x is at most $\lfloor \frac{3}{v} N_{k,v}(3) \rfloor$. So the maximum number of monochromatic triangles in the K_{v+1} is

$$\left\lfloor \frac{n-1}{k} \right\rfloor + \left\lfloor \left(1 + \frac{3}{v}\right) N_{k,v}(3) \right\rfloor. \qquad \square$$

This theorem provides the upper bounds 2 and 5 for $N_{2,6}(3)$ and $N_{3,17}(3)$, both of which can be met [47, 96, 120].

4.5 Sum-Free Sets

For this section you need to remember that $\mathbb{Z}_n$ is the set of integers modulo n. The residue class of x, the set of all integers leaving remainder x on division by n, is denoted by $[x]$. Addition is defined on $\mathbb{Z}_8$ by the rule: $[x] + [y] = [x+y]$. Two important properties of $\mathbb{Z}_n$ is that it has an identity element $[0]$—$[x] + [0] = [x]$ for any x—and each element $[x]$ has an *inverse* $[-x]$— $[x] + [-x]$ always equals $[0]$. For example, in $\mathbb{Z}_8$, the inverse of $[5]$ is $[3]$. For convenience we shall usually just write x instead of $[x]$. The set of non-zero elements of $\mathbb{Z}_n$ is written $\mathbb{Z}_n^*$.

As an introduction we derive the construction of the red graph of Figure 4.2. The vertices of the graph are labeled with the elements of $\mathbb{Z}_8$. The set $\mathbb{Z}_8^*$ is partitioned into two sets:

$$\mathbb{Z}_8^* = \{3, 4, 5\} \cup \{1, 2, 6, 7\}.$$

Call the two sets R and B, respectively. Then $x \sim y$ in G if and only if $x - y \in R$. It follows that two vertices are joined in $\overline{G}$ if and only if their difference is in B. Observe that both R and B contain the additive inverses of all their elements; this is important because the differences $x - y$ and $y - x$ both correspond to the same edge xy. (This property might be relaxed for some applications to directed graphs.)

Notice that R contains no solution to the equation

$$a + b = c;$$

no element of R equals the sum of two elements of R. We say R is a *sum-free set*. By contrast, B is not sum-free; not only is $1 + 6 = 7$, but also $1 + 1 = 2$ (a, b and c need not be distinct). If xyz were a triangle in G, then $x - y$, $y - z$ and $x - z$ would all be members of R; but

$$x - y + y - z = x - z,$$

so R would not be sum-free.

In general, a non-empty subset S of $\mathbb{Z}_n$ is a *sum-free set* if there never exist elements a, b of S such that $a + b \in S$. (This means that 0 cannot belong to S, since $0 + 0 = 0$.) S is *symmetric* if $-x \in S$ whenever x is in S. A *symmetric sum-free partition* of G is a partition of $\mathbb{Z}_n^*$ into symmetric sum-free sets. As examples,

$$\begin{array}{rcl}
\mathbb{Z}_5^* & = & \{1, 4\} \cup \{2, 3\} \\
\mathbb{Z}_9^* & = & \{3, 4, 5\} \cup \{1, 7\} \cup \{2, 6\}
\end{array}$$

are symmetric sum-free partitions.

If S is a symmetric sum-free set in $\mathbb{Z}_n$, the *graph of* S is the graph with vertex-set G, where x and y are adjacent if and only if $x - y \in S$. From our earlier discussion, it follows that

THEOREM 4.6 (Graphs of Sum-Free Sets)
If S is a symmetric sum-free set of order s in $\mathbb{Z}_n$, then the graph of S is a triangle-free regular graph of degree s on n vertices.

If there is a symmetric sum-free partition $S_1 \cup S_2 \cup \cdots \cup S_k$ of $\mathbb{Z}_n^*$, then one obtains a k-coloring of K_n that contains no monochromatic triangle, by applying c_i to all the edges of the graph of S_i for $i = 1, 2, \ldots, k$. So

COROLLARY 4.6.1
If there exists a sum-free partition of an n-element group into k parts, then $R_k(3) > n$.

For example, the partition

$$\mathbb{Z}_5^* = \{1, 4\} \cup \{2, 3\}$$

provides the well-known partition of $\mathbb{Z}_5$ into two 5-cycles that is used in proving that $R_2(3) = 6$. The partition

$$\mathbb{Z}_8^* = \{3, 4, 5\} \cup \{1, 7\} \cup \{2, 6\}$$

yields a (not very interesting) triangle-free three-coloring of K_8.

Students who have studied a little abstract algebra will realize that all of this can be applied to abelian groups in general. (If you don't know what this means, skip to the end of this section.) There are two abelian groups of order 16 that have symmetric sum-free partitions. The group $\mathbb{Z}_4 \times \mathbb{Z}_4$ is the set of all ordered pairs xy where both x and y come from $\{0, 1, 2, 3\}$ and

$$xy + zt = (x + z)(y + t)$$

(additions modulo 4). Then

$$(\mathbb{Z}_4 \times \mathbb{Z}_4)^* = R \cup B \cup G$$

where

$$
\begin{aligned}
R &= \{02, 10, 30, 11, 33\}, \\
B &= \{20, 01, 03, 13, 31\}, \\
G &= \{22, 21, 23, 12, 32\}.
\end{aligned}
\tag{4.2}
$$

$(\mathbb{Z}_2 \times \mathbb{Z}_2 \times \mathbb{Z}_2 \times \mathbb{Z}_2)^*$ has a similar partition

$$
\begin{aligned}
R &= \{1000, 1100, 1010, 1111, 0001\}, \\
B &= \{0010, 0011, 1011, 0111, 1101\}, \\
G &= \{0100, 0110, 0101, 1110, 1001\}.
\end{aligned}
\tag{4.3}
$$

The existence of these partitions proves of course that $R_3(3) \geq 17$. To see that $R_3(3) = 17$, we use the following argument. Suppose K_{17} could be colored in three colors. Select any vertex x. Since there are 16 edges incident with x, there must be at least six in one color, red say. Consider the subgraph generated by the other endpoints of those edges. If it has a red edge, then there is a red triangle; if not, the subgraph is a K_6 colored in the two remaining colors, and it must contain a monochromatic triangle.

The above argument can be used to show that $R_4(3) \leq 66$, but there is no sum-free partition of a 65-element group into four parts. In fact, we know that $51 \leq R_4(3) \leq 65$ ([36, 125]). The lower bound was proven by exhibiting a triangle-free coloring of K_{50}, while the upper bound comes from a lengthy argument proving that if a triangle-free four-coloring of K_{65} existed, then the adjacency matrices of the monochromatic subgraphs would have eigenvalues of irrational multiplicities.

The method of sum-free sets can be generalized to avoid larger complete subgraphs. For example, consider the subset $B = \{1, 2, 6, 7\}$ of K_8 that arose in discussing $R(3, 4)$. This set is not sum-free, and its graph will contain triangles. However, suppose there were a K_4 in the graph, with vertices a, b, c and d. Then B would contain a solution to the following system of three simultaneous equations in six unknowns:

$$\begin{aligned}
x_{ab} \quad + \quad x_{bc} \quad\quad\quad &= \quad x_{ac} \\
x_{ac} \quad\quad\quad + \quad x_{cd} &= \quad x_{ad} \\
x_{bc} \quad + \quad x_{cd} &= \quad x_{bd}
\end{aligned}$$

(in each case, x_{ij} will be either $i - j$ or $j - i$). But a complete search shows that B contains no solution to these equations. So the graph contains no K_4. (The graph is $\overline{G}$ in Figure 4.2.)

4.6   Bounds on Ramsey Numbers

Very few Ramsey numbers are known. Consequently, much effort has gone into proving upper and lower bounds.

LEMMA 4.7
If p and q are integers greater than 2, then

$$R(p,q) \le R(p-1,q) + R(p,q-1).$$

Proof. Write $m = R(p-1,q) + R(p,q-1)$. Suppose the edges of K_m are colored in red and blue. We shall prove that K_m contains either a red K_p or a blue K_q. Two cases arise.

(i) Suppose that one of the vertices x of K_m has at least $s = R(p-1,q)$ red edges incident with it, connecting it to vertices $x_1, x_2, \ldots, x_s$. Consider the K_s on these vertices. Since its edges are colored red or blue, it contains either a blue K_q, in which case the lemma is proved, or a red K_{p-1}. Let the set of vertices of the red K_{p-1} be $\{y_1, y_2, \ldots, y_{p-1}\}$. Then the vertices $x, y_1, \ldots, y_{p-1}$ are those of a red K_p and again the lemma holds.

(ii) Suppose that no vertex of K_m has $R(p-1,q)$ red edges incident with it. Then every vertex must be incident with at least $m - 1 - [R(p-1,q)-1] = R(p,q-1)$ blue edges. The argument is then analogous to that of part (i). $\square$

THEOREM 4.8 (Upper Bound)
For all integers, $p, q \ge 2$,

$$R(p,q) \le \binom{p+q-2}{p-1}.$$

Proof. Write $n = p + q$. The proof proceeds by induction on n. Clearly, $R(2,2) = 2 = \binom{2+2-2}{2-1}$. Since $p, q \ge 2$, we can have $n = 4$ only if $p = q = 2$.

Hence the given bound is valid for $n = 4$. Also for any value of q, $R(2, q) = q = \binom{2+q-2}{2-1}$, and similarly for any value of p, $R(p, 2) = p = \binom{p+2-2}{p-1}$, so the bound is valid if $p = 2$ or $q = 2$.

Without loss of generality assume that $p \geq 3$, $q \geq 3$ and that

$$R(p', q') \leq \binom{p' + q' - 2}{p' - 1}$$

for all integers p', q' and n satisfying $p' \geq 2$, $q' \geq 2$, $p' + q' < n$ and $n > 4$. Suppose the integers p and q satisfy $p + q = n$.

We apply the induction hypothesis to the case $p' = p - 1$, $q' = q$, obtaining

$$R(p - 1, q) \leq \binom{p + q - 3}{p - 2},$$

and to $p' = p$, $q' = q - 1$, obtaining

$$R(p, q - 1) \leq \binom{p + q - 3}{p - 1}.$$

But by the properties of binomial coefficients,

$$\binom{p + q - 3}{p - 2} + \binom{p + q - 3}{p - 1} = \binom{p + q - 2}{p - 1},$$

and from Lemma 4.7

$$R(p, q) \leq R(p - 1, q) + R(p, q - 1),$$

so

$$R(p, q) \leq \binom{p + q - 3}{p - 2} + \binom{p + q - 3}{p - 1} = \binom{p + q - 2}{p - 1}. \qquad \square$$

If $p = 2$ or $q = 2$ or if $p = q = 3$, this bound is exact. But suppose $p = 3$, $q = 4$. Then $\binom{p+q-2}{p-1} = \binom{5}{2} = 10$, and the exact value of $R(3, 4)$ is 9. Again if $p = 3$, $q = 5$, then $\binom{p+q-2}{p-1} = \binom{6}{2} = 15$, whereas the exact value of $R(3, 5)$ is 14. In general, Theorem 4.8 shows that

$$R(3, q) \leq \binom{q + 1}{2} \frac{q(q + 1)}{2} = \frac{q^2 + q}{2}.$$

But for the case $p = 3$, this result can be improved. It may be shown that for every integer $q \geq 2$,

$$R(3, q) \leq \frac{q^2 + 3}{2}.$$

The following lower bound for $R_n(k)$ was proved by Abbott [1].

THEOREM 4.9 (Lower Bound)
For integers $s, t \geq 2$,

$$R_n(st - s - t + 2) \geq (R_n(s) - 1)(R_n(t) - 1) + 1.$$

Proof. Write $p = R_n(s) - 1$ and $q = R_n(t) - 1$. Consider a K_p on vertices $x_1, x_2, \ldots, x_p$ and a K_q on vertices $y_1, y_2, \ldots, y_q$. Color the edges of K_p and K_q in n colors $c_1, c_2, \ldots, c_n$ in such a way that K_p contains no monochromatic K_s and K_q contains no monochromatic K_t (such colorings must be possible by the definitions of p and q).

Now let K_{pq} be the complete graph on the vertices w_{ij}, where $i \in 1, 2, \ldots, p$ and $j \in 1, 2, \ldots, q$. Color the edges of K_{pq} as follows:

(i) edge $w_{gj} w_{gh}$ is given the color $y_j y_h$ received in K_q;

(ii) if $i \neq g$, $w_{ij} w_{gh}$ is given the color $y_j y_h$ received in K_p.

Now write $r = st - s - t + 2$ and let G be any copy of K_r contained in K_{pq}. Suppose G is monochromatic, with all its edges colored c_1. Two cases arise:

(i) There are s distinct values of i for which the vertex w_{ij} belongs to G. Then from the coloring scheme K_p contains a monochromatic K_s, which is a contradiction;

(ii) There are at most $s - 1$ distinct values of i for which w_{ij} belongs to G. Suppose there are at most $t - 1$ distinct values of j such that w_{ij} belongs to G. Then G has at most $(s-1)(t-1) = st - s - t + 1 = r - 1$ vertices, which is a contradiction. So there is at least one value of i such that at least t of the vertices w_{ij} belong to G. Applying the argument of case (i), K_q contains a monochromatic K_t, which is again a contradiction.

Thus, K_{pq} contains no monochromatic K_r. □

In order to develop the ideas of sum-free sets and obtain some bounds for $R_n(3)$, we define the *Schur function*, $f(n)$, to be the largest integer such that the set

$$1, 2, \ldots, f(n)$$

can be partitioned into n mutually disjoint non-empty sets $S_1, S_2, \ldots, S_n$, each of which is sum-free. Obviously $f(1) = 1$, and $f(2) = 4$ where $\{1, 2, 3, 4\} = \{1, 4\} \cup \{2, 3\}$ is the appropriate partition. Computations have shown that $f(3) = 13$ with

$$\{1, 2, \ldots, 13\} = \{3, 2, 12, 11\} \cup \{6, 5, 9, 8\} \cup \{1, 4, 7, 10, 13\}$$

as one possible partition, that $f(4) = 44$ and $f(5) \geq 138$.

LEMMA 4.10

For any positive integer n

$$f(n+1) \geq 3f(n) + 1,$$

and since $f(1) = 1$,

$$f(n) \geq \frac{3^n - 1}{2}.$$

Proof. Suppose that the set $S = \{1, 2, \ldots, f(n)\}$ can be partitioned into the n sum-free sets $S_1 = \{x_{11}, x_{12}, \ldots, x_{1\ell_1}\}, \ldots, S_n = \{x_{n1}, x_{n2}, \ldots, x_{n\ell_n}\}$. Then the sets

$$
\begin{aligned}
T_1 &= \{x_{11}, 3x_{11} - 1, 3x_{12}, 3x_{12} - 1, \ldots, 3x_{1\ell_1}, 3x_{1\ell_1} - 1\}, \\
T_2 &= \{x_{21}, 3x_{21} - 1, 3x_{22}, 3x_{22} - 1, \ldots, 3x_{2\ell_1}, 3x_{2\ell_1} - 1\}, \\
&\cdots \\
T_n &= \{3x_{n1}, 3x_{n1} - 1, 3x_{n2}, 3x_{n2} - 1, \ldots, 3x_{n\ell_n}, 3x_{n\ell_n} - 1\}, \\
T_{n+1} &= \{1, 4, 7, \ldots, 3f(n) + 1\}
\end{aligned}
$$

form a partition of $\{1, 2, \ldots, 3f(n) + 1\}$ into $n + 1$ sum-free sets. So

$$f(n+1) \geq 3f(n) + 1.$$

Now $f(1) = 1$, so the above equation implies that

$$f(n) \geq 1 + 3 + 3^2 + \cdots + 3^{n-1} = \frac{3^n - 1}{2}. \qquad \square$$

THEOREM 4.11 (Bounds for Triangles)

For any positive integer

$$\frac{3^n + 3}{2} \leq R_n(3) \leq n(R_{n-1}(3) - 1) + 2.$$

Proof. (i) The proof of the upper bound is a generalization of the method used to establish $R_3(3) \leq 17$.

(ii) Let $K_{f(n)+1}$ be the complete graph on the $f(n) + 1$ vertices $x_0, x_1, \ldots, x_{f(n)}$. Color the edges of $K_{f(n)+1}$ in n colors by coloring $x_i x_j$ in the kth color if and only if $|i - j| \in S_k$.

Suppose the graph contains a monochromatic triangle. This must have vertices x_a, x_b, x_c with a.b.c, such that $a - b$, $b - c$, $a - c \in S_k$. But now $(a - b) + (b - c) = a - c$, contradicting the fact that S_k is sum-free. Hence

$$f(n) + 1 \leq R_n(3) - 1,$$

so that

$$\frac{3^n - 1}{2} + 2 = \frac{3^n - 3}{2} \leq R_n(3),$$

which proves the lower bound. $\qquad \square$

4.7 The General Form of Ramsey's Theorem

THEOREM 4.12 (Ramsey's Theorem)
Suppose S is an s-element set. Write $\Pi_r(S)$ for the collection of all r-element subsets of S, $r \geq 1$. Suppose the partition

$$\Pi_r(S) = A_1 \cup A_2 \cup \cdots \cup A_n$$

is such that each r-subset of S belongs to exactly one of the A_i, and no A_i is empty. If the integers $p_1, p_2, \ldots, p_n$ satisfy $r \leq p_i \leq s$, for $i = 1, 2, \ldots, n$, then there exists an integer

$$R(p_1, p_2, \ldots, p_n; r),$$

depending only on $n, p_1, p_2, \ldots, p_n$ and r, such that if $s \geq R(p_1, p_2, \ldots, p_n; r)$, then for at least one i, $1 \leq i \leq n$, there exists a p_i-element subset T of S, all of whose r-subsets belong to A_i.

The proof is by induction.

Obviously the pigeonhole principle is case $r = 1$ of Ramsey's Theorem. The case $r = 2$ is the graphical case that we have been discussing; instead of $R(p_1, p_2, \ldots, p_n, 2)$ we just write $R(p_1, p_2, \ldots, p_n)$.

Exercises 4A

1. How many times must we throw a die in order to be sure that we get the same score at least twice?

2. Let $a_1, a_2, \ldots, a_n$ be n (not necessarily distinct) integers. Show that there exist integers s and t such that $1 \leq s \leq t \leq n$ and $a_s + a_{s+1} \cdots + a_t \equiv 0 \pmod{n}$.

3. A night club's patron parking lot contains 25 cars, and there are 101 patrons in the night club. Assume all of the patrons arrived by car; find the greatest number k such that some car must have contained at least k people.

4. An accountant returns from vacation with exactly 10 calendar weeks to go before the April 15th deadline to file her clients' income tax statements. She decides to complete at least one tax return per day during

the five-day work week, but no more than eight tax returns in any calendar week. Show that there is some set of consecutive work days during which she completes exactly 19 tax returns.

5. A student in an English class is expected to turn in journal entries. The class meets four times a week, and the student wants to turn in at least one journal entry per class day, but no more than six journal entries during a calendar week for the entire 15-week semester. Show that there exists some sequence of consecutive days during which the student turns in exactly 29 journal entries.

6. Let $S = \{1, 2, \ldots, n\}$, let r be an integer not belonging to S and let $T = S \cup \{r\}$. Show that there exists $x \in S$ such that the sum of all the elements of $T \backslash \{x\}$ is divisible by n.

7. Let p be a prime, $p \neq 2$, $p \neq 5$. Show that there exists a positive power of p which, written in the decimal system, ends $\ldots 01$.

8. Let x be any positive real number and let n be any positive integer. Show that there exist positive integers h and k such that

$$h - \frac{1}{n} \leq kx \leq h + \frac{1}{n}.$$

(That is, show that some integral multiple of x is arbitrarily close to an integer.)

9. What is the relation between $R(p_1, p_2, p_3, \ldots, p_n; r)$ and $R(p_2, p_1, p_3, \ldots, p_n; r)$?

10. Prove that $R(3, 5) \leq 14$.

11. Prove that $R(3, 5) > 13$ by choosing $R = \{4, 6, 7, 9\}$, $B = \{1, 2, 3, 5, 8, 10, 11, 12\}$ in Z_{13}. (*Note:* Together with the preceding exercise, this proves that $R(3, 5) = 14$.)

12. Verify that the partition in (4.2) is in fact a symmetric sum-free partition.

13. Is it possible to 2-color the edges of K_{35} so that no red K_4 or blue K_5 occurs?

Exercises 4B

1. How many times must we throw two dice in order to be sure that we get the same total score at least six times?

2. Let n and B be given positive integers. Show that there exists an integer divisible by n that contains only the digits 0 and 1 when it is written in base B.

3. Let $A = \{a_1, a_2, \ldots, a_5\}$ be a set of five distinct positive integers, none of which exceeds 9. Show that the sums of the non-empty subsets of A cannot all be distinct.

4. Let $a_1, a_2, \ldots, a_t, t > n^2$, be a sequence of distinct real numbers. Show that this sequence contains a monotonic subsequence of more than n terms.

5. Suppose we have a 6×6 chessboard and 18 dominoes, such that each domino can just cover two squares of the chessboard with an edge in common. The 10 lines (five horizontal, five vertical) that divide the board into squares are called *rulings*. Show that in any arrangement of dominoes that precisely covers the chessboard, there is at least one ruling crossed by no domino.

6. In September, October and November, a student solves at least one problem per day but not more than 13 problems per week. Show that she solves exactly 12 problems on some set of consecutive whole days.

7. Let $a_1, a_2, \ldots, a_{n+1}$ be $n+1$ distinct positive integers. Show that there exist two of them, a_i and a_j, such that n divides $a_i - a_j$.

8. Verify that, for any $p \geq 2$ and any $q \geq 2$,

$$R(2, q) = q, \qquad R(p, 2) = p.$$

9. Prove by construction (not using Theorem 4.4) that any edge-coloring of K_6 in two colors must contain at least two monochromatic triangles. (*Hint:* Look at the monochromatic subgraphs.)

10. Prove that any edge-coloring of K_7 in two colors must contain at least three monochromatic triangles.

11. Prove that $R(4, 4) \leq 18$.

12. Suppose there exists a painting of K_{17} in the three colors red, blue and green that contains two or fewer monochromatic triangles. If v is any vertex, write $R(v)$, $B(v)$ and $G(v)$ for the sets of vertices joined to v by red, blue and green edges respectively, and write $r(v) = |R(v)|$, and so on.

   (i) Select a vertex x that lies in no monochromatic triangle. Prove that one of $\{r(x), b(x), g(x)\}$ is equal to 6 and the other two each equal 5.

(ii) Without loss of generality, say $r(x) = 6$. Let S be the set of all vertices of K_{17} that lie in monochromatic triangles. Prove that $S \subseteq R(x)$.

(iii) If y is any member of $B(x)$, it is clear that S lies completely within $R(y)$, $B(y)$ or $G(y)$. Prove that, in fact, $S \subseteq R(y)$.

(iv) Prove that there must exist two vertices y_1 and y_2 in $B(x)$ such that $y_1 y_2$ is red.

(v) Use the fact that $S \subseteq R(y_1) \cap R(y_2)$ to prove that K_{17} contains more than two red triangles. So $N_{3,17}(3) \geq 3$.

13. Verify that the partition in (4.3) is in fact a symmetric sum-free partition.

14. Show that $R(4, 4) > 17$, by choosing

$$R = \{1, 2, 4, 8, 9, 13, 15, 16\}, \quad B = \{3, 5, 6, 7, 10, 11, 12, 14\}$$

in Z_{17}. (*Note:* Together with Exercise 11, this proves that $R(4, 4) = 18$.)

15. Is it possible to 2-color the edges of K_{25} so that no monochromatic K_5 occurs?

Problems

Problem 1:

(i) Let S be an s-set, where $s > nm$. Show that if S is partitioned into n subsets, then at least one of the subsets contains more than m elements.

(ii) Let $a_1, a_2, \ldots, a_n$ be real numbers such that $\sum_{i=1}^{n} a_i > b$. Show that for at least one value of i, $i = 1, \ldots, n$, $a_i \geq b/n$.

(iii) Let $a_1, a_2, \ldots, a_n$ be real numbers such that $\sum_{i=1}^{n} a_i \leq b$. Show that for at least one value of i, $i = 1, \ldots, n$, $a_i \leq b/n$.

(iv) Let $a_1, a_2, \ldots, a_n$ be real numbers such that $\sum_{i=1}^{n} a_i < b$. Show that for at least one value of i, $i = 1, \ldots, n$, $a_i < b/n$.

Problem 2: Suppose N objects are distributed between B boxes in such a way that no box is empty. How many objects must we choose in order to be sure that we have chosen the entire contents of at least one box?

Problem 3: There are 12 signs of the (Western) Zodiac, corresponding roughly to which month a person is born in. Similarly, there are twelve animal signs of the so-called "Chinese Zodiac," (Rat, Ox, Tiger, Rabbit, Dragon, Snake, Horse, Goat, Monkey, Rooster, Dog, and Boar) corresponding roughly to which year a person is born in. Suppose there are 145 people in a room. Show that:

(i) There must be 13 people who share the same sign of the Western Zodiac;

(ii) There must be 13 people who share the same sign of the Chinese Zodiac;

(iii) There must be at least two people who have the same sign in both Zodiacs.

Problem 4: Let p be a given prime and let $R_0, R_1, \ldots, R_{p-1}$ be the residue classes modulo p. Let $A = \{a_1, a_2, \ldots, a_n\}$ be a set of integers, and define a partition of A into sets $A_0, A_1, \ldots, A_{p-1}$ by

$$A_i = \{a_{i1}, a_{i2}, \ldots, a_{ik_i}\} = A \cap R_i, 0 \le i \le p-1.$$

(Obviously, $k_i = |A_i|$.) What is a necessary and sufficient condition on n and p that $k_i = k_j$ for at least one pair of distinct integers i and j, where $0 \le i, j \le p-1$?

Problem 5: (For those who have studied a little abstract algebra.) Let $a_1, a_2, \ldots, a_n$ be n (not necessarily distinct) elements of a group of order n written multiplicatively. Show that there exist integers s and t such that $1 \le s \le t \le n$ and $a_s a_{s+1} \ldots a_t = 1$.

Problem 6: Choose $n^2 + 1$ closed intervals on the real line. Prove that at least one of the following is true:

(i) $n + 1$ of the intervals have a common point;

(ii) $n + 1$ of the intervals are pairwise disjoint.

Problem 7: Find a three-coloring of the edges of $K_{2R(m,n)-2}$ so that there is no K_3 of color c_1, no K_m of color c_2, and no K_n of color c_3; find the bound on $R(3, m, n)$ obtained from this coloring.

Problem 8: Theorem 4.4 uses the quantity

$$\left\lfloor \frac{n}{2} \left\lfloor \left(\frac{n-1}{2} \right)^2 \right\rfloor \right\rfloor.$$

For which n does this not simply equal $\left\lfloor \frac{n(n-1)^2}{8} \right\rfloor$?

Problem 9: Prove Theorem 4.12.

Chapter 5

The Principle of Inclusion and Exclusion

5.1 Unions of Events

The addition principle could be restated as follows: if X and Y are two disjoint sets then
$$|X \cup Y| = |X| + |Y|.$$

But suppose X and Y are *any* two sets, not necessarily disjoint, and you need to list all members of $X \cup Y$. If you list all members of X, then list all the members of Y, you certainly cover all of $X \cup Y$, but the members of $X \cap Y$ are listed twice. To fix this, you could count all members of both lists, then subtract the number of duplicates:

$$|X \cup Y| = |X| + |Y| - |X \cap Y|. \tag{5.1}$$

A Venn diagram illustrating this equation is shown in Figure 5.1. To enumerate $X \cup Y$, first count all elements in the shaded areas and then subtract the number of elements in the heavily shaded area.

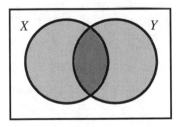

FIGURE 5.1: Enumerating $X \cup Y$ using (5.1).

If X is the collection of all the objects of type I in some universal set and Y consists of all objects of type II, then (5.1) expresses the way to count the objects that are either of type I or type II:

(i) count the objects of type I;

(ii) count the objects of type II;

(iii) count the objects that are of both types;

(iv) add answers (i) and (ii) and then subtract answer (iii).

Example 5.1: The makers of Red Hat licorice put coupons in packets of their product. Each coupon is an anagram of REDHAT, and if your anagram contains either of the words RED or HAT embedded, you win a free pack of red licorice hats. For example, HREDTA is a winner; RHEADT is not. The number of different coupons possible is 6!, or 720. How many winning anagrams are there?

To calculate the number of winners, we count two sets, R and H. The collection R of anagrams containing RED can be enumerated as follows: they are equivalent to ways of organizing the four symbols H, A, T and RED (in other words, treat "RED" as one symbol). So $|R| = 4! = 24$. Similarly, H, the set of those containing "HAT," has 24 elements. The intersection, $R \cap H$, has only two elements, REDHAT and HATRED. (Unfortunate, that second one.) From (5.1),

$$|R \cap H| = |R| + |H| - |R \cap H| = 24 + 24 - 2 = 46.$$

So there are 46 winning anagrams. The percentage of winning anagrams is about 6.4%. □

Another rule, sometimes called the *rule of sum*, says that the number of objects of type I equals the number that are both of type I and of type II, plus the number that are of type I but not of type II; in terms of the sets X and Y, this is

$$|X| = |X \cap Y| + |X \backslash Y|. \tag{5.2}$$

For example, suppose Dr. John Walters teaches a course on C^{++} and a course on operating systems. There are 26 students who take both courses, and 20 who take the C^{++} course but not the other. Obviously there are 46 students taking C^{++}. This illustrates Equation (5.2): if X is the C^{++} class list and Y is the OS class list, then

$$\begin{aligned} 46 &= |X| &= |X \cap Y| &+ |X \backslash Y| \\ &= 26 &+ 20. \end{aligned}$$

The two rules (5.1) and (5.2) can be combined to give

$$|X \cup Y| = |Y| + |X \backslash Y|. \tag{5.3}$$

Figure 5.2 illustrates the last two rules. In the left-hand diagram, $X \backslash Y$ is dark and $X \cap Y$ light, and (5.2) tells us to enumerate T by adding those two parts. In the right-hand diagram, Y is dark and $X \backslash Y$ light, and from (5.3) we count the contents of the shaded set $X \cup Y$ by adding those two parts.

We now extend the above work to three sets. In Equation (5.1), replace Y by a union, $Y \cup Z$. Then

$$|X \cup Y \cup Z| = |X| + |Y \cup Z| - |X \cap (Y \cup Z)|.$$

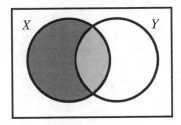 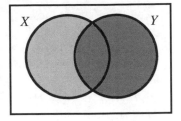

FIGURE 5.2: Illustration of (5.2) (left) and (5.3) (right).

The term $|Y \cup Z|$ can be replaced by

$$|Y| + |Z| - |Y \cap Z|;$$

as a consequence of the distributive laws for intersections and unions of sets, $X \cap (Y \cup Z) = (X \cap Y) \cup (X \cap Z)$, and the last term can be written as

$$|X \cap Y| + |X \cap Z| - |X \cap Y \cap Z|.$$

So the whole expression is

$$
\begin{aligned}
|X \cup Y \cup Z| \;=\; & |X| + |Y| + |Z| \\
& -|X \cap Y| - |Y \cap X| - |X \cap Z| \\
& +|X \cap Y \cap Z|.
\end{aligned}
$$

5.2   The Principle

This formula can be generalized:

THEOREM 5.1 (Principle of Inclusion and Exclusion)
The number of elements in $X_1 \cup X_2 \cup \cdots \cup X_n$ is

$$
\begin{aligned}
& |X_1 \cup X_2 \cup \cdots \cup X_n| \\
=\; & \sum_{i_1=1}^{n} |X_{i_1}| - \sum_{i_1=1}^{n} \sum_{i_2=i_1+1}^{n} |X_{i_1} \cap X_{i_2}| + \cdots \\
& +(-1)^{k-1} \sum_{i_1=1}^{n} \sum_{i_2=i_1+1}^{n} \cdots \sum_{i_k=i_{k-1}+1}^{n} |X_{i_1} \cap X_{i_2} \cap \cdots \cap X_{i_k}| \\
& + \cdots + (-1)^{n-1} |X_{i_1} \cap X_{i_2} \cap \cdots \cap X_{i_n}|.
\end{aligned}
\tag{5.4}
$$

Proof. We can think of the process as:
1. Count the elements of all the sets (with repetitions), $\sum |X_{i_1}|$.
2. Subtract all elements that were counted at least twice
   (that is, count intersections of two sets), $\sum\sum |X_{i_1} \cap X_{i_2}|$.
3. Add in all elements that were counted at least three times
   (that is, count intersections of three sets), $\sum\sum\sum |X_{i_1} \cap X_{i_2} \cap X_{i_3}|$.
   $\cdots$

Now suppose an object belongs to precisely r of the sets, say sets $X_{j_1}, X_{j_2}, \ldots,$ X_{j_r}. The element is counted r times in calculating $\sum |X_{i_1}|$, and contributes r to that part of the sum (5.4). It appears $\binom{r}{2}$ times in subsets of the form $X_{i_1} \cap X_{i_2}$, since it appears in precisely those for which $\{i_1, i_2\}$ is a 2-set of $\{j_1, j_2, \ldots, j_r\}$; so it contributes $-\binom{r}{2}$ to that part of (5.4). Continuing, we obtain total contribution

$$\sum_{k=1}^{r} (-1)^{k-1} \binom{r}{k}, \tag{5.5}$$

which equals 1 by Corollary 2.3.3 (see Exercise 2A.14). So (5.4) counts 1 for every member of $X_1 \cup X_2 \cup \cdots \cup X_n$; therefore its total is $|X_1 \cup X_2 \cup \cdots \cup X_n|$. $\square$

Example 5.2: The Red Hat licorice makers, following the success of their first contest, introduce "Red Hat 21," a casino-type game played with licorice hats. Again, they run a promotion; this time, the tickets are anagrams of REDHAT21, and the winning tickets must either contain RED or HAT or 21. For example, RH21EDTA and HA2RED1A are winners. How many winners are there this time? What percentage of tickets are winners?

We proceed as we did in Example 5.1, but this time we count three sets, R, H and T (for "twenty-one"). The anagrams containing RED are equivalent to ways of organizing the six symbols H, A, T, 2, 1 and RED, so $|R| = 6!$. Similarly H, the set of those containing "HAT," has 6! elements and T has 7! elements. The intersections are counted similarly: $R \cap H$, has 4! elements (RED, HAT, 2 and 1) while $|R \cap T| = |H \cap T| = 5!$. Finally, there are 3! elements in $R \cap H \cap T$, corresponding to the 3! ways of ordering the three words RED, HAT and 21. From (5.4), the number of elements of $R \cup H \cup T$ is

$$
\begin{aligned}
R \cup H \cup T \ = \quad & (|R| + |H| + |T|) \\
- \ & (|R \cap H| + |R \cap T| + |H \cap T|) \\
+ \ & (|R \cap H \cap T|) \\
= \quad & (6! + 6! + 7!) \\
- \ & (4! + 5! + 5!) \\
+ \ & (3!) \\
= \quad & (720 + 720 + 5040) - (24 + 120 + 120) - 6 \\
= \quad & 6480 - 264 + 6 \ = \ 6222.
\end{aligned}
$$

So there are 6,222 winners. There are $8! = 40320$ tickets, so about 15.4% are winners. □

Example 5.3: The following diagram represents a set of streets. How many ways can we walk from the southwest corner to the northeast corner of the following if all three of the streets labeled *b1*, *b2*, and *b3* are blocked due to construction?

Let X be the set of all walks from southwest to northeast. As in Example 2.2, we see that there are $|X| = C(12,5) = 792$ total walks. If we let X_i be the walks that pass through block *bi* for $i = 1, 2, 3$, we can compute $|X_1|$ (there are 210 of these, as discussed before), $|X_2|$ ($\binom{6}{3} \cdot \binom{5}{2} = 200$ walks), and $|X_3|$ ($C(8,4) \cdot C(3,1) = 210$ of these). Now, we count the intersections; there are $C(4,2) \cdot C(5,2)$ walks that pass through both *b1* and *b2*, or $|X_1 \cap X_2| = 60$ such walks. We similarly compute $|X_1 \cap X_3| = C(4,2) \cdot C(3,2) \cdot C(3,1)$ walks, or 54. This comes from counting walks from the southwest corner to the southwest endpoint of *b1*, from the northeast endpoint of *b1* to the southwest endpoint of *b3*, and from the northeast endpoint of *b3* to the northeast corner. Next, $|X_2 \cap X_3| = C(6,3) \cdot C(3,1)$ or 60 walks that pass through *b2* and *b3*.

Finally, we ask how many walks include all three forbidden blocks; there are $C(4,2) \cdot C(3,1)$ or 18, because we only count walks to *b1* and from *b3*. Then the number of walks that include none of these is $792 - (210 + 200 + 210) + (60 + 54 + 60) - 18 = 328$ distinct walks. □

Suppose all the sets X_i are subsets of some set S (that is, S is a universal set for the discussion). Using the general form of de Morgan's Laws, we get the following alternative form of (5.4):

$$
\begin{aligned}
&|\overline{X_1} \cap \overline{X_2} \cap \cdots \cap \overline{X_n}| \\
=\ &|S| - |X_1 \cup X_2 \cup \cdots \cup X_n| \\
=\ &|S| - \sum_{i_1=1}^{n} |X_{i_1}| + \sum_{i_1=1}^{n} \sum_{i_2=i_1+1}^{n} |X_{i_1} \cap X_{i_2}| + \cdots \\
&+ (-1)^k \sum_{i_1=1}^{n} \sum_{i_2=i_1+1}^{n} \cdots \sum_{i_k=i_{k-1}+1}^{n} |X_{i_1} \cap X_{i_2} \cap \cdots \cap X_{i_k}| \\
&+ \cdots + (-1)^n |X_{i_1} \cap X_{i_2} \cap \cdots \cap X_{i_n}|.
\end{aligned}
\tag{5.6}
$$

Suppose we want to know how many of the positive integers smaller than n are perfect squares, perfect cubes, or perfect higher powers. As an example we shall find out how many of the positive integers less than 200 are perfect powers higher than the first. Trivially, 1 is a perfect kth power for every k, so we omit 1 from our calculations, and add 1 to the answer. In this case we need only consider up to seventh powers: $2^8 = 256 > 200$, so every eighth or higher power of x is greater than 200 whenever $x \geq 2$. For $i = 2, 3, \ldots, 6$ let us write P_k for the set of all kth powers of integers that lie between 2 and 199 (inclusive). Then, from (5.4), the number of integers that belong to at least one of these sets is

$$
\begin{aligned}
& |P_2| + |P_3| + |P_4| + |P_5| + |P_6| + |P_7| \\
- \quad & (|P_2 \cap P_3| + |P_2 \cap P_4| + \ldots + |P_5 \cap P_7| + |P_6 \cap P_7|) \\
+ \quad & (|P_2 \cap P_3 \cap P_4| + |P_2 \cap P_3 \cap P_5| + \ldots + |P_5 \cap P_6 \cap P_7|) \\
- \quad & (|P_2 \cap P_3 \cap P_4 \cap P_5| + \ldots + |P_4 \cap P_5 \cap P_6|) \\
+ \quad & (|P_2 \cap P_3 \cap P_4 \cap P_5 \cap P_6| + \ldots + |P_3 \cap P_4 \cap P_5 \cap P_6 \cap P_7|) \\
- \quad & |P_2 \cap P_3 \cap P_4 \cap P_5 \cap P_6 \cap P_7|.
\end{aligned}
$$

All of these sets and their sizes are easy to calculate. The largest square up to 199 is 14^2, so $P_2 = \{2, 3, 4, \ldots, 14\}$ and $|P_2| = 13$, $P_3 = \{2, 3, 4, 5\}$ so $|P_3| = 4$, $|P_4| = 3$, $|P_5| = |P_6| = |P_7| = 1$. If $k > 7$ then P_k is empty. In general $P_i \cap P_j = P_l$, where l is the least common multiple of i and j, and similarly for intersections of three or more, so $P_2 \cap P_3 = P_2 \cap P_6 = P_3 \cap P_6 = P_2 \cap P_3 \cap P_4 = P_2 \cap P_3 \cap P_6 = P_6$, $P_2 \cap P_4 = P_4$, and all the others are empty. Therefore the total is

$$
\begin{aligned}
& |P_2| + |P_3| + |P_4| + |P_5| + |P_6| + |P_7| \\
- \quad & |P_2 \cap P_3| - |P_2 \cap P_4| - |P_2 \cap P_6| - |P_3 \cap P_6| \\
+ \quad & |P_2 \cap P_3 \cap P_4| + |P_2 \cap P_3 \cap P_6|
\end{aligned}
$$

which equals $13 + 4 + 3 + 1 + 1 + 1 - 1 - 3 - 1 - 1 + 1 + 1 = 19$. So the answer to the original question is 20 (adding 1 for the integer 1).

In some applications of Inclusion/Exclusion, all the sets X_i are the same size, as are all the intersections of two sets (like $X_{i_1} \cap X_{i_2}$), all the intersections of three sets, and so on. If the intersection of any k of the sets X_i always has s_k elements, then (5.4) becomes

$$
n s_1 - \binom{n}{2} s_2 + \binom{n}{3} s_3 - \cdots + (-1)^{n-1} s_n. \tag{5.7}
$$

This is called the *uniform form* of the Principle of Inclusion and Exclusion.

Example 5.4: A bank routing number is a string of nine digits. For the purposes of this example, let us assume that any digit can be used in any position, so that there are 10^9 possible routing numbers (the number of sequences length 9 on 10 symbols). How many routing numbers contain all the odd digits?

To calculate how many routing numbers contain all the odd digits, we find out how many *do not* contain all the odd digits, then subtract that number from 10^9. Write R_x for the set of all routing numbers that *do not* contain x. Then we want to calculate $|R_1 \cup R_3 \cup R_5 \cup R_7 \cup R_9|$. For any x, the number of routing numbers not containing x equals the number of sequences of length 9 on 9 symbols, so $|R_x| = 9^9$; similarly, if $x \neq y$ then $|R_x \cap R_y| = 8^9$; if x, y, z are all different then $|R_x \cap R_y \cap R_z| = 7^9$, and so on. So, from (5.7), the sum is

$$5 \times 9^9 - \binom{5}{2} \times 8^9 + \binom{5}{3} \times 7^9 - \binom{5}{4} \times 6^9 + 5^9$$

and the answer we require is

$$10^9 - 5 \times 9^9 + 10 \times 8^9 - 10 \times 7^9 + 5 \times 6^9 - 5^9. \qquad \square$$

5.3 Combinations with Limited Repetitions

Suppose we have a set S with n elements. Theorem 2.2 tells us that, if no repetitions are allowed, the number of ways to choose a collection of size k is

$$\binom{n}{k} = \frac{n!}{k!(n-k)!}. \tag{5.8}$$

Theorem 2.8 tells us that, if any object can be repeated as often as we wish, the number of ways to choose a collection of size k is

$$\binom{n+k-1}{k} = \binom{n+k-1}{n-1} = \frac{(n+k-1)!}{k!(n-1)!}. \tag{5.9}$$

Instead of repetitions, we could discuss *distinguishable* and *indistinguishable* elements. Suppose we have a set with infinitely many elements of type A and infinitely many elements of type B. We do not distinguish between elements of the same type. Then we would say there are three distinguishable subsets of size 2: AA (that is, a set containing two elements, both of type A), AB and BB. This is equivalent to saying there are three different ways to choose a 2-subset of $\{A, B\}$, with repetitions allowed. Then (5.9) tells us the number of distinguishable subsets of size k that can be chosen from a set with n types of element, and infinitely many of each type.

We now discuss the question, what if there are some indistinguishable elements, but only finitely many of some or all types?

Suppose a set contains a_1 elements of type A_1, a_2 elements of type A_2, $\ldots$, a_n elements of type A_n. We shall denote such a set

$$\{a_1 \circ A_1, a_2 \circ A_2, \ldots, a_n \circ A_n\}.$$

If there are infinitely many elements of type A_i, we write $\infty \circ A_i$. It is clear that, in determining the number of subsets of size k, there is no difference between an element that occurs k times and one that appears infinitely often.

Suppose you want to count all subsets of size k of the set $\{a_1 \circ A_1, a_2 \circ A_2, \ldots, a_n \circ A_n\}$, where $\sum a_i > k$. First, consider the collection X of all k-subsets of $S = \{\infty \circ A_1, \infty \circ A_2, \ldots, \infty \circ A_n\}$. From (5.9) there are $\binom{n+k-1}{n-1}$ such sets. Now write X_1 for the collection of such sets with more than a_1 elements A_1, X_2 for the collection of such sets with more than a_2 elements A_2, and so on. Then you want to count all sets belonging to none of the X_i, so the answer is

$$|\overline{X_1} \cap \overline{X_2} \cap \cdots \cap \overline{X_n}|.$$

From (5.6), this equals

$$|X| - \sum_{i_1=1}^{n} |X_{i_1}| + \sum_{i_1=1}^{n} \sum_{i_2=i_1+1}^{n} |X_{i_1} \cap X_{i_2}| + \cdots$$
$$+ (-1)^k \sum_{i_1=1}^{n} \sum_{i_2=i_1+1}^{n} \cdots \sum_{i_k=i_{k-1}+1}^{n} |X_{i_1} \cap X_{i_2} \cap \cdots \cap X_{i_k}|$$
$$+ \cdots + (-1)^n |X_{i_1} \cap X_{i_2} \cap \cdots \cap X_{i_n}|.$$

From (5.9), $|X| = \binom{n+k-1}{n-1}$. The set X_1 consists of all k-subsets of S with at least $a_1 + 1$ members of type A_1; these are equivalent to subsets of size $(k - a_1 - 1)$ (simply add $(a_1 + 1)$ more A_1s to each set), so they number $\binom{n+(k-a_1-1)-1}{n-1} = \binom{n+k-a_1-2}{n-1}$. The sets $X_2, \ldots, X_n$ are enumerated in the same way. Similarly, $X_1 \cap X_2$ consists of all the subsets with at least $(a_1 + 1)$ A_1s and $(a_2 + 1)$ A_2s, and these are derived by adding a_1 A_1s and a_2 A_2s to subsets of size $(k - a_1 - a_2 - 2)$, so $|X_1 \cap X_2| = \binom{n+k-a_1-a_2-3}{n-1}$, and so on.

Example 5.5: How many subsets of size 10 can be selected from the set

$$\{4 \circ A_1, 3 \circ A_2, 6 \circ A_3\}?$$

Following the preceding notation, $|X| = \binom{3+10-1}{2} = \binom{12}{2} = 66$, $|X_1| = \binom{7}{2} = 21$, $|X_2| = \binom{8}{2} = 28$, $|X_3| = \binom{5}{2} = 10$, $|X_1 \cap X_2| = \binom{3}{2} = 3$, and $|X_1 \cap X_3| = |X_2 \cap X_3| = |X_1 \cap X_2 \cap X_3| = 0$. So the answer is $66 - (21 + 28 + 10) + 3 - 0 = 10$. $\square$

We chose a small example so that the answer could be checked by hand. If pqr denotes the subset $\{p \circ A_1, q \circ A_2, r \circ A_3\}$, the solutions are

$$433, \ 424, \ 415, \ 406, \ 334, \ 325, \ 316, \ 235, \ 226, \ 136.$$

The same technique applies when there are infinitely many elements of some or all types.

Example 5.6: Solve Example 4.5 when you also have unlimited elements of type A_4.

Now $n = 4$, so $|X| = \binom{13}{3} = 286$, $|X_1| = \binom{8}{3} = 56$, $|X_2| = \binom{9}{3} = 84$, $|X_3| = \binom{6}{3} = 20$, $|X_1 \cap X_2| = \binom{4}{3} = 4$, and $|X_1 \cap X_3| = |X_2 \cap X_3| = |X_1 \cap X_2 \cap X_3| = 0$. (As type A_4 is unrestricted, there is no set X_4.) So the answer to this new question is $286 - (56 + 84 + 20) + 4 - 0 = 122$. Even in this small case, hand checking is already tedious. □

Just as in the restricted case, there is an application to integer equations.

Example 5.7: How many integer solutions are there to the equation $x + y + z = 14$, subject to $3 \le x \le 7, 1 \le y \le 4, 0 \le z \le 6$?

Define $p = x - 3, q = y - 1, r = z$. Then the conditions become $p + q + r = 10$, $0 \le p \le 4, 0 \le q \le 3, 0 \le r \le 6$, and the problem is equivalent to Example 4.5; the answer is 10.

Example 5.8: Enumerate the solutions to $x + y + z \le 14$ under the same restrictions.

With the same substitutions, the equation becomes $p + q + r \le 10$, or equivalently $p + q + r + s = 10$, where s is any non-negative integer. The answer, 122, follows from Example 4.6. □

5.4 Derangements

A *derangement* of an ordered set of objects is a way of rearranging the objects so that none appears in its original position. A derangement of $(1, 2, \ldots, n)$, for example, is an arrangement $(x_1, x_2, \ldots, x_n)$ of the first n integers in which $x_i = i$ never occurs. We write D_n for the number of possible derangements of n objects; clearly D_n does not depend on the type of objects in the set.

For consistency, we define $D_0 = 1$. The first few values of D_n are easy to calculate. For example, $D_1 = 0$ (how can you put the members of a 1-element set in a different order?) and $D_2 = 1$. The six permutations of 123 are 123, 132, 213, 231, 312, 321, and the only derangements are 231 and 312, so $D_3 = 2$. But there are nine derangements of four objects, and the number grows rapidly.

One expression for D_n is derived from the Principle of Inclusion and Exclusion.

THEOREM 5.2 (The Derangement Series)

For every positive integer n,

$$D_n = n![1 - \frac{1}{1!} + \frac{1}{2!} - \frac{1}{3!} + \cdots + (-1)^n \frac{1}{n!}]. \qquad (5.10)$$

Proof. We write X_i for the set of all arrangements $(x_{1,2}, \ldots, n)$ of $(1, 2, \ldots, n)$ in which $\alpha_i = i$. In order to use Inclusion/Exclusion we enumerate X_i for any i, then $X_i \cap X_J$, and so on.

Each arrangement with $x_1 = 1, x_2 = 2, \ldots,$ and $x_i = i$ can be derived from an arrangement of $(i + 1`, i + 2, \ldots, n)$ by putting $(1, 2, \ldots, i)$ in front of it, and each arrangement of $(i + 1`, i + 2, \ldots, n)$ gives precisely one member of X_1 in this way, so

$$|X_1 \cap X_2 \cap \cdots \cap X_i| = (n - i)!.$$

Obviously the intersection of any i sets has the same size, and there are $\binom{n}{j}$ ways of choosing j sets. So

$$\sum \sum \cdots \sum |X_{j_1} \cap X_{j_2} \cap \cdots \cap X_{j_i}| = \binom{n}{i}(n - i)! = \frac{n!}{i!}.$$

The theorem follows upon substituting this formula into (5.6). □

The series in (5.10) is the expansion of e^{-1}, truncated after $n + 1$ terms. So, by choosing n to be sufficiently large, we can make $D_n/n!$ as close as we wish to e^{-1}. In fact, $D_n/n!$ differs from e^{-1} by less than 1% provided n is at least 6.

While Theorem 5.2 is useful in theory, it would be very tedious to use it to calculate large derangement numbers. The following recursive result is much easier to use.

THEOREM 5.3 (Recursive Derangement Formula)

$$D_n = (n - 1)(D_{n-1} + D_{n-2}).$$

Proof. The D_n derangements of $\{1, 2, \ldots, n\}$ can be partitioned into $n - 1$ classes $X_1, X - 2, \ldots, X_{n-1}$, where X_k consists of all the derangements with last element k. (No derangement will have last element n.) To prove the theorem, it is sufficient to show that each set X_k has $D_{n-1} + D_{n-2}$ elements.

The elements of X_k can be classified further. Write Y_k for the subset of X_k consisting of all the derangements in X_k with kth element n and Z_k for the set of those whose kth element is not n.

Of the members of Y_k, if entries n and k are deleted, the remaining elements must be a derangement of the $(n-2)$-set $T = \{1, 2, \ldots, k-1, k+1, \ldots, n-1\}$, and different arrangements give rise to different members of Y_k. So the number of elements of Y_k equals the number of derangements of T, or D_{n-2}. The members of Z_k are just the derangements of $\{1, 2, \ldots, k-1, n, k+1, \ldots, n-1\}$, so $|Z_k| = D_{n-1}$.

So $|X_k| = |Y_k| + |Z_k| = D_{n-2} + D_{n-1}$, and $D_n = \sum_{i=1}^{n-1} |S_k|$, giving the result. $\qquad\square$

Example 5.9: Find D_5.

We know $D_3 = 2$ and $D_2 = 1$. So $D_4 = 3(D_3 + D_2) = 3(2 + 1) = 9$. Then
$$D_5 = 4(D_4 + D_3) = 4(9 + 2) = 44. \qquad\square$$

Example 5.10: Five people attend a party and all check their coats. When they leave, they are given the wrong coats; no one gets their own coat. If the coats were returned at random, what is the chance of this happening?

The coats could have been returned in 5! ways. The chance of a derangement is $D_5/5! = 44/120 = 11/30$, or approximately 37%. $\qquad\square$

Example 5.11: Five couples, man and woman, attend a party and all check their coats. When they leave, they are given the wrong coats; each man is given a man's coat, but no one gets their own coat. In how many ways can this happen?

Label the couples 1, 2, 3, 4, 5. Say man i gets the coat belonging to man x_i. Then $x_1 x_2 x_3 x_4 x_5$ is a derangement of 12345, so there are D_5 ways in which this can happen. As the men's coats are all accounted for, each woman gets a woman's coat, but the wrong one. So this can also happen in D_5 ways. By the multiplication principle, there are $D_5^2 = 44^2 = 1936$ ways. $\qquad\square$

If we subtract nD_{n-1} from each side of the result of Theorem 5.3, we obtain
$$D_n - nD_{n-1} = -[D_{n-1} - (n-1)D_{n-2}],$$

provided $n \geq 3$. The expression on the right-hand side is derived from the left-hand side by subtracting 1 from each expression of the variable, and then negating. Repeating this process, we get

$$
\begin{aligned}
D_n - nD_{n-1} &= (-1)[D_{n-1} - (n-1)D_{n-2}] \\
&= (-1)^2[D_{n-2} - (n-2)D_{n-3}] \\
&= (-1)^3[D_{n-3} - (n-3)D_{n-4}] \\
&= \cdots \\
&= (-1)^{n-2}[D_2 - 2D_1].
\end{aligned}
$$

We know $D_2 = 1$ and $D_1 = 0$. So
$$D_n - nD_{n-1} = (-1)^{n-2}(1 - 0),$$

or more simply
$$D_n = nD_{n-1} + (-1)^n. \qquad (5.11)$$

Exercises 5A

1. You need to choose a set of four members from $\{a, b, c, d, e, f, g, h\}$ that contains either a or b.

    (i) In how many ways can this be done?

    (ii) In how many ways can it be done if you may not include both a and b?

2. Find the number of positive integers that are smaller than 200 and are relatively prime to 21.

3. Find the number of integers between 1 and 5,000 inclusive that are not divisible by 4, 5 or 6.

4. How many elements are there in the set $A \cap B$, if $|A| = 28$, $|B| = 34$, and:

    (i) A and B are disjoint?

    (ii) $|A \cup B| = 47$?

    (iii) $A \subseteq B$?

5. 2,200 voters were surveyed about a new federal building. Of them, 703 were in favor of financing it with a new sales tax, 525 favored a new property tax, while 336 said they would vote in favor of either measure.

    (i) How many favor property tax but not sales tax?

    (ii) How many favor sales tax but not property tax?

    (iii) How many would vote against either form of funding?

6. Eighty schoolchildren were surveyed about their reading habits. It was found that 52 had read most or all *Harry Potter* books, 25 had read *Lemony Snicket* and 20 had read *Artemis Fowl*. Nine read both *Harry Potter* and *Artemis Fowl*, 11 read both *Lemony Snicket* and *Harry Potter*, and eight read both *Lemony Snicket* and *Artemis Fowl*. Eight of the children do not read any of the series.

    (i) How many have read all three series?

    (ii) How many have read exactly one of the three?

7. Inspired by the makers of Red Hat licorice, the distributors of Red Hat Linux decide to start a contest. Each contestant receives an anagram of REDHATLINUX, and is a winner if *none* of the words RED, HAT, LINUX appears in its correct order. For example, HLXREDATINU is *not* allowed, because of RED. Of the 11! anagrams, how many are winners?

8. Find the number of distinguishable 12-subsets of

$$\{5 \circ A, 6 \circ B, 6 \circ C\}.$$

9. Find the number of distinguishable 12-subsets of

$$\{5 \circ A, 7 \circ B, 2 \circ C, \infty \circ D\}.$$

10. Determine the number of integer solutions to

$$x + y + z + t = 18$$

that satisfy $2 \leq x \leq 4, 1 \leq y \leq 6, 3 \leq z \leq 6, 4 \leq t \leq 7$.

11. Find the number of integers between 1 and 6,000 inclusive that are neither perfect squares nor perfect cubes.

12. How many elements are there in the union of the four sets X_1, X_2, X_3, X_4 if each of the sets has precisely 16 elements, each intersection $X_i \cap X_j$ has precisely eight elements, each triple of sets has exactly one element in common, and there is no element common to all four sets?

13. Each of the sets X_1, X_2, X_3, X_4 has 24 elements. X_1 and X_3 are disjoint; every other intersection of two of the sets has 12 elements. $|X_1 \cap X_2 \cap X_4| = 7$, $|X_2 \cap X_3 \cap X_4| = 1$. What is the size of $X_1 \cup X_2 \cup X_3 \cup X_4$?

14. Calculate D_6, D_7 and D_8.

15. Six women at a party check their coats. In how many ways can the following occur?
    (i) No one gets her own coat back.
    (ii) Exactly one woman gets her own coat back.
    (iii) At least one woman gets her own coat back.
    (iv) Every woman but one gets her own coat back.

16. What is the number of permutations of $\{1, 2, 3, 4, 5, 6, 7\}$ in which exactly three of the integers are in their original position?

Exercises 5B

1. A survey was carried out. It was found that 20 of the people surveyed followed football on television, while 15 followed baseball; 12 watched both football and baseball, but 14 watched neither. Use the formula (5.1) to find out how many people were interviewed.

2. Find the number of positive integers that are smaller than 100 and are relatively prime to 10.

3. Find the number of integers that are not divisible by 4, 5 or 6:
   (i) between 1 and 10,000 inclusive;
   (ii) between 5,001 and 10,000 inclusive.

4. How many elements are there in the set $S \cup T$, if $|S| = 14$, $|T| = 17$, and:
   (i) S and T are disjoint?
   (ii) $|S \cap T| = 11$?
   (iii) $S \cap T$ is empty?

5. A survey was carried out. It was found that 185 of the people surveyed drank tea regularly, while 344 were coffee drinkers; 126 regularly drank both beverages, while 84 drank neither. Use formula (5.1) to find out how many people were interviewed.

6. Find the number of distinguishable 10-subsets of

$$\{3 \circ A, 5 \circ B, 6 \circ C\}.$$

7. Find the number of distinguishable 9-subsets of

$$\{4 \circ A, 5 \circ B, 6 \circ C, \infty \circ D\}.$$

8. Determine the number of integer solutions to

$$x + y + z + t = 24$$

that satisfy $0 \le x \le 4, 2 \le y \le 8, 5 \le x \le 10, 2 \le x \le 5$.

9. Find the number of integers between 1 and 10,000 inclusive that are not perfect squares, perfect cubes or perfect fifth powers.

10. A shopping precinct contains Staples, Office Depot and Best Buy stores. In a survey of 150 shoppers, it is found that
    (a) 87 shopped at Office Depot;
    (b) 101 shopped at Staples;
    (c) 92 shopped at Best Buy;
    (d) 72 shopped at both Office Depot and Staples;
    (e) 75 shopped at both Staples and Best Buy;
    (f) 62 shopped at both Office Depot and Best Buy;
    (g) 52 shopped at all three stores.
   (i) How many people shopped only at Office Depot?

(ii) How many people shopped at Office Depot and Staples, but not at Best Buy?

(iii) How many people shopped at exactly one of the stores?

(iv) How many people shopped at none of the three stores?

11. You have five sets, X_1, X_2, X_3, X_4, X_5, each of size 12. X_1, X_3 and X_5 are pairwise disjoint; $|X - 2 \cap X_4| = 6$; every other intersection of two of the sets has three elements. What is the size of the following sets: $X_1 \cup X_2 \cup X_3 \cup X_4$, $X_2 \cup X_4 \cup X_5$, $X_1 \cup X_2 \cup X_3 \cup X_4 \cup X_5$?

12. Eight job applicants are to be interviewed by eight members of a hiring committee. Each will be interviewed twice. To avoid bias, each candidate is assigned their first interviewer at random; the second interviewer is also chosen at random, but no candidate is to get the same interviewer twice.

(i) How many different ways are there to schedule the first round of interviews?

(ii) How many different ways are there to schedule the whole process?

Problems

Problem 1: Derive the following formula, for subsets A, B, C of any universal set:

$$|A \cup C| = |A| + |B \cap C| + |\overline{B} \cap C| - |A \cap B \cap C| - |A \cap \overline{B} \cap C|.$$

Problem 2: Prove that D_{n+1} is even if and only if D_n is odd.

Problem 3: At a party, n people are seated in a room with $n + 1$ chairs. They go to another room for supper; when they return and sit down again, it is found that no one occupies the same chair as before. Show that this can occur in precisely $D_n + D_{n+1}$ ways.

Problem 4: Prove

$$n! = \sum_{k=0}^{n} \binom{n}{k} D_k.$$

(Remember, D_0 is defined to be 1.)

Problem 5: Use Equation (5.11) to prove Equation (5.10).

Chapter 6

Generating Functions and Recurrence Relations

6.1 Generating Functions

Some kinds of combinatorial calculations are modeled by the behavior of infinite power series or even finite power series (i.e., polynomials). For instance, we considered in Chapter 2 how to determine the number of ways of selecting k items from a set of n; for simplicity, we will consider $n-4$, $k=2$. Suppose we wish to multiply $(1 + x_1)(1 + x_2)(1 + x_3)(1 + x_4)$. In the product, we ask how many terms of the form $x_i x_j$ there will be; clearly, $\binom{4}{2} = 6$ of them, since that is how many ways we can choose two of the factors to contribute an x_i. Now, if we set $x_1 = x_2 = x_3 = x_4 = x$, the product is the polynomial $(1 + x)^4$. The coefficient of x^2 in this product is $\binom{4}{2} = 6$. (The reader will recall that this subscripting technique is how we proved Theorem 2.3, the Binomial Theorem.) In the same way, some numbers of combinatorial interest (such as a sequence of integers) have properties that become clear or easier to prove when they are used as coefficients of a polynomial or an infinite series. These considerations motivate us to define the *generating function* of a sequence $a_0, a_1, \ldots$ to be the formal power series $\sum_{i \geq 0} a_i x^i$.

The expression *formal power series*indexformal power series is understood to mean that we do not evaluate the power series at any point, or concern ourselves with the question of convergence necessarily. Indeed, some uses of generating functions do not require infinite series at all; in this case we shall assume that $a_n = 0$ for all n greater than some specific integer N.

Example 6.1: Suppose we wish to buy a dozen donuts from a shop with varieties v_1, v_2, v_3, and v_4. We want between three and five donuts of variety v_1 (because the shop has only 5 available, and at least three people will want one); similarly, we require $1 \leq |v_2| \leq 10$, $4 \leq |v_3| \leq 6$, and $0 \leq |v_4| \leq 7$.

In Chapter 5 we introduced the Principle of Inclusion and Exclusion, and it could be used to eliminate the cases that violate the given constraints, but in this case the process would be enormously difficult. Instead, consider the product

$$(x^3 + x^4 + x^5) \times (x + x^2 + x^3 + x^4 + x^5 + x^6 + x^7 + x^8 + x^9 + x^{10}) \times$$
$$(x^4 + x^5 + x^6) \times (1 + x + x^2 + x^3 + x^4 + x^5 + x^6 + x^7) \qquad (6.1)$$

and ask where terms involving x^{12} originate. If we multiply out the given polynomials and do not collect like powers of x, a given term of x^{12} might come from the x^4 in the first factor, the x^2 in the second factor, the x^6 in the third, and the 1 in the final factor. This would correspond to a dozen donuts consisting of 4 of type v_1, 2 of type v_2, 6 of type v_3, and none of type v_4. Similarly, any other way of making up a dozen donuts satisfying the given constraints must correspond to exactly one term x^{12} of the product. Thus we see that in general the coefficient of x^n in the product above determines the number of ways of making up a box of n donuts satisfying the constraints. The product is

$$x^8 + 4x^9 + 10x^{10} + 18x^{11} + 27x^{12} + 36x^{13} + 45x^{14} + 54x^{15} +$$
$$62x^{16} + 68x^{17} + 70x^{18} + 68x^{19} + 62x^{20} + 54x^{21} + 45x^{22} + 36x^{23} +$$
$$27x^{24} + 18x^{25} + 10x^{26} + 4x^{27} + x^{28} \qquad (6.2)$$

$\square$

Now, multiplying out the product (6.1) above may seem only slightly easier than enumerating possible solutions by hand; but there are reasons why the product might be preferable. For instance, we might be faced with the same problem for 18 donuts, or 10. The polynomial will give us the answers for those problems with no additional work. The relative prevalence of computer algebra systems, from *Mathematica* and *Maple* on down to some sophisticated pocket calculators, means that multiplication need no longer be terribly time-consuming. And finally, algebraic identities can in some cases reduce the effort of calculating the product.

Example 6.2: Assume that we again want a dozen donuts from the same shop, but now the shop has an unlimited supply (i.e., at least 12) of each variety; the lower bound constraints still apply.

Rather than limiting each factor to x^{12}, we allow each factor to represent an infinite series. We have

$$(x^3 + x^4 + \dots)(x + x^2 + \dots)(x^4 + x^5 + \dots)(1 + x + x^2 + \dots),$$

and since each factor is a geometric series we can use the familiar formula for the sum of such a series to see that it is equivalent to the expression

$$\frac{x^8}{(1-x)^4} = x^8 \frac{1}{1 - 4x + 6x^2 - 4x^3 + x^4}.$$

Now, polynomial long division (done with the coefficients in increasing order) will allow us to compute a Maclaurin polynomial, and the coefficient of x^{12} will be the solution to our problem. We get

$$x^8 + 4\,x^9 + 10\,x^{10} + 20\,x^{11} + 35\,x^{12} + 56\,x^{13} + 84\,x^{14} + 120\,x^{15} + 165\,x^{16} + 220\,x^{17} + \cdots$$

These figures conform well to our intuition that the numbers should be larger, as there are fewer constraints (so there are more ways to make up a box of donuts). □

Just as we can calculate the values for the donut problem above, we can use infinite series in a formal way to describe any infinite sequence of integers. Consider the most basic example of this sort, the sequence $\alpha_n = 1$. As we saw in Example 5.2, it will be useful to recall $f_\alpha(x) = \sum_{i=1}^\infty 1 x^i = x/(1-x)$. In calculus, we were concerned with the values of x for which the series converged; in combinatorics, we ignore the question.

Another sequence of combinatorial import has a generating function that may be simply computed. Suppose $\beta_n = n$, $n \geq 0$. Then $f_\beta(x) = \sum_1^\infty a_i x^i = x/(1-x)^2$. This formula is easily derived by differentiating the ordinary infinite geometric series, and multiplying by x. Note that since $a_0 = 0$ we begin the summation at $i = 1$; if we begin at $i = 0$, we get the same series. Differentiating again, with another multiplication by x, gives us another formula: $\sum_1^\infty i^2 x^i = x(1+x)/(1-x)^3$. With these formulas, we are able to find the generating function for any integer sequence produced by a quadratic polynomial.

Example 6.3: Find the generating function for $a_n = \binom{n}{2}$.

We may write $a_n = \frac{1}{2}n^2 - \frac{1}{2}n$, so that the generating function will be $f(x) = \frac{1}{2}\left(\sum_2^\infty n^2 - \sum_1^\infty n\right)$. We sum from 2 because $\binom{1}{2} = 0$; we could as easily sum from 0 or 1. Using the values of these sums given above, we see $f(x) = x^2/(1-x)^3$. □

To the elementary manipulations shown above may be added a few others; for instance, consider the triangular numbers $T_n = 1 + 2 + \cdots + n$ $n \geq 1$. The well-known formula is just $T_n = \binom{n+1}{2}$. We may re-index the series given in the foregoing example by letting $j = i + 1$. So:

$$f_T(x) = \sum_1^\infty T_i x^i = \sum_2^\infty T_{j-1} x^{j-1} = \sum_2^\infty \binom{j}{2} x^{j-1}.$$

Multiplying the right-hand side by x must yield, by the previous example, $x^2/(1-x)^3$; so the desired result is $f_T(x) = \sum_1^\infty T_i x^i = x/(1-x)^3$.

Now, we pause to point out a pattern in some of the foregoing results. The sequence α_n is related to β_n by $\beta_n = \sum_{i=1}^n \alpha_i$. We also notice $f_\alpha(x) = (1-x)f_\beta(x)$. Upon noting that $T_n = \sum_{i=1}^n \beta_i$, we look at the generating functions; $f_T(x) = (1-x)f_\beta(x)$. The pattern holds, and we conjecture and prove a result.

THEOREM 6.1 (Partial Sums)
Given sequences a_n and b_n with $b_n = \sum_{i \leq n} a_i$, their generating functions are related by $f_a(x) = (1 - x)f_b(x)$.

The proof is left as Problem 6.1.

Other series manipulations will be convenient from time to time; we summarize a few of these in theorem form.

THEOREM 6.2 (Incomplete Sums)
Given $f_a(x) = \sum_{i=0}^{\infty} a_i x^i$, we have $\sum_{i=k}^{\infty} a_i x^i = f_a(x) - a_0 - a_1 x - \cdots - a_{k-1} x^{k-1}$.

THEOREM 6.3 (Series for Derivative)
For a_n and $f_a(x)$ as above, we have $x f_a'(x) = \sum_{i \leq \infty} i a_i x^i$.

Proof. Differentiate both sides of $f_a(x) = \sum a_i x^i$ and multiply by x. □

We used the next theorem as part of doing Example 5.3, so we ought to at least state it.

THEOREM 6.4 (Series for Addition)
If the sequences a_n and b_n have generating functions f_a and f_b, respectively, then the sequence $c_1 a_n + c_2 b_n$ has generating function $c_1 f_a + c_2 f_b$.

We leave the proof to Exercise 6B.9.

There are other identities involving generating functions. We refer the interested reader to the Problems at the end of this chapter, or to [139].

Example 6.4: Let $a_n = 3^n - 2^n$, for $n \geq 0$. Then $f_a(x) = \sum_{i=0}^{\infty} a_i x^i$. Evaluate this sum.

This may be summed using the geometric series:

$$\sum_{i=0}^{\infty}(3^i - 2^i)x^i = \sum_{i=0}^{\infty}(3x)^i - \sum_{i=0}^{\infty}(2x)^i = \frac{1}{1 - 3x} - \frac{1}{1 - 2x} = \frac{x}{(1 - 3x)(1 - 2x)}$$

so $f_a(x) = \frac{x}{(1-3x)(1-2x)}$. □

More generally, it is true that if a_n is any sequence defined as a linear combination of nth powers, the generating function will be rational. Furthermore, its denominator will have roots equal to the reciprocals of the bases of the nth powers. In the last example we had a linear combination of 3^n and 2^n, and the denominator of the generating function has roots $1/3$ and $1/2$. We will explore other properties of the denominator later.

6.2   Recurrence Relations

Frequently in studying a sequence of numbers, we discover that although we may not have a simple, "closed-form" expression for each number, we may have some idea of a relationship by which a given term depends upon those terms that have gone before. We have seen this sort of formula in Chapter 2.

A recurrence relation is said to be *linear* if the value for a_n is given as a finite sum, each summand of which is a multiple of the first power of some previous term, or a constant or function of n; so, for example, $a_n = na_{n-1}+2n$ is linear, and $a_n = (a_{n-1})^2 - 2n$ is nonlinear.

Similarly, we say that the recurrence relation is *homogeneous* if the value for a_n is given as a finite sum, each summand of which is a multiple of some previous terms. So, $a_n = a_{n-1} + 2n + 1$ is *nonhomogeneous* (but linear) and $a_n = (a_{n-1})^2$ is homogeneous (but nonlinear). The coefficients may be non-constant, so that $a_n = na_{n-1}$ is both linear and homogeneous.

The *degree* of a recurrence is the largest i such that the recurrence depends upon a_{n-i}. We give $a_n = na_{n-3}-1$ as an example of a linear nonhomogeneous recurrence relation of degree 3. The degree is important because the higher the degree, the more initial values we need to define a sequence precisely.

If we wish to write down the first several values of a_n given by $a_n = (a_{n-1})^2 - a_{n-2}$, we must know two initial values, say $a_0 = 1$ and $a_1 = 2$. From these, we may compute $a_2 = a_1^2 - a_0 = 3$, $a_3 = a_2^2 - a_1 = 7$, and so on. These initial values are called the *initial conditions*. The reader who has taken a course in differential equations will notice the similarity of the terminology with that subject. Any sequence of integers may satisfy many recurrence relations, including linear and nonlinear, homogeneous and nonhomogeneous, of many different degrees.

It will occasionally be useful to prove that two sequences that appear to be the same really are the same. One way to do this is to show that the two sequences satisfy the same recurrence relation with the same initial conditions. This is a short form of what more formally could be a proof by induction of the equality of the two sequences.

Example 6.5: Consider the formula $a_n = a_{n-1}+2a_{n-2}$, for $n \geq 2$, $a_0 = 0$, and $a_1 = 3$. Clearly, we can compute any a_n with this formula by repeatedly applying the equation to the initial conditions. So $a_2 = a_1 + 2a_0 = 3$, and $a_3 = a_2 + 2a_1 = 9$. Can we find a closed-form formula for a_n?

The generating function is one of several ways to solve this problem. Consider $f_a(x) = \sum_{i \geq 0} a_i x^i$. We can use the recurrence to evaluate this function. Multiply both sides of the equation by x^n and sum over n to obtain:

$$\sum_{n=2}^{\infty} a_n x^n = \sum_{n=2}^{\infty} a_{n-1}x^n + 2\sum_{n=2}^{\infty} a_{n-2}x^n.$$

Each summation may be interpreted in terms of $f_a(x)$, using Theorem 6.2. We get $f_a(x) - a_0 - a_1 x = x(f_a(x) - a_0) + 2x^2 f_a(x)$. Now we can solve for $f_a(x)$. We get $f_a(x)(1 - x - 2x^2) = a_0 + a_1 x - a_0 x$ or $f(x) = 3x/(1 - x - 2x^2)$. Now, in Example 5.4, we saw that the reciprocals of the roots of the denominator of $f(x)$ were used as bases of exponentials in the formula of a_n. That is, a_n involved 2^n and 3^n, and the denominator had roots $1/2$ and $1/3$. In this case, the denominator is $(1 - 2x)(1 + x)$. We simplify using partial fractions.

$$\frac{3x}{(1 - 2x)(1 + x)} = \frac{A}{1 - 2x} + \frac{B}{1 + x} \Rightarrow A = 1, \quad B = -1.$$

This yields $f_a(x) = \frac{1}{1 - 2x} - \frac{1}{1 + x}$; we can use the geometric series formula here. We know $\frac{1}{1 - 2x} = \sum_{i \geq 0} 2^i x^i$ and $\frac{1}{1 + x} = \sum_{i \geq 0} (-1)^i x^i$, so $f_a(x) = \sum_{i \geq 0} (2^i - (-1)^i) x^i = \sum_{i \geq 0} a_i x^i$. It follows that $a_n = 2^n - (-1)^n$. □

We could compute this formula in another way. Suppose we were to assume that the sequence a_n had a representation as a linear combination of exponential functions. Then $a_n = r^n$ would give us, via the recurrence, an equation $r^{n+1} = r^n + 2r^{n-1}$. Dividing by r^{n-1} yields a quadratic equation, $r^2 - r - 2 = 0$. This is called the *characteristic equation* of the recurrence. Finding the roots gives us $r = 2$ or $r = -1$, the bases of the exponential functions. We are now able to surmise that $a_n = c_1 2^n + c_2 (-1)^n$. Using the given values for a_0 and a_1 we are able to solve for c_1 and c_2 to get the same formula as above.

That method works for distinct roots. If there are repeated roots, we must work a little harder.

Example 6.6: Let us consider $a_{n+1} = 2a_n - a_{n-1}$, with $a_0 = 1$, $a_1 = 2$.

It is easy to find a closed expression for a_n using a guess and a proof by induction, but we will try generating functions. As with our first example, we multiply by x^{n+1} and sum over n to arrive at:

$$f_a(x) - a_0 - a_1 x = 2x f_a(x) - 2a_0 x - x^2 f_a(x) \text{ or } f_a(x) = (x - 1)^{-2}.$$

Partial fractions will not assist us in this case. Instead, we turn for inspiration to an earlier example, the sequence β_n, which had as its generating function $f_\beta(x) = x(x - 1)^{-2}$. We write this sequence as $\beta_n = (n + 0)(1)^n$ and ask whether a formula of the type $a_n = (c_1 n + c_2)1^n$ will work. A little work reveals $a_n = n + 1$. □

THEOREM 6.5 (Repeated Roots)
If the characteristic equation has a repeated root $(x - r)^k$, this corresponds to a term of the form $(c_{k-1} n^{k-1} + \cdots + c_1 n + c_0) r^n$ in the formula for a_n.

We omit the proof, but see Problem 6.3.

Example 6.7: Solve the recurrence $a_{n+1} = 4a_n - 4a_{n-1}$, where $a_0 = -2$, $a_1 = 2$.

The characteristic equation is $x^2 - 4x + 4 = 0$, so the solution is $a_n = (c_1 n + c_2)2^n$ for some c_1 and c_2. Using the given values of a_0 and a_1 we get $-2 = c_2$ and $2 = 2c_1 + 2c_2$. Solving for c_1 gives $a_n = (3n - 2)2^n$. □

As a concrete example, we introduce the reader to a sequence of numbers involved with a problem that is over 800 years old, about rabbits.

The *Fibonacci numbers* $f_1, f_2, f_3, \ldots$ are defined as follows. $f_1 = f_2 = 1$, and if n is any integer greater than 2, $f_n = f_{n-1} + f_{n-2}$. This famous sequence is the solution to a problem posed by Leonardo of Pisa, or Leonardo Fibonacci (Fibonacci means *son of Bonacci*) in 1202:

> A newly born pair of rabbits of opposite sexes is placed in an enclosure at the beginning of the year. Beginning with the second month, the female gives birth to a pair of rabbits of opposite sexes each month. Each new pair also gives birth to a pair of rabbits of opposite sexes each month, beginning with their second month.

The number of pairs of rabbits in the enclosure at the beginning of month n is f_n. An English translation of Fibonacci's work *Liber Abaci* is available (see [102]) for those interested.

Some interesting properties of the Fibonacci numbers involve the idea of *congruence* modulo a positive integer. We say a is *congruent to* b modulo n, written "$a \equiv b \pmod n$," if and only if a and b leave the same remainder on division by n. In other words n is a divisor of $a - b$, or in symbols $n \mid (a - b)$.

Example 6.8: Prove by induction that the Fibonacci number f_n is even if and only if n is divisible by 3.

Assume n is at least 4. $f_n = f_{n-1} + f_{n-2} = (f_{n-2} + f_{n-3}) + f_{n-2} = f_{n-3} + 2f_{n-2}$, so $f_n \equiv f_{n-3} \pmod 2$.

We first prove that, for $k > 0, f_{3k}$ is even. Call this proposition $P(k)$. Then $P(1)$ is true because $f_3 = 2$. Now suppose k is any positive integer, and $P(k)$ is true: $f_{3k} \equiv 0 \pmod 2$. Then (putting $n = 3k + 3$) $f_{3(k+1)} \equiv f_{3k} \pmod 2 \equiv 0 \pmod 2$ by the induction hypothesis. So $P(k+1)$ is true; the result follows by induction. To prove that, for $k > 0, f_{3k-1}$ is odd—call this proposition $Q(k)$—we note that $Q(1)$ is true because $f_1 = 1$ is odd, and if $Q(k)$ is true, then f_{3k-1} is odd, and $f_{3(k+1)-1} \equiv f_{3k-2} \pmod 2 \equiv 1 \pmod 2$. We have $Q(k+1)$ and again the result follows by induction. The proof for $k \equiv 1 \pmod 3$ is similar. □

This sequence is one of the most-studied sequences of integers in all of mathematics. Generating functions make it possible for us to understand many properties of the sequence. Suppose $f(x)$ is the generating function for the Fibonacci numbers. We multiply both sides of the recurrence by x^{i+1} and sum as in Example 5.5.

$$\sum_{i=2}^{\infty} f_{i+1}x^{i+1} = \sum_{i=2}^{\infty} f_i x^{i+1} + \sum_{i=2}^{\infty} f_{i-1}x^{i+1}$$

Careful use of the formula for $f(x)$ will help us to represent this equation in more understandable terms. So, using Theorem 6.2, the first term in the equation is $\sum_{i=2}^{\infty} f_{i+1}x^{i+1} = (f(x) - f_1 x - f_2 x^2)$. We substitute similarly for the other two terms to get

$$f(x) - f_1 x - f_2 x^2 = xf(x) - f_1 x^2 + x^2 f(x)$$

which we may solve for $f(x)$, to get

$$x = f(x)(1 - x - x^2) \text{ or } f(x) = x/(1 - x - x^2)$$

We may confirm that polynomial long division, as in the earlier example, produces the Fibonacci numbers (see Problem 6.4). But there is a more useful and fascinating consequence of our calculation of $f(x)$. Recall that for the earlier case of the sequence $a_n = 3^n - 2^n$ in Example 6.4, the generating function was also a rational function, and that the bases of the exponents were reciprocals of the roots of the denominator. Alternatively, as in Example 6.5, we may use partial fractions or the characteristic equation.

First, we assume that the Fibonacci numbers may be represented by an expression of the form $f_n = r^n$, $r \geq 1$. Then the recurrence gives us $r^{n+1} = r^n + r^{n-1}$, and dividing by r^{n-1} gives us a quadratic $r^2 - r - 1$ with two real roots, traditionally denoted $\phi_1 = (1 + \sqrt{5})/2$ and $\phi_2 = (1 - \sqrt{5})/2$. To find f_n we simply assume $f_n = c_1 \phi_1^n + c_2 \phi_2^n$ and use the initial values to solve for c_1 and c_2.

$$f_1 = 1 = c_1 \frac{1 + \sqrt{5}}{2} + c_2 \frac{1 - \sqrt{5}}{2}, \qquad f_2 = 1 = c_1 \frac{6 + 2\sqrt{5}}{4} + c_2 \frac{6 - 2\sqrt{5}}{4}$$

This gives us $c_1 = 1/\sqrt{5}$ and $c_2 = -1/\sqrt{5}$. This formula for the Fibonacci numbers is sometimes called *Binet's formula*, although Binet was not its first discoverer. Different starting values yield a different sequence, but it is not hard to see that the values $f_1 = 0$, $f_2 = 1$ will give the same numbers with different subscripts, and likewise $f_1 = 1$, $f_2 = 2$. So, the "first" (in some sense) set of initial values that gives us a different sequence is $f_1 = 1$, $f_2 = 3$. This sequence has also been studied extensively; it is more commonly called the sequence of *Lucas numbers*, after the mathematician Edouard Lucas. There is a formula for the nth Lucas number L_n similar to the Binet formula for Fibonacci numbers; see Problem 6.5.

The Fibonacci numbers represent the number of ways of writing an integer as an ordered sum of 1s and 2s. More precisely, f_n is the number of ways of writing $n - 1$ as such a sum, for $n \geq 2$. The proof of this statement comes from asking how the sum begins. There are f_n ways to write $n - 1$, and we place a 1 at the beginning of such sequences to create some of the ways of writing n as such a sum. Then there are f_{n-1} ways to write $n - 2$; we may place a 2 at the beginning of any of these sequences to obtain ways of writing n. It follows that $f_{n+1} = f_n + f_{n-1}$. Now, $f_1 = 1$ (which counts the empty sum only) and $f_2 = 1$ since we can only write 1 as a sum of a single 1 and no 2s. Thus we have the same recurrence and initial conditions as the Fibonacci sequence; it follows by mathematical induction that the sequences are the same. The Fibonacci numbers arise in other contexts as well; see Problem 6.7 for an example.

Our next example will concern a construction by the Greek mathematician Theon of Smyrna. At the time when Theon lived, through the early second century AD, it was already known that the square root of two was not rational. However, there was work done on finding a rational sequence of approximations to the value, and it was Theon who found (see [44]) the process that has become known as *Theon's ladder*. We begin with the first "rung" of the ladder by setting $a_0 = b_0 = 1$. Each successive rung is created by setting $a_n = a_{n-1} + 2b_{n-1}$ and $b_n = a_{n-1} + b_{n-1}$. Then the rational number $r_n = a_n/b_n$ is our nth approximation. This gives us $a_1 = 3$ and $b_1 = 2$, so that our second approximation to $\sqrt{2}$ is $r_1 = 1.5$. Continuing, we will find $a_2 = 7$, $b_2 = 5$, and $r_2 = 7/5 = 1.4$.

Example 6.9: Find a closed form for the quantities a_n and b_n and use this closed form to show that the ratio r_n indeed approaches $\sqrt{2}$.

We clearly cannot deal with these two intertwined recurrences as they stand, so we must try to eliminate one of the sequences. We observe that the second recurrence may be rewritten as $a_{n-1} = b_n - b_{n-1}$ and therefore $a_n = b_{n+1} - b_n$. We substitute these values into the first recurrence to get

$$b_{n+1} - b_n = b_n - b_{n-1} + 2b_{n-1} \quad \text{or} \quad b_{n+1} = 2b_n + b_{n-1}.$$

Similar calculations will give us $a_{n+1} = 2a_n + a_{n-1}$, the same recurrence. Using the characteristic equation and the values of a_1, a_2, b_1, and b_2 as given above, we arrive at the formulas for a_n and b_n.

$$a_n = \frac{(1 + \sqrt{2})^{n+1} + (1 - \sqrt{2})^{n+1}}{2} \qquad b_n = \frac{(1 + \sqrt{2})^{n+1} - (1 - \sqrt{2})^{n+1}}{2\sqrt{2}}$$

Now we look at the ratio $r_n = a_n/b_n$ and take the limit as n goes to ∞. We note that $(1 - \sqrt{2})^{n+1}$ is less than 1 in absolute value and goes rapidly to 0 as n increases.

$$r_n = \sqrt{2} \, \frac{(1 + \sqrt{2})^{n+1} + (1 - \sqrt{2})^{n+1}}{(1 + \sqrt{2})^{n+1} - (1 - \sqrt{2})^{n+1}}$$

When we replace the $(1 - \sqrt{2})^{n+1}$ with 0, we get the result. □

6.3 From Generating Function to Recurrence

We have seen that we can use a recurrence for a sequence to develop a generating function. This process works both ways. Recall that for a sequence defined by a linear combination of exponential functions we discovered that the bases were the roots of the characteristic function, and were the reciprocals of the roots of the polynomial in the denominator of the rational function. This means that the denominator of the generating function is related to the recurrence relation.

THEOREM 6.6 (Rational Generating Function)

If a sequence a_n has a rational generating function with denominator $q(x) = 1 + c_1 x + \cdots + c_n x^n$ then it satisfies the recurrence $a_n + c_1 a_{n-1} + \cdots + c_n a_0 = 0$.

Proof. We will prove explicitly the case in which the denominator is a quadratic with distinct non-zero roots, say $1/r_1$ and $1/r_2$. This implies the denominator is $(1 - r_1 x)(1 - r_2 x)$, and $a_n = d_1 r_1^n + d_2 r_2^n$ for some coefficients d_1 and d_2. Since the denominator looks like $1 - (r_1 + r_2)x + r_1 r_2 x^2$, the recurrence we are to satisfy is $a_n - (r_1 + r_2)a_{n-1} + r_1 r_2 a_{n-2} = 0$. We easily confirm this by replacing each a_i in the recurrence by $d_1 r_1^i + d_2 r_2^i$. Other cases are done similarly. □

Example 6.10: Find a recurrence satisfied by the numbers of boxes of donuts from Example 5.1.

Since the denominator is $1 - 4x + 6x^2 - 4x^3 + x^4$, we see that the number of ways a_n of filling a box of n donuts with at least three of variety v_1, at least one of variety v_2 and at least four of variety v_3 satisfies $a_n - 4a_{n-1} + 6a_{n-2} - 4a_{n-3} + a_{n-4} = 0$, or $a_n = 4a_{n-1} - 6a_{n-2} + 4a_{n-3} - a_{n-4}$. □

Example 6.11: In the last section, we discussed the Fibonacci numbers, and noted that they enumerate ways to write n as an ordered sum of 1s and 2s. Now, let us ask about unordered sums. That is, in how many ways can we write an integer n as a sum of 1s and 2s where we consider sums equivalent if they contain the same numbers of 1s and 2s?

Let this quantity be a_n. It is not too hard to see that this is merely the coefficient of x^n in the product $(1 + x^2 + x^4 + \ldots)(1 + x + x^2 + \ldots) = (\sum x^{2i})(\sum x^i)$. We get a term of x^n for each solution to the equation $n = 2m + k$, which means for each possible factor $x^{2m} x^k$, where the x^{2m} comes from the first sum and the x^k from the second. The generating function, found via geometric series, is just $\left[(1 - x^2)(1 - x)\right]^{-1}$, with denominator $1 - x - x^2 + x^3$.

This says that our recurrence is $a_n = a_{n-1} + a_{n-2} - a_{n-3}$. We can compute the value of a_n for small n; doing so for several values of n will give us initial conditions for the recurrence, and allow us to check our results.

n	0	1	2	3	4	5	6	7	8	9	10
a_n	1	1	2	2	3	3	4	4	5	5	6

The recurrence works nicely, and we suspect that the formula for a_n is just $a_n = \lceil \frac{n+1}{2} \rceil$. In fact, this formula is clear when we think about it for a moment; to write n as a sum of 1s and 2s, we may write it as a sum of n 1s or $n-2$ 1s and a single 2, or (up to) $\lfloor n/2 \rfloor$ 2s and a 1 or not depending on the parity of n. □

We finish with an example that does not use real numbers. Just as Binet's formula for the Fibonacci numbers always produces integers despite irrational quantities, in the same way some recurrences always produce integers despite complex quantities.

Example 6.12: Solve the recurrence $a_{n+1} = 2a_n - a_{n-1} + 2a_{n-2}$ with initial values $a_0 = a_1 = 6$, $a_2 = 4$.

We get a characteristic equation of $x^3 - 2x^2 + x - 2 = 0$, which factors nicely to get $x = \pm i$ and $x = 2$. We conclude that a general solution is of the form $c_1 i^n + c_2(-i)^n + c_3 \cdot 2^n$. We get three equations in three unknowns using the given initial values; so that $c_1 + c_2 + c_3 = a_0 = 6$, and $(c_1 - c_2)i + 2c_3 = 6$, and $-c_1 - c_2 + 4c_3 = 4$. This yields $c_1 = 2 - i$, $c_2 = 2 + i$, and $c_3 = 2$. Our final formula is $a_n = (2-i)i^n + (2+i)(-i)^n + 2^{n+1}$. □

6.4 Exponential Generating Functions

Suppose we are dealing with sequences that grow very quickly, such as $a_n = n!$ or $b_n = 2(n-1)!$. One difficulty with the kinds of generating functions discussed in the earlier sections is that these sequences will both give us the function that is 0 at $x = 0$ and is undefined for all other values of x. For sequences that grow faster than exponentials, another kind of generating function is desirable. Accordingly, we define the *exponential generating function* *(EGF)* of a sequence $a_0, a_1, \ldots$ to be the formal power series $\sum_{i \geq 0} a_i \frac{x^i}{i!}$.

Example 6.13: Find the exponential generating function of $a_n = 1$.

From the definition, we have $f_a(x) = \sum_{n \geq 0} x^i / i! = e^x$. Similarly, if $b_n = r^n$ for some constant r, we have $f_b(x) = e^{rx}$. □

There is a formula comparable to that of Theorem 6.2.

THEOREM 6.7 (Incomplete Sums—EGFs)
Given $f_a(x) = \sum_{i=0}^{\infty} a_i \frac{x^i}{i!}$, we have $\sum_{i=0}^{\infty} a_{i+k} \frac{x^i}{i!} = f_a^{(k)}(x)$.

Proof. Given $f_a(x) = a_0 + a_1x/1! + a_2x^2/2! + a_3x^3/3! + \ldots$, we differentiate both sides. This gives us $f'(x) = a_1 + a_2x/1! + a_3x^2/2! + \ldots$. This is the case $k = 1$ of the theorem; induction on k completes the proof. □

Roughly speaking, exponential generating functions enumerate some kinds of problems in which order is important, where the other, "ordinary" generating functions enumerate problems in which order is not important.

Example 6.14: We wish to eat k donuts, and there are an unlimited supply of donuts of each of three varieties v_1, v_2, and v_3. Order is important; so in the case $k = 4$ we distinguish between, for instance, eating two of variety v_1 first and then two of variety v_2, as opposed to eating the two v_2 donuts first and then the two v_1 types. In how many ways can we do this? In how many ways can we do this if we require at least two donuts of variety v_1?

The solution to the first problem is easy enough; each possible way of eating four donuts corresponds to a string of length k consisting of the symbols 1, 2, or 3 (denoting variety). The answer is just 3^k or for $k = 4$, 81. Alternatively, consider the product

$$\left(1 + x + \frac{x^2}{2!} + \frac{x^3}{3!} + \ldots\right) \times \left(1 + x + \frac{x^2}{2!} + \frac{x^3}{3!} + \ldots\right) \times \left(1 + x + \frac{x^2}{2!} + \frac{x^3}{3!} + \ldots\right).$$

The coefficient of $x^k/k!$ in this product will be the solution. Suppose we eat four donuts, including two of variety 1 and one of each other variety. The number of ways of doing this is $4!/(2!1!1!)$, as we saw in Chapter 2. This corresponds to the term $x^2/2! \times x/1! \times x/1! = \left(\dfrac{x^4}{4!}\right)\dfrac{4!}{2!1!1!}$. The coefficient of $x^4/4!$ is thus $4!/(2!1!1!)$, corresponding to the number of distinct ways of ordering the donuts. The factor of $x^2/2!$ comes from the first factor of the product, corresponding to donuts of variety v_1, and the factors of $x/1!$ come from the second and third factors of the product, corresponding to varieties v_2 and v_3. This turns out to be just $(e^x)^3 = e^{3x}$. The coefficient of $x^k/k!$ is clearly 3^k.

Now, how do we restrict ourselves to the case of at least two donuts of type v_1? In the first factor of the product above, the terms $1 + x/1!$ correspond to 0 or 1 donut of that variety. So we replace the first factor with $(x^2/2! + x^3/3! + \ldots)$ This gives us the product $(e^x - 1 - x)(e^{2x})$. Similarly, we may find the solution for any other restrictions on donut types. □

Example 6.15: How many k-letter words can be made from a, b, e, or g if we must use at least one of a and e, and no more than 3 of b or g?

We form the product $(x + x^2/2! + x^3/3! + \ldots)^2(1 + x/1! + x^2/2! + x^3/3!)^2$. The first factor corresponds to a and e, and the second to b and g.

So, if we want the solution for $k = 6$ (for example) we form the product $(e^x - 1)^2(1 + x + x^2/2 + x^3/6)^2 = (e^{2x} - 2e^x + 1)(x^6/36 + x^5/6 + 7x^4/12 +$

$4x^3/3 + 2x^2 + 2x + 1)$. Expanding the first factor out in series as far as x^6, we get

$$x^2 + 3x^3 + 55x^4/12 + 19x^5/4 + 1321x^6/360 + \ldots$$

This makes it clear that the coefficient of $x^6/6!$ is 2642, the solution to the given problem. □

Computer algebra systems make these computations much more convenient. The student with access to *Mathematica* may enter a problem as follows:

```
Series[(E^x-1)^2*(1+x+x^2/2+x^3/6)^2, {x, 0, 9}]
```

Following this with Shift+Enter (to evaluate the expression) produces:

$$x^2 + 3x^3 + \frac{55x^4}{12} + \frac{19x^5}{4} + \frac{1321x^6}{360} + \frac{89x^7}{40} + \frac{7369x^8}{6720} + \frac{259x^9}{576} + O[x]^{10}.$$

In *Maple*, essentially the same result comes from this input:

```
series((exp(x)-1)^2*(1+x+(1/2)*x^2+(1/6)*x^3)^2, x = 0, 10);
```

Exercises 6A

1. Find the generating function for the sequences:
   (i) $a_n = n^3$.
   (ii) $b_n = 3n^2 - 2n - 1$.
   (iii) $c_n = \binom{n}{3}$.
   (iv) $d_n = 3^n - \frac{3}{2}2^n$.

2. A bakery sells plain, chocolate chip, and banana-walnut muffins; on a particular day, the bakery has three of each kind.
   (i) How many ways are there to purchase a selection of n muffins?
   (ii) How many ways are there if we wish at least one of each kind?
   (iii) How many ways are there if we wish at least two banana-walnut muffins, but have no restrictions on the numbers of the other kinds of muffins?
   (iv) How many ways are there to purchase n muffins if the bakery has an unlimited supply of each kind? Find a recurrence relation for this sequence.

3. Find the solution for the recurrence with given initial values:

(i) $a_{n+1} = a_n + 12a_{n-1}$, $a_0 = 1, a_1 = 4$.

(ii) $a_{n+1} = 4a_n - 4a_{n-1}$, $a_0 = -2, a_1 = -2$.

(iii) $a_{n+1} = a_n + 12a_{n-1}$, $a_0 = 1, a_1 = 3$.

(iv) $a_{n+1} = 4a_n - 4a_{n-1}$, $a_0 = 1, a_1 = 2$.

(v) $a_{n+1} = a_n + 12a_{n-1}$, $a_0 = 0, a_1 = 1$.

(vi) $a_{n+1} = 4a_n - 4a_{n-1}$, $a_0 = 0, a_1 = 1$.

4. Compute the first 12 Fibonacci numbers.

5. Find a generating function for a_n, the number of ways to represent the integer n as an unordered sum of 1s and 3s. Use this generating function to find a recurrence relation satisfied by a_n. Can you find a closed form for this quantity?

6. Find the generating function for $a_n = 2^n - n$; use this function to find a recurrence that a_n satisfies.

7. Find exponential generating functions for the following sequences:

(i) $a_n = \binom{n}{2}$.

(ii) $b_n = 3^n - 2^{n-1}$.

(iii) $c_n = 2$.

(iv) $d_n = 2^n + 2$.

8. Find a sequence whose exponential generating function is $\cosh(x) = (e^x + e^{-x})/2$.

9. To celebrate a holiday, a child is permitted (under supervision in a safe place) to play with sparklers that when ignited emit showers of sparks in one of four colors. The child is permitted only one sparkler at a time, and may have no more than k sparklers total.

(i) Find the exponential generating function for the total number of ways (counting order) for the child to play with the sparklers.

(ii) As above, find the exponential generating function if the child is to use at least one of each color (so $k \geq 4$).

(iii) Assume that the child may use between one and three of each color of sparkler, and find the exponential generating function.

Exercises 6B

1. Find the generating function for the sequences:
   (i) $a_n = n^4$.
   (ii) $b_n = 6n^4 - 4n^2$.
   (iii) $c_n = \binom{n}{4}$.
   (iv) $d_n = 2^n + 2(-1)^n$.

2. A supply cabinet contains four red pens, three blue pens, and three black pens.
   (i) How many ways are there for a staff member to take n pens?
   (ii) How many ways if we wish one of each color?
   (iii) How many ways are there if we want at least two red pens but have no other restrictions?
   (iv) How many ways are there to take n pens if the supply cabinet has an unlimited quantity of each color? Find a recurrence relation for this sequence.

3. Find the solution for the recurrence with given initial values:
   (i) $a_{n+1} = a_n + 2a_{n-1}, a_0 = 1, a_1 = 2$.
   (ii) $a_{n+1} = 2a_n - a_{n-1}, a_0 = 0, a_1 = 1$.
   (iii) $a_{n+1} = a_n + 2a_{n-1}, a_0 = 1, a_1 = -1$.
   (iv) $a_{n+1} = 2a_n - a_{n-1}, a_0 = 2, a_1 = 1$.
   (v) $a_{n+1} = a_n + 2a_{n-1}, a_0 = 5, a_1 = 4$.
   (vi) $a_{n+1} = 2a_n - a_{n-1}, a_0 = 1, a_1 = 1$.

4. Prove the given result about the Fibonacci numbers, for all positive integers n.
   (i) f_n is divisible by 3 if and only if n is divisible by 4.
   (ii) f_n is divisible by 4 if and only if n is divisible by 6.
   (iii) f_n is divisible by 5 if and only if n is divisible by 5.
   (iv) $f_1 + f_2 + \cdots + f_n = f_{n+2} - 1$.
   (v) $f_1 + f_3 + \cdots + f_{2n-1} = f_{2n}$.
   (vi) $f_n^2 + f_{n+1}^2 = f_{2n+1}$.

5. Find the generating function for a_n, the number of ways to represent the integer n as an unordered sum of 1s and 4s. Use this generating function to find a recurrence relation satisfied by a_n. Can you find a closed form for this quantity?

6. Find the generating function for the sequence $a_n = 3^n - 2n$; use this function to find a recurrence relation that a_n satisfies.

7. Find exponential generating functions for the following sequences:
   (i) $a_n = n^2 + n$.
   (ii) $b_n = 4^n - 3^{n-1}$.
   (iii) $c_n = -1$.
   (iv) $d_n = (-1)^n + 3$.

8. Find a sequence whose exponential generating function is $\sinh(x) = (e^x - e^{-x})/2$.

9. To celebrate a holiday, a child is permitted (under supervision in a safe place) to play with sparklers that when ignited emit showers of sparks in one of three colors. The child is permitted only one sparkler at a time, and may have no more than k sparklers total.
   (i) Find the exponential generating function for the total number of ways (counting order) for the child to play with the sparklers.
   (ii) As above, find the exponential generating function if the child is to use at least one of each color (so $k \geq 3$).
   (iii) Assume that the child may use between one and four of each color of sparkler, and find the exponential generating function.

10. Prove Theorem 6.4.

Problems

Problem 1: Prove Theorem 6.1.

Problem 2: Let $a_n = (c_1 n + c_0)r^n$, where $c_1 \neq 0$. Without using Theorem 6.5, prove that $f_a(x) = \sum_{i \geq 0} a_i x^i$ implies that $f_a(x)$ is a rational function with denominator $(1 - rx)^2$.

Problem 3: Compute the quotient $1/(1 - x - x^2)$ by polynomial long division and confirm that the Fibonacci numbers are the result.

Problem 4: Find a Binet-type formula for the Lucas numbers using the same method as was used with the Binet formula for the Fibonacci numbers.

Problem 5: Find the number of ways a_n to cover a $2 \times n$ checkerboard with n rectangles (2×1 or 1×2, also known as *dominos*).

Problem 6: How many binary strings of length n are there that contain no two consecutive 1s?

Problem 7: Prove the *convolution identity:* if $f_a(x) = \sum_{n \geq 0} a_n x^n$ and $f_b(x) = \sum_{n \geq 0} b_n x^n$, then $f_a(x) f_b(x)$ is the ordinary generating function for the sequence $c_n = a_0 b_n + a_1 b_{n-1} + \cdots + a_i b_{n-i} + \cdots + a_n b_0$.

Problem 8: Suppose that $h(n) = an^2 + bn + c$ for some real numbers a, b, c; the generating function $\sum_{n=0}^{\infty} h(n) x^n$ is a rational function $p(x)/(1-x)^3$ for some polynomial $p(x)$ of degree 2 or less. Show that the coefficient of x^2 in $p(x)$ is $h(-1)$; show that $p(1) = 2a$.

Problem 9: Prove that if $f_a(x) = \sum_{n \geq 0} a_n x^n/n!$ and $f_b(x) = \sum_{n \geq 0} b_n x^n/n!$ then $f_a(x) f_b(x)$ is the exponential generating function for the sequence $c_n = \binom{n}{0} a_0 b_n + \binom{n}{1} a_1 b_{n-1} + \cdots + \binom{n}{i} a_i b_{n-i} + \cdots + \binom{n}{n} a_n b_0$.

Problem 10: We saw the *derangement numbers* $D_n = n!(1/0! - 1/1! + \cdots + (-1)^i/i! + \cdots + (-1)^n/n!)$ in Chapter 5. Use the recurrence $D_n = nD_{n-1} + (-1)^n, n \geq 2$, to find the exponential generating function for D_n.

Chapter 7

Catalan, Bell, and Stirling Numbers

7.1 Introduction

One of the basic purposes of combinatorics is enumeration, and we have seen many ways to enumerate arrangements of objects. The binomial coefficients enumerate ways of choosing a subset of distinct objects from a set; factorials and permutation numbers enumerate ways of choosing an ordered subset; and so on. In general, we distinguish several classes of problems involving enumeration of arrangements of objects. The objects may be distinct or identical; the arrangements may be considered distinct when order is changed (*order important*) or not (*order unimportant*); and we may distinguish among the classes in which they are arranged, or not. Also, we may ask whether repetition is allowed, or not (and if so, whether the repetition may be unlimited); whether all objects must be arranged; and whether all the classes into which they are to be arranged must receive at least one object.

For example, consider the number of ways of putting Halloween candies into two bags. If the candies are all identical, there are clearly fewer ways of arranging them than if all the candies are distinct. Again, if the bags are considered identical, there are fewer ways than if we have already assigned one bag to a particular child. Of course, if the candies are distinct it is not likely that the child will be terribly concerned about whether the chocolate bar went into the bag before or after the lollipop; if this makes a difference, then clearly there are more ways to arrange the candies. If we decide that repetition is allowed, then placing the chocolate bar into one bag does not mean that the other bag cannot also get a chocolate bar; we might have a limited number of chocolate bars available, or not. Are we allowed not to distribute (say) the lemon drop, or must we use all the available candy? Is it fair to leave one of the bags empty? These questions also affect the number of ways in which the distribution might be done. We have already seen ways to enumerate arrangements for some of these cases; the purpose of this chapter is to discuss enumeration in some cases that we have not before. This discussion will include quantities for which (unlike the binomial coefficients, for instance) no easy formula exists. These quantities are given the names of the mathematicians with whom they are chiefly associated.

7.2 Catalan Numbers

In the late eighteenth century, the extraordinarily prolific Swiss mathematician Leonhard Euler (about whom more is written in Appendix 3) investigated the question: How many ways are there of dividing up a regular polygon into triangles using non-intersecting diagonals? He produced, after great effort, a sequence that today is denoted C_n, representing the number of ways of so dividing an $(n+2)$-gon. By convention, we will define $C_0 = C_1 = 1$. This sequence turned out to arise in a remarkable number of different places, and (for reasons we will see later) was named eventually after the Belgian mathematician Eugène Catalan. We provide a table below of the Catalan numbers.

n	0	1	2	3	4	5	6	7
C_n	1	1	2	5	14	42	132	429

To investigate the Catalan numbers, we begin by considering an $(n+2)$-gon and we choose one side of this polygon. We can make this chosen side the side of a triangle in some triangulation of the polygon in exactly n ways; just choose one of the n corners that do not touch this side and draw lines from this corner to the two corners at either end of the chosen side. Each such triangle is the start of a triangulation, so our expression for C_n will involve a sum over n terms. Now suppose that the triangle we have drawn that includes our chosen edge leaves a $(k+2)$-gon on one side and an $(n-k+1)$-gon on the other side. Notice that this is valid for $0 \leq k \leq n-1$; a "2-gon" is one side of the polygon, and this means that our chosen triangle shares *two* sides with the original polygon. Now to finish the triangulation, there are $C_k \times C_{n-k-1}$ ways, because we can triangulate the $(k+2)$-gon in C_k ways, and the $(n-k+1)$-gon in C_{n-k-1} ways. Summing, we arrive at the *Catalan recurrence*: $C_n = \sum_{0 \leq k \leq n-1} C_k \times C_{n-k-1}$. We illustrate with Figure 7.1, using a hexagon (so $n + 2 = 6$). It should be clear that C_0, the number of ways of triangulating a "two-cycle," is 1 by convention just to make the recurrence work (some authors simply say that C_0 is undefined).

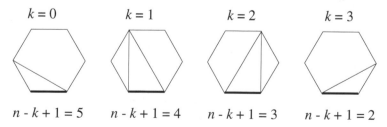

$k = 0$	$k = 1$	$k = 2$	$k = 3$
$n - k + 1 = 5$	$n - k + 1 = 4$	$n - k + 1 = 3$	$n - k + 1 = 2$

FIGURE 7.1:   Triangulating the hexagon.

It was not until the 1830s that Catalan investigated the question of how many ways one can correctly arrange n pairs of parentheses. Here "correctly" means that the expression makes sense, i.e., that each right parenthesis is matched by a preceding unique corresponding left parenthesis and vice versa. Thus, for $n = 0$ and $n = 1$ we get 1; for $n = 2$ we get 2, enumerating the set $\{(()), ()()\}$; and for $n = 3$ we find five ways, namely $\{((())), (())(), ()(()), (()()), ()()()\}$. Because the sequence begins $1, 1, 2, 5, \ldots$ we naturally wonder whether it is the same sequence as Euler's.

If we let C_n denote the number of ways to correctly arrange n pairs of parentheses, we can find a recurrence that relates each value to the preceding values. Take any correct arrangement of n pairs of parentheses, $n \geq 1$, and look at the leftmost parenthesis. It is matched by a right parenthesis elsewhere in the expression, either immediately following or at the end or somewhere in the middle of the expression. Because the expression is a correct arrangement, there must be a correct arrangement of 0 or more pairs of parentheses between this first parenthesis and the corresponding right parenthesis, and a correct arrangement of 0 or more pairs following the corresponding right parenthesis. Suppose there are k pairs of parentheses between it and its "mate." Then there are $n - k - 1$ pairs of parentheses following the mate. How many ways can this be accomplished for a given value of k? Clearly, there are $C_k \times C_{n-k-1}$ ways. Summing over k, we see that $C_n = \sum_{0 \leq k \leq n-1} C_k \times C_{n-k-1}$. Because this sequence has the same initial values and satisfies the same recurrence as the sequence of polygon triangulations, they are identical.

THEOREM 7.1 (Values of Catalan Numbers)

$$C_n = \binom{2n}{n} / (n + 1).$$

Proof. In how many ways may we arrange n pairs of parentheses in any order whatsoever? A string of n pairs of parentheses has length $2n$; we select n of these positions to be the left parentheses. Thus there are exactly $\binom{2n}{n}$ ways to do so. Write $C_n = \binom{2n}{n} - I_n$, where I_n denotes the number of *incorrect* arrangements of parentheses. We shall count I_n and the theorem will follow as a result. Consider an incorrect sequence of parentheses, denoted $a_1 a_2 \ldots a_{2n}$, where each a_i is either a left or a right parenthesis, and there are n of each. Because this sequence is incorrect, there must be a smallest subscript k such that a_k is a right parenthesis that is not matched by a preceding left parenthesis. Because k is the least such subscript, a_k is preceded by matching pairs of left and right parentheses, so k must be odd.

Now, for the trick: we reverse each of the first k parentheses and leave the others unchanged. This gives us a string of $n + 1$ left parentheses and $n - 1$ right parentheses. Clearly, there are $\binom{2n}{n+1}$ such strings. This process is reversible. Given such a string, we can find the first point k at which the number of left parentheses exceeds the number of right parentheses, and by reversing the parentheses up to that point, recover precisely the same incorrect expression we started with. This process provides a one-to-one and onto correspondence between the incorrect strings of n left and n right parentheses and the strings of $n+1$ left and $n-1$ right parentheses. This establishes the equation $I_n = \binom{2n}{n+1}$. We now calculate $C_n = \binom{2n}{n} - \binom{2n}{n+1} = \binom{2n}{n}/(n+1)$. □

The Catalan numbers occur in numerous places throughout all of mathematics, and we will provide examples here, in the exercises, and in the problems.

Example 7.1: We wish to walk to a point that is n blocks north and n blocks east of our current position, but we never want to be more blocks north of our current position than we are east. In how many ways can we do such a walk if we stick to the streets?

We model this problem by starting at the origin $(0,0)$ and take a total of $2n$ steps, each of which consists of adding 1 to either the first or the second coordinate (but not both). The requirement that we may not go further north than east means that the allowable points (x, y) at each step must satisfy $x \leq y$. We will remember from Chapter 2 that the total number of paths without this restriction is $\binom{2n}{n}$. We might try to count paths for which y becomes greater than x and subtract, but there is an easier way.

Observe that a walk from $(0,0)$ to (n,n) may be described by a string of directions, where each direction is N for north or E for east. The condition equates to saying that at any point we have no more Ns than Es. Therefore, take any walk, and change N to a right parenthesis, ")" and E to a left parenthesis, "(". We see that we have an expression involving nested parentheses, and this must be a correct arrangement because as we move from left to right we never will have encountered more right parentheses symbols than left parentheses. It is not hard to see that this will provide a bijection between walks of the sort we are asked for and correct arrangements of n pairs of parentheses. It follows that there are exactly C_n walks from $(0,0)$ to (n,n) that meet the condition that we are never further north than east. □

Example 7.2: A *rooted binary tree* is a kind of graph (as discussed in Section 1.5); a collection of points (also called *nodes* or *vertices*) where one distinguished vertex, called the *root*, is placed at the top of the diagram. Each vertex has either zero or two edges descending down to any further vertices. A vertex is said to be *internal* if it has two edges descending. How many rooted binary trees with n internal vertices are there?

Figure 7.2 shows the cases for $n = 0$, 1, and 2.

If we denote by a_n the number of rooted binary trees with n internal points, we see from the diagram that $a_0 = a_1 = 1$ and $a_2 = 2$. To prove generally the equality $a_n = C_n$, we derive a recurrence. Observe that for $n \geq 1$, removal of the root leaves us with two rooted binary trees. Since the root

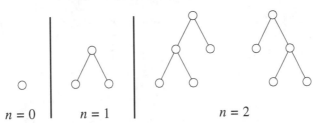

$n = 0$ $n = 1$ $n = 2$

FIGURE 7.2: Rooted binary trees with n internal points.

is an internal point of the original rooted binary tree, the two trees we get after removing it have k and $n - k - 1$ internal points, for some value k with $0 \leq k \leq n - 1$. Now, as with the problem of triangulations of the polygon, we may write $a_n = \sum_{k=0}^{n-1} a_k a_{n-k-1}$. The sequence a_n has the same initial terms and satisfies the same recurrence as the Catalan numbers so it is the same sequence. $\qquad \square$

We spent all of Chapter 6 discussing generating functions. The Catalan numbers will provide us with a challenge.

THEOREM 7.2 (The Catalan Number Generating Function)

$$C(x) = \sum_{n=0}^{\infty} C_n x^n = \frac{1 - \sqrt{1 - 4(x)(1)}}{2x}$$

Proof. We will use the convolution identity from Problem 6.8. Let $C(x) = \sum_{i \geq 0} C_n x^n = 1 + x + 2x^2 + 5x^3 + \dots$. This identity gives us:

$$C^2(x) = \sum_{i \geq 0} (C_0 C_i + C_1 C_{i-1} + \dots + C_i C_0)\, x^i.$$

The Catalan recurrence with which we began this section tells us that the parenthesized expression on the right side of this equation is just C_{i+1}, so $C^2(x) = \sum_{i \geq 0} C_{i+1} x^i$ or $C^2(x) = 1 + 2x + 5x^2 + 14x^3 + \dots$. (The reader should confirm this by squaring the polynomial $1 + x + 2x^2 + 5x^3 + 14x^4$, neglecting terms past x^4.) Now we multiply both sides by x and add 1: $xC^2(x) + 1 = 1 + x + 2x^2 + 5x^3 + 14x^4 + \dots = C(x)$. Now what are we to make of the equation $xC^2(x) - C(x) + 1 = 0$? This is just a quadratic equation in C, for any particular value of x. We may apply the quadratic formula with $a = x$, $b = -1$, and $c = 1$. We arrive at two possibilities:

$$C(x) = \frac{1 \pm \sqrt{1 - 4x}}{2x}.$$

To determine whether to use the plus or the minus, look at $C(0)$; we get $C(0) = 1 + 1 \cdot 0 + 2 \cdot 0^2 + \dots = 1$. Now, setting $x = 0$ in the formula above means division by 0 regardless of whether we use the plus or the minus. A closer look is in order.

If we choose the plus sign, then as x approaches 0 we arrive at the expression $2/0$, which clearly is undefined. If we choose the minus, we arrive at the indeterminate form $0/0$, which is susceptible to L'Hôpital's rule. This yields $\lim_{x \to 0} C(x) = 1$, as required. If there is any doubt that we have found the correct generating function, we may use Newton's binomial theorem or Maclaurin series to expand the radical and confirm that Catalan numbers are the result.   □

We have only touched very lightly upon the subject of Catalan numbers, a subject about which a great deal is known. For more information, the interested reader is encouraged to see ([40], ch. 20), reprinted in [43] for a very readable exposition; much more technical detail is available in [66]. Another source of good information on the Catalan numbers is [108] and the online addendum [109].

7.3   Stirling Numbers of the Second Kind

Consider the number of ways of distributing n distinct candies to k distinct children in such a way that each candy is distributed and no child fails to get a candy ($n \geq k$). Our first impulse might be to attach to each of the candies a number from 1 to k indicating which child gets that particular candy; by putting the candies in line in some order, we have an n-digit number in base k, and can easily enumerate the possibilities. However, this does not work, for we have not assured ourselves that each child gets at least one candy. The k^n possible n-digit numbers in base k include many possibilities that omit one or more base-k digits.

Likewise, we cannot say (as we might if the candies were identical) that we will simply distribute one candy to each child in the beginning, and then arrange any remaining candies afterwards; it makes a difference which candy is distributed to which child, and a distribution in which (for instance) child 1 gets candies A and B might be counted as two distributions, depending on whether candy A was handed out first (we are assuming that order is not important here). Instead we approach this problem from a different point of view.

We still associate with each candy a number denoting which child receives it; regard this association as a function from an n-set (the candies) to a k-set (the children) and we require that it be *onto*. This means we have rephrased our original problem as follows. *How many surjections are there from an n-set onto a k-set?* (Recall *surjection* just means *onto function*.)

There is an ancient mathematical tradition that says that, in order to solve a problem, you merely invent a symbol, define the symbol to be the solution to the problem, and then announce that you've solved the problem, giving the symbol as the solution. In this tradition, we define the *Stirling number of the second kind*, $\left\{ {n \atop k} \right\}$, to be the number of ways to divide n distinct objects into k identical classes. Because the children are distinct, we must multiply by $k!$ to take into account the number of possible ways of ordering the children. Thus the number of surjections from an n-set onto a k-set is just $k!\left\{ {n \atop k} \right\}$. We have now solved the problem; if we are given nine candies and six children, we may distribute the candies in $6!\left\{ {9 \atop 6} \right\}$ ways. Other notations for the Stirling numbers of the second kind include $S_n^{(k)}$ and $S(n,k)$; our notation was popularized by Knuth ([48], [64]), who attributes it to Karamata ([55]). Knuth also suggests that the notation $\left\{ {n \atop k} \right\}$ be read as "n subset k."

The irritated demands of those unmoved by the tradition discussed in the previous paragraph suggest we look more deeply into the properties of the numbers $\left\{ {n \atop k} \right\}$ in order to produce numerical values. To this end, we establish a few conventions. For instance, we will define $\left\{ {0 \atop 0} \right\} = 1$ for much the same reason that we define $0! = 1$; it is convenient for formulas that we will develop soon. If $n < k$ or if n and k are not both non-negative integers, then we will say $\left\{ {n \atop k} \right\} = 0$. It is easy to see that $\left\{ {n \atop 1} \right\} = 1$, since the only way to put n objects into one class is for the class to have all the objects. Similarly, if $n = k$ we clearly have $n!$ ways to arrange the classes so that each gets exactly one distinct object, so that $\left\{ {n \atop n} \right\} = 1$.

We now seek a recurrence relation that will let us calculate $\left\{ {n \atop k} \right\}$ from smaller Stirling numbers of the second kind.

THEOREM 7.3 (A Stirling Number Recurrence)

$$\left\{ {n \atop k} \right\} = \left\{ {n-1 \atop k-1} \right\} + k \left\{ {n-1 \atop k} \right\}$$

Proof. Suppose we assign objects $\{o_1, o_2, \ldots o_n\}$ to k classes. When we come to the last object o_n, we may place it in a class by itself; if we do this, there are $\left\{ {n-1 \atop k-1} \right\}$ ways in which we might have arranged the $n-1$ preceding objects into $k-1$ classes. Otherwise, we already have k non-empty classes, and we must place the new object into one of the pre-existing classes; there are k ways to do this, and $\left\{ {n-1 \atop k} \right\}$ ways in which the $n-1$ earlier objects were placed. The result follows. $\qquad\square$

Example 7.3: Calculate the value $\left\{ {4 \atop 2} \right\}$ by applying the recurrence above.

$$\left\{ {4 \atop 2} \right\} = \left\{ {3 \atop 1} \right\} + 2\left\{ {3 \atop 2} \right\} = \left\{ {2 \atop 0} \right\} + 1\left\{ {2 \atop 1} \right\} + 2\left(\left\{ {2 \atop 1} \right\} + 2\left\{ {2 \atop 2} \right\} \right)$$

We use $S(2,0) = 0$ and $S(2,1) = S(2,2) = 1$ to simplify further.

$$\left\{ {4 \atop 2} \right\} = 0 + \left\{ {2 \atop 1} \right\} + 2\left\{ {2 \atop 1} \right\} + 4(1) = 3\left\{ {2 \atop 1} \right\} + 4$$

So, $\left\{ {4 \atop 2} \right\} = 7$. □

It is somewhat tedious to find the Stirling numbers of the second kind by use of the recurrence. One approach is to create a table similar to Pascal's triangle and refer to the table, shown below.

To extend the table or verify any entry in it, choose an element. Multiply that element by the number at the top of its column, and add to it the element on its left (or 0 if there is no element to the left). The result should be the number below the chosen element.

$n\backslash k$	0	1	2	3	4	5	6
0	1	0	0	0	0	0	0
1	0	1	0	0	0	0	0
2	0	1	1	0	0	0	0
3	0	1	3	1	0	0	0
4	0	1	7	6	1	0	0
5	0	1	15	25	10	1	0
6	0	1	31	90	65	15	1

Example 7.4: How many ways are there to factor the number $2 \cdot 3 \cdot 5 \cdot 11 \cdot 23 = 7590$ into a product of two factors each greater than 1? How many for three such factors?

It is clear that each of these factors needs to use some of the distinct primes in the unique factorization of 7590. Accordingly, we take two boxes and put prime factors into them. As there are five prime factors and two boxes, we find $\left\{ {5 \atop 2} \right\} = 15$. Note that we count the factorization 2×3795 the same as 3795×2, for instance, because the boxes are identical. In the same way, $\left\{ {5 \atop 3} \right\} = 25$, so we have 25 ways of writing 7590 as a product of three factors, each bigger than 1. □

Example 7.5: Angered by all the candies being given to children at the beginning of this section, a group of indignant parents take five distinct pamphlets on the dangers of childhood obesity to give to two textbook authors. In how many ways may this be done if the two authors are considered distinct? In how many ways if the authors are considered identical? In either case, no author escapes being given at least one pamphlet, and there is only one copy of each of the five pamphlets.

Here $n = 5$ refers to the pamphlets, and $k = 2$ refers to the authors. From the table, $\left\{ {5 \atop 2} \right\} = 15$, so if the authors are distinct the answer is $2! \times 15$ or 30, and if the authors are identical we get 15. □

Example 7.6: Properly chastened, the authors meet with three indignant parents and offer them a collection of six distinct healthful snacks. (The authors never offer *healthy* snacks, as by definition *healthy* implies that the snacks are still alive and potentially able to fight back. Ethical, dietary, and legal considerations forbid.) Assuming the three parents are distinct, in how many ways can this be done if each parent is to receive at least one snack? In how many ways if a parent may refuse a snack (all snacks are distributed)? In how many ways if we do not require that all snacks be distributed, and any parent may refuse a snack?

The first problem is just a matter of looking at the table; $n = 6$, $k = 3$, so we get $3! \times \left\{ {6 \atop 3} \right\} = 6 \times 90 = 540$.

The second problem must be done in pieces. If all three parents accept a snack, 540 is still the answer; if one parent refuses, there are three ways to choose a parent to refuse and $2! \times \left\{ {6 \atop 2} \right\} = 62$ ways to distribute the snacks to the other two parents, so we get another 186 possibilities. Then if two parents refuse, we get another three possibilities (three choices of the parent to accept all six snacks, or if you like $\binom{3}{1} \times \left\{ {6 \atop 1} \right\} = 3$). Adding these, we find a total of 729 possible distributions of snacks.

The third problem is the easiest or most difficult, and may be solved by considering the trash can as a fourth potential snack recipient. If all four possible recipients get a snack, we have $4! \times \left\{ {6 \atop 4} \right\} = 1560$. If one recipient (parent or trash can) gets no snack, we count $\binom{4}{1} \times 3! \times \left\{ {6 \atop 3} \right\} = 2160$. Two snackless recipients yields $\binom{4}{2} \times 2! \times \left\{ {6 \atop 2} \right\} = 372$. Finally, if all the snacks go to one recipient, there are four possibilities. Adding these numbers yields 4096 ways. Realizing that this answer is just 4^6, we see that we could have saved ourselves much trouble by saying: Associate with each snack a number from $\{0, 1, 2, 3\}$ according to which recipient receives it. Placing these numbers in order gives us a six-digit base-4 number, of which there are 4^6. □

This example leads us to seek a generalization. A classic theorem concerning Stirling numbers of the second kind gives a formula that has been used as the definition. We use the notation $[r]_k = r!/(r-k)! = r(r-1)\ldots(r-k+1)$, the *falling factorial* notation.

THEOREM 7.4 (A Stirling Number Formula)

$$r^n = \sum_{k=0}^{r} \left\{ {n \atop k} \right\} [r]_k$$

Proof. Suppose we have r recipients of n objects, and, although all the objects must be distributed, some recipients may get no object. If all r recipients get an object, we have $r! \times \left\{ {n \atop r} \right\}$, which is the $k = r$ summand of the formula. For the case in which one recipient gets no object, we count $\binom{r}{1} \times (r-1)! \times \left\{ {n \atop r-1} \right\}$, which is equivalent to the $k = r - 1$ summand. In general, if i recipients get no objects, we have $\binom{r}{i} \times (r-i)! \times \left\{ {n \atop r-i} \right\} = \frac{r!}{i!} \left\{ {n \atop r-i} \right\}$

which is the $r - i$th summand. On the other hand, the quantity we are counting is the number of n-digit base-r integers, of which there are r^n.   □

Example 7.7: Write the polynomial x^3 as a linear combination of the polynomials 1, x, $x(x - 1)$, and $x(x - 1)(x - 2)$.

In the preceding theorem, substitute 3 for n and x for r; this gives us

$$x^3 = \sum_{k=0}^{3} \left\{ {3 \atop k} \right\} \frac{x!}{(x - k)!}$$

and we simplify this to get $x^3 = 0 \cdot 1 + 1 \cdot x + 3 \cdot x(x - 1) + 1 \cdot x(x - 1)(x - 2)$, or $x^3 = x + 3x(x - 1) + x(x - 1)(x - 2)$. Even though the recurrence requires that x be an integer, the identity holds for any value of x.   □

A number of identities exist for particular classes of Stirling numbers of the second kind; for instance, in [48] it is shown that

$$\left\{ {n \atop 2} \right\} = 2^{n-1} - 1.$$

This, as we have seen with other combinatorial identities, may be proved in a variety of ways. The simplest but least enlightening is the use of mathematical induction using the recurrence of Theorem 7.3. We adapt the proof from [48], which is a combinatorial proof giving more insight into the formula. Consider a set of n objects $o_1, o_2, \ldots o_n$ and two identical empty boxes. We toss object o_n into one of the boxes; they are no longer identical, as one is not empty. Now, for the remaining $n - 1$ objects, we have a choice; toss the object into the box with o_n, or into the other box. There are 2^{n-1} ways to do this. One of the resulting distributions consists of all n objects in one box, and the other box is empty. Because we are counting only distributions in which neither box is empty, we subtract 1 to arrive at $2^{n-1} - 1 = \left\{ {n \atop 2} \right\}$.

This formula is one of a number of identities for $\left\{ {n \atop k} \right\}$ for specific k; see the problems for more examples. Each may be proved by induction or combinatorial techniques. There is another recurrence relation for $\left\{ {n \atop k} \right\}$ that may be useful in computing the entries in a table such as the one given above.

THEOREM 7.5 (Another Stirling Number Recurrence)

$$\left\{ {n \atop k} \right\} = \sum_{i=0}^{n-1} \binom{n-1}{i} \left\{ {i \atop k-1} \right\} \quad for \ n > 0$$

Proof. Notice first of all that many of these terms will be 0 (when $i < k - 1$) and recall $\left\{ {0 \atop 0} \right\} = 1$. Assume we have chosen a way of arranging all n objects into k boxes, and look at the box that contains the nth object o_n. That box will contain $n - i$ total objects, where $0 \leq i \leq n - 1$. If we eliminate this box, we have a way of arranging i objects into $k - 1$ boxes, and there are clearly

$\left\{ {i \atop k-1} \right\}$ ways of doing this. There are also $\binom{n-1}{i}$ ways of selecting the objects that do not go into the box with o_n. Summing over all values of i gives all $\left\{ {n \atop k} \right\}$ ways of arranging the objects. □

Example 7.8: Prove the identity $\left\{ {n \atop 2} \right\} = 2^{n-1} - 1$ using Theorem 7.5.

Substituting 2 for k, we get $\left\{ {n \atop 2} \right\} = \sum_{i=0}^{n-1} \binom{n-1}{i} \left\{ {i \atop 1} \right\}$; since $\left\{ {i \atop 1} \right\} = 1$ for $i > 0$, we have $\left\{ {n \atop 2} \right\} = \sum_{i=1}^{n-1} \binom{n-1}{i} = 2^{n-1} - 1$ by the binomial theorem. □

7.4 Bell Numbers

How many ways can we assign n students to work in groups? We do not specify the number of students in any group (except that a group must be non-empty), and do not specify the number of groups (except that there must be at least one and no more than n groups). We easily see this to be the sum of Stirling numbers of the second kind. We define the *Bell number* B_n to be the number of ways to partition a set of n distinct objects. We may also think of it as the number of ways to place a set of n elements into identical boxes, where the number of boxes is not specified, and write $B_n = \sum_{i=0}^{n} \left\{ {n \atop i} \right\}$.

Example 7.9: We have three distinct snacks and wish to put them into bags. How many ways can this be done?

There is exactly one way to do this with three bags (each snack into its own bag); three ways to do this with two bags (depending on which snack goes into a bag by itself); and one way with one bag. Thus $B_3 = 5$. □

Example 7.10: How many ways are there of writing $2 \times 3 \times 5 \times 7 = 210$ as a product of distinct integers?

Because 210 is (as shown) a product of distinct primes, we may think of a factor as a subset of the set $\{2, 3, 5, 7\}$. Thus the number of ways of writing 210 as a product of integers is the number of ways of partitioning that set, or $B_4 = 15$. Here, the product 1×210 represents the partition $\{2\,3\,5\,7\}$; the product 6×35 represents the partition $\{2\,3\}, \{5\,7\}$, and so forth. Notice that writing 6×35 is considered the same as 35×6. □

Summing a row of the Stirling triangle will work for smaller Bell numbers, but it would be worthwhile to find a better way to compute, say, B_{25} that would not involve so much work. Such a formula is given in the following theorem.

THEOREM 7.6 (Bell Number Recurrence)

$$B_{n+1} = \sum_{i=0}^{n} \binom{n}{i} B_i$$

The proof is left to the reader (see Problem 6.7). Note that B_0 is 1.

Example 7.11: Find B_6.

The recurrence gives $B_6 = B_0 + 5B_1 + 10B_2 + 10B_3 + 5B_4 + B_5$. We may compute these by summing rows in the triangle, or by recursively applying Theorem 7.6 to get $B_1 = 1$, $B_2 = 2$, $B_3 = 5$, $B_4 = 15$, and $B_5 = 52$. This gives us $B_6 = 1 + 5 + 20 + 50 + 75 + 52 = 203$. This corresponds to the sum of the sixth row of the triangle in Section 7.3.   □

Another approach is called the Bell triangle. Just as binomial coefficients may be computed easily by means of Pascal's triangle, the Bell numbers may be computed by a triangular array. We create the array by means of the following rules:

1. The first row consists of the number 1 by itself.

2. Each row is one entry longer than the row before it.

3. The first item in each row is equal to the last item in the preceding row.

4. Each entry after the first is the sum of the entry to its left and the entry directly above that entry.

Applying these rules, we get the following table.

1					
1	2				
2	3	5			
5	7	10	15		
15	20	27	37	52	
52	67	87	114	151	203

The rightmost entries give us a list of Bell numbers; $B_0 = 1$, $B_1 = 1$, $B_2 = 2$, $B_3 = 5$, $B_4 = 15$, $B_5 = 52$, and $B_6 = 203$. This is only a small selection of the properties of the Bell numbers; we refer the interested reader to ([42]), also found in the CD-ROM collection ([43]) for a very readable discussion. More technical details and identities may be found in [48], [63], and [139].

7.5   Stirling Numbers of the First Kind

Recall that Theorem 7.4 tells us that the Stirling numbers of the second kind may be used to write monomials x^n as linear combinations of falling factorials, so we may write $x^3 = \left\{ {3 \atop 0} \right\} + \left\{ {3 \atop 1} \right\} x + \left\{ {3 \atop 2} \right\} x(x-1) + \left\{ {3 \atop 3} \right\} x(x-1)(x-2)$. Indeed, the original definition of $\left\{ {n \atop k} \right\}$ was as the coefficient in that equation. It is easier to write the falling factorial in terms of sums of monomials; so we can figure that $[n]_3 = n(n-1)(n-2) = n^3 - 3n^2 + 2n$. These coefficients are referred to as the *Stirling numbers of the first kind*, or sometimes as *signed Stirling numbers of the first kind*, and we denote them $\left[{n \atop k} \right]$. Thus we have $[x]_k = \sum_{i=0}^{k} \left[{k \atop i} \right] x^i$.

We refer to the numbers $\left| \left[{n \atop k} \right] \right|$ as *unsigned Stirling numbers of the first kind*. The unsigned Stirling numbers of the first kind may also be viewed as the coefficients of the polynomial $[x]^k = x(x+1) \ldots (x+k-1)$ (the "rising factorial"). For example, where $[x]_3 = x(x-1)(x-2) = x^3 - 3x^2 + 2x$, we have $[x]^3 = x(x+1)(x+2) = x^3 + 3x^2 + 2x$.

These numbers are more easily calculated than Stirling numbers of the second kind; we simply multiply out the polynomial $[x]_k = x(x-2) \ldots (x - k+1)$. It is also easy to find a recurrence for signed or unsigned Stirling numbers of the first kind.

THEOREM 7.7 (Another Stirling Number Recurrence)

$$\left[{m \atop n} \right] = \left[{m-1 \atop n-1} \right] - (m-1) \left[{m-1 \atop n} \right]$$

Proof. We begin by using the definition and applying a bit of algebra.

$$\sum_{j=0}^{m} \left[{m \atop j} \right] x^j = x(x-1) \ldots (x - m + 1) = [x]_m = (x - m + 1)[x]_{m-1}$$

Replacing the falling factorial on the right with the equivalent sum gives us:

$$\sum_{j=0}^{m} \left[{m \atop j} \right] x^j = (x - m + 1) \sum_{j=0}^{m-1} \left[{m-1 \atop j} \right] x^j$$

and with the distributive law we obtain

$$\sum_{j=0}^{m} \left[{m \atop j} \right] x^j = x \sum_{j=0}^{m-1} \left[{m-1 \atop j} \right] x^j - (m-1) \sum_{j=0}^{m-1} \left[{m-1 \atop j} \right] x^j.$$

What is the coefficient of x^n on each side of the equation above? We get $\left[{m \atop n} \right]$ on the left and $\left[{m-1 \atop n-1} \right] - (m-1) \left[{m-1 \atop n} \right]$ on the right. Since the two polynomials are equal, the coefficients are equal for each n, establishing the theorem. □

A similar result holds for the unsigned Stirling numbers of the first kind.

THEOREM 7.8 (Unsigned Stirling Number Recurrence)

$$\left| \left[{m \atop n} \right] \right| = \left| \left[{m-1 \atop n-1} \right] \right| + (m-1) \left| \left[{m-1 \atop n} \right] \right|$$

The proof is essentially the same as that of Theorem 7.7.

By convention, we take $[x]_0 = 1$, so $\left[\begin{smallmatrix} 0 \\ 0 \end{smallmatrix}\right] = 1$. It is not hard to see that for $n > 0$ we have $\left[\begin{smallmatrix} n \\ 0 \end{smallmatrix}\right] = 0$ because $[x]_m$ is a multiple of x and so has no x^0 term. In the same way, since the leading coefficient of the polynomial $[x]_m$ is 1, we see $\left[\begin{smallmatrix} m \\ m \end{smallmatrix}\right] = 1$. Also, we set $\left[\begin{smallmatrix} n \\ m \end{smallmatrix}\right] = 0$ whenever $m > n$. It is not quite as easy to find other coefficients; some examples are given in the problems.

The unsigned Stirling numbers of the first kind have a combinatorial interpretation like that of the Stirling numbers of the second kind. Suppose we have m people and wish to seat them at n identical circular tables, leaving no table empty. In how many ways can we do this, assuming that the arrangement around the table makes a difference? In particular, if there are one or two people at a table, there is only one way to seat them; with three people at a table, there are two ways; with four people, there are six ways. In general, if there are k people at a circular table, then there are $(k-1)!$ ways to arrange them; we can line them up in any of $k!$ ways, and then we divide by k because we do not distinguish between two seatings that differ only by which of the k chairs the first person sat in. But this does not help us compute the number of ways of seating the m people, because different seatings may use different values of k. Let us denote the number of ways to seat m people at the n tables as $s(m, n)$, where clearly $n \leq m$. It should be clear, for example, that $s(m, m) = 1$; each table has one person seated at it, and as the tables are identical this can be done in just one way. Also, $s(m, 0) = 0$ for $m > 0$ and $s(0, 0) = 1$ by convention.

THEOREM 7.9 (A Combinatorial Interpretation)

$$s(m, n) = \left| \left[\begin{array}{c} m \\ n \end{array} \right] \right|$$

Proof. We first establish the recurrence relation of Theorem 7.8. Suppose we have m people labeled 1 through m. We may seat person m at a table alone and seat the other people in $s(m-1, n-1)$ ways, or we may seat the other $m-1$ people at n tables, and squeeze person m into a table in $m-1$ ways, according to which of the other $m-1$ people is seated to person m's right. There are $(m-1)s(m-1, n)$ to do this. Adding these two gives us $s(m, n) = s(m-1, n-1) + (m-1)s(m-1, n)$. Now, since the numbers $s(m, n)$ and $\left| \left[\begin{smallmatrix} m \\ n \end{smallmatrix} \right] \right|$ satisfy the same recurrence relation and (as we have already seen) $s(m, 0) = \left| \left[\begin{smallmatrix} m \\ 0 \end{smallmatrix} \right] \right|$, it follows that $s(m, n) = \left| \left[\begin{smallmatrix} m \\ n \end{smallmatrix} \right] \right|$ for any m and n. □

The next result requires a definition from the theory of permutation groups. A *permutation* of a set of n elements is a bijection from that set to itself. A *cycle* of a permutation is a subset of the elements with the property that each member of the subset is mapped to any other member of the subset by repeated applications of the permutation. For example, consider the permutation σ on the set $\{1, 2, 3, 4, 5, 6\}$ given by the table below.

n	1	2	3	4	5	6
$\sigma(n)$	3	4	5	2	3	6

By applying σ repeatedly to the element 1, we see that $\sigma(1) = 3$, $\sigma^2(1) = \sigma(3) = 5$, and $\sigma^3(1) = \sigma(5) = 1$; so the set $\{1, 3, 5\}$ is a cycle. The other cycles are $\{2, 4\}$ and $\{6\}$. Groups of permutations are discussed in Chapter 8.

COROLLARY 7.9.1

There are $\left|\left[\begin{smallmatrix} n \\ m \end{smallmatrix}\right]\right|$ permutations on n letters with m cycles.

Proof. We establish a one-to-one correspondence between ways to seat n people at m tables and the specified permutations. Given a seating, let σ be the permutation that maps i to j if and only if person i is sitting immediately to the left of person j. Conversely, given a permutation with m cycles, take the people whose labels correspond to the elements of a cycle, and seat them at a table in the order they appear in the cycle; repeat this procedure until all m cycles have been turned into seats at all m tables. This process gives us a bijection, so the two sets have the same number of elements. $\qquad\square$

As with Stirling numbers of the second kind, we may present them in the form of a table and refer to it, shown below. To extend the table or verify any entry in it, choose an element. Multiply that element by the number n at the left of its row, and subtract it from the element on its left (or 0 if there is no element to the left). The result should be the number below the chosen element.

$n\backslash k$	0	1	2	3	4	5	6
0	1	0	0	0	0	0	0
1	0	1	0	0	0	0	0
2	0	−1	1	0	0	0	0
3	0	2	−3	1	0	0	0
4	0	−6	11	−6	1	0	0
5	0	24	−50	35	−10	1	0
6	0	−120	274	−225	85	−15	1

7.6 Computer Algebra and Other Electronic Systems

Although it is important to be able to manipulate the identities and recurrences involved in the Catalan, Bell, and Stirling numbers, we can quickly find the values of these quantities with a variety of software packages. Each of the major packages has some ability to calculate these quantities rapidly.

Even a spreadsheet can produce the Catalan numbers quickly and efficiently. As an example, consider Microsoft Excel. We use the function COM-

BIN, which produces binomial coefficients. To generate the Catalan numbers, open Excel, and place the focus in cell A1. Enter 1 and go to A2. In A2, place the cursor in the formula bar above the cells and type =A1+1 and press Enter. At this point, cell A2 should read 2. We will arrange for the cells in column A to display the number of the row as follows: Select all of the cells A1 down to A25 (say), by placing the cursor in A1, holding down the Shift key, and pressing the down arrow until all the cells you wish are highlighted. Then use the Fill command and Fill Down. (The location of the Fill command may vary according to the version you are using.)

To compute the Catalan numbers, we place the cursor in cell B1 and type = COMBIN(2*A1, A1)/(A1+1). This command should insert the formula for the first Catalan number C_1 in that cell. Now select cells B1 to B25 and again select Fill Down. In less than half a minute you have calculated the values of all the Catalan numbers up to C_{25}, and will probably find that the computer must use exponential notation for those past about C_{22}.

Both *Mathematica* and *Maple* also have the capability of producing Catalan numbers. For *Mathematica*, we first type <<DiscreteMath`Combinatorica` to load the package of combinatorial functions. The backwards apostrophe before and after the name Combinatorica is found to the left of the 1 in the top row of most keyboards. Now the name in *Mathematica* of the function we want is CatalanNumber, with capital letters as shown. This instruction:

Table[CatalanNumber[i], {i, 1, 25}]

produces a list of the first 25 Catalan numbers more quickly than the similar exercise with the Excel spreadsheet. In addition, we get the larger numbers such as $C_{25} = 4861946401452$ without exponential notation and truncation as with Excel.

The Combinatorica package is not required for the Stirling numbers of the first and second kinds, which are available by default. To generate the triangle of Stirling numbers of the second kind, for instance, we may enter this:

Table[StirlingS2[i,j], {i,0,6},{j,0,6}]//TableForm

The first part of the command, the Table command, produces a two-dimensional array indexed by the variables i and j. The StirlingS2 function produces a Stirling number of the second kind (just as the StirlingS1 function produces those of the first kind). The two sections in braces indicate the ranges of the indices, and the TableForm command arranges the results in a square grid.

There is no built-in function to calculate the Bell numbers in *Mathematica*, but that causes no difficulty.

Sum[StirlingS2[6, i], {i, 0, 6}]

If we wish to generate Bell numbers frequently, it is better to define a function to do this for us. The syntax is the same as for other *Mathematica* functions.

bell[n_] := Sum[StirlingS2[n, i], {i, 0, n}]

After entering that line, for example, bell[10] yields 115975.

The *Maple* package, like *Mathematica*, requires a command to load the combinatorial functions. In *Maple*, the command `with(combinat);` loads this package and enables one to calculate Catalan and the various Stirling numbers. The Catalan numbers are not built-in, but are easily accessed by means of the binomial coefficients. These are found using the `numbcomb(n,m)` function that calculates $\binom{n}{m}$. Entering the definition requires the use of the right-arrow key.

`seq(numbcomb(i)/i+1`$\rightarrow$`, i=0..25);`

Entering the character / tells *Maple* to begin the denominator of a fraction; the right arrow key means that subsequent entries will not appear in the denominator. It is important to distinguish between this and the right-arrow symbol in *Maple* that is generated by the two keys - (hyphen) and > (greater-than) and which indicates function definition.

This command produces the first 25 Catalan numbers. As with *Mathematica*, no exponential notation or truncation occurs.

Stirling numbers are produced in *Maple* using the commands `stirling1` and `stirling2`. Thus, `stirling1(5, 3);` produces 35. Again, *Maple*'s `seq` command will produce a sequence, but we can get a two-dimensional array with `Matrix()`. Thus the table of Stirling numbers of the first kind could be produced by this command:

`Matrix[7, (m,n)-> stirling1(`$m-1, n-1$`));`

In this line, the combination of the hyphen and the greater-than sign produce an arrow that indicates a function taking the pair (m, n) as input and producing $\begin{bmatrix} m-1 \\ n-1 \end{bmatrix}$ as output. The numeral 7 indicates a 7×7 matrix.

Maple also has the function `bell` that produces the Bell numbers; after loading the combinatorial functions as above, we may type `bell(10);` to get 115975.

Exercises 7A

1. A ticket-taker sells tickets to a tacky tatting show. Each tatting ticket is ten dollars. If there are $2n$ customers lined up to take tickets to the tatting show, and n of them have \$20 bills and the other half have \$10 bills, in how many ways can they line up so that the ticket-taker will always have correct change (she begins with no change)? It may help to represent the line as "(" for a customer with a \$20 bill and ")" for a customer with a \$10 bill.

2. Draw each of the distinct ways of triangulating a regular hexagon.

3. Place $2n$ dots in a row along the x-axis, and imagine that we connect pairs of dots with arcs that curve up into the first quadrant. We are to

do this so that each dot is connected with exactly one other dot, and with no arcs crossing one another; show that C_n is the number of ways this can be done.

4. Use the Catalan recurrence to verify the table value for C_5.

5. Use Theorem 7.1 to calculate C_9.

6. Make a 4×4 matrix whose (i, j)-entry is $\left\{ {i \atop j} \right\}$. Calculate the inverse of this matrix (by hand, with a calculator, or using a computer algebra system such as *Mathematica* or *Maple*). Compare the entries in this inverse with the Stirling numbers of the first kind.

7. Use the recurrence to calculate $\left\{ {7 \atop k} \right\}$ for $k = 2, 3, 4$.

8. How many ways are there to factor the number $2 \times 3 \times 7 \times 19 = 798$ into a product of two factors each greater than 1?

9. How many ways are there to distribute six distinct chocolate rabbits into three identical Easter baskets, where each basket is to receive at least one chocolate rabbit? In how many ways can this be done if the baskets are distinct?

10. Repeat the previous exercise if we do not have to distribute all six rabbits.

11. Five dishonest lobbyists set out to bribe three corrupt politicians. If each politician receives at least one bribe, and each lobbyist offers a bribe to exactly one politician, in how many ways can we assign lobbyists to politicians, assuming the politicians are distinct?

12. Five dishonest lobbyists are caught engaging in bribery by federal authorities, and thrown into two jail cells. If each jail cell is to hold at least one lobbyist, in how many ways can we arrange them? Assume the cells are identical.

13. Six investors wish to invest their savings into four mutual funds; they have eight funds from which to choose, and each investor will put his or her savings into just one fund. In how many ways can four funds be chosen and the money invested? Assume the funds are distinct and each of the four chosen funds receives some money.

14. How many ways are there to write $154 = 2 \times 7 \times 11$ as a product of distinct integers? Write out each product.

15. How many partitions of a set of eight elements are there?

16. Use the recurrence of Theorem 7.7 to find $\left[{7 \atop 4} \right]$, $\left[{7 \atop 5} \right]$, and $\left[{7 \atop 6} \right]$.

17. We have six people who will be seated at two circular tables. How many ways can we do this if the tables are identical? How many ways if one table is distinguished?

Exercises 7B

1. An election takes place in which each of two candidates A and B will receive n votes; there are $2n$ voters. In how many ways can the voters line up so that candidate B will never be in the lead? It may help to represent the line with a ")" for a voter for B and "(" for a voter for A.

2. Show each of the distinct ways of properly pairing up four pairs of parentheses.

3. Use the Catalan recurrence to verify the table value of C_6.

4. Use Theorem 7.1 to calculate C_8.

5. Make a 5×5 matrix whose (i, j)-entry is $\left\{ {i \atop j} \right\}$. Calculate the inverse of this matrix (by hand, with a calculator, or using a computer algebra system such as *Mathematica* or *Maple*). Compare the entries in this inverse with the Stirling numbers of the first kind.

6. Use the recurrence to calculate $\left\{ {7 \atop k} \right\}$ for $k = 5, 6, 7$.

7. How many ways are there to factor the number $2 \times 3 \times 5 \times 11 \times 23 = 7,590$ into a product of two factors each greater than 1?

8. In how many ways can we distribute our last seven (distinct) candy bars on Halloween night to two identical trick-or-treaters if we ensure each child gets at least one candy bar? How many ways are there if the trick-or-treaters are distinct (different costumes)?

9. Repeat the previous exercise in the case where we do not need to distribute all the candy bars.

10. Seven federal agents set out to investigate three politicians suspected of being corrupt. In how many ways can this be done, assuming that each politician is investigated by at least one agent, and the politicians are considered distinct?

11. Three corrupt politicians are sent to two luxurious minimum-security penal institutions. If each institution receives at least one politician, and the institutions are identical, in how many ways can we do this?

12. Seven investors wish to invest their savings into four mutual funds; they have six funds from which to choose, and each investor will put his or her savings into just one fund. In how many ways can four funds be chosen and the money invested? Assume the funds are distinct and each of the four chosen funds receives some money.

13. How many ways are there to write $385 = 5 \times 7 \times 11$ as a product of distinct integers? Write out each product.

14. How many partitions of a set of seven elements are there?

15. Use the recurrence of Theorem 7.7 to find $\begin{bmatrix} 7 \\ 1 \end{bmatrix}$, $\begin{bmatrix} 7 \\ 2 \end{bmatrix}$, and $\begin{bmatrix} 7 \\ 3 \end{bmatrix}$.

16. We have six people who will be seated at four circular tables. In how many ways can we do this if the tables are identical? In how many ways if one table is distinct from the other two?

Problems

Problem 1: A group of $2n$ people are seated around a circular table. In how many ways can they shake hands simultaneously so that every participant shakes hands with another and no handshakes cross other handshakes?

Problem 2: Show that $\sum_{i \leq n} \left| \begin{bmatrix} n \\ i \end{bmatrix} \right| = n!$.

Problem 3: Find the reciprocal of the generating function $C(x) = \frac{1 - \sqrt{1 - 4(x)}}{2x}$ and rationalize the denominator. Now use Newton's binomial theorem, an infinite series from calculus, or a computer algebra system such as *Mathematica* or *Maple* to expand $1/C(x)$ in an infinite series. What sequence does this reciprocal function generate?

Problem 4:

(i) Use the formula of Theorem 7.1 to show $C_{n+1} = C_n \dfrac{2(2n+1)}{n+2}$.

(ii) Show $C_{n+1} = \displaystyle\prod_{i=0}^{n} \dfrac{2(2i+1)}{i+2}$.

(iii) Find $\displaystyle\lim_{n \to \infty} \dfrac{C_{n+1}}{C_n}$.

Problem 5: Show that if p is an odd prime, then $\left\{ \begin{matrix} p \\ p-1 \end{matrix} \right\}$ is a multiple of p.

Problem 6: Show that if n is odd then $\left\{ \begin{matrix} n \\ 2 \end{matrix} \right\}$ is divisible by 3.

Problem 7: Prove Theorem 7.6.

Problem 8: Prove that $\left[{n \atop n-1} \right] = -\binom{n}{2}$.

Problem 9: Confirm that $\left[{n \atop n-2} \right] = n(n+1)(n+2)(3n+5)/24$ for $2 \leq n \leq 6$.

Chapter 8

Symmetries and the Pólya–Redfield Method

We have seen how to enumerate circular permutations in Chapter 2; the idea there is that two configurations are considered identical if one may be transformed into another by means of a rotation. There are other possible ways for two configurations to be equivalent, and in this chapter we look at the ways to enumerate configurations that are distinct with respect to some collection of symmetries. We will refer to these configurations as "colorings," even though what is meant by "color" might be nothing that an artist or decorator would recognize. For instance, we may seat politicians at a circular table, and only care which of (say) three political leanings (conservative, moderate, or liberal) the politician has. Thus, we might label seats with the "colors" C, M, and L. If we are concerned with gender, we might use "colors" M and F; and with religion or ethnicity, there would be many possible labels. For class-scheduling purposes, a classroom might get a label indicating that it can accommodate large classes or that it is equipped with a computer projector.

8.1   Introduction

In this chapter we discuss how to find the number of ways of coloring objects (such as seats at a circular table, or charms on a charm bracelet) that are distinct with respect to certain operations. Thus, we say two colorings of seats at a circular table are the same (or equivalent, in the sense of an equivalence relation) if one coloring may be transformed into another by a rotation. In the same way, colorings of beads on a necklace or charm bracelet may be considered equivalent under a rotation, or when the necklace is flipped over.

Suppose we wish to color each corner of a square one of two colors. In how many ways can this be accomplished? We can start off by naïvely dividing the question into cases. There are two ways (one for each color) if we are to use exactly one of the colors, so that the square is all one color. If we are to use both colors, there are more possibilities. So we might do three corners in color c_1, and the other in color c_2; or we might do two corners in color

143

c_1 and the other two in color c_2. In the second case, we realize that there is a difference if adjacent corners are to get the same color, or not. At this point, the problem appears to be tedious and simply a matter of eliminating possibilities by hand. For a large problem (as, say, a dozen seats and four colors), this approach is wholly inadequate. A more effective approach uses some basic facts of symmetry groups; we begin by discussing groups.

8.2   Basics of Groups

For our purposes, a *group* is a collection of bijective (one-to-one and onto) functions (called *permutations*) from a (usually finite) set S to itself that satisfies the *group axioms*, which follow.

- The *identity*, commonly denoted e, is the function that maps an element $x \in S$ to itself; $e(x) = x$ $\forall x \in S$. Every group must contain the identity.

- The group is closed under composition of functions; for any σ, τ in the group G, the function $\sigma \circ \tau$ defined by $\sigma \circ \tau(x) = \sigma(\tau(x))$ is also in G.

- The group contains inverses; that is, for any $\sigma \in G$ there is an element σ^{-1} in G with the property that $\sigma \circ \tau = \tau \circ \sigma = e$.

The reader should be familiar with the set of integers modulo n, denoted $\mathbb{Z}_n$, under addition, which we saw in Chapter 4. For instance, $\mathbb{Z}_4 = \{[0], [1], [2], [3]\}$; the group operation $+$ behaves like ordinary addition for small numbers, and gives us the remainder upon division by 4, so for example $[3] + [2] = [1]$ in the group. The group axioms are satisfied; the identity element is $[0]$, each element has an inverse (so $[1]^{-1} = [3]$, for instance), and the operation is clearly closed since the only remainders possible upon division by 4 are 0 through 3. We may think of these elements as being functions from the set $\{0, 1, 2, 3\}$ to itself as well. So we may consider $[1]$ to be the function $f(n) = n + 1 (\bmod 8)$; this means that $[1]$ is the function that takes 1 to 2, 2 to 3, and 3 to 0. It is also common to represent these functions as acting on the set $\{1, 2, 3, 4\}$, and we will use this convention when we are looking at symmetries of objects.

For the most part, we will consider the elements of a group to be functions acting on the set $S = \{1, 2, \ldots n\}$ but may also use, for example, a set of different ways to color an object. More abstract definitions of the group include the axiom that the operation $\circ$ is *associative*, meaning that $\tau \circ (\sigma \circ \pi) = (\tau \circ \sigma) \circ \pi$. This clearly holds for composition of functions, and for addition modulo n. The student should notice that our notation differs significantly from the notation used in many abstract or modern algebra texts. In many texts, the notation $\tau\pi$ is used to mean $\pi(\tau(x))$; we write multiplication "backwards" by this standard.

For any $n > 0$, there is at least one group of order n (that is, containing n elements). The only group of order 1 is the group consisting only of the identity. We will describe the elements of a group using the *two-line notation,* or *cycle notation.* (We looked at cycles of permutations in Section 7.5.) So consider the function that maps each corner of a square to the next corner clockwise from the center, corresponding to a 90° rotation clockwise, as in Figure 8.1.

FIGURE 8.1: Square with numbered corners.

This function ρ will rotate the circle one turn clockwise; in two-line form, this is denoted
$$\begin{pmatrix} 1 & 2 & 3 & 4 \\ 2 & 3 & 4 & 1 \end{pmatrix}.$$

The same permutation, written in cycle notation, is (1234). In the two-line notation, we evaluate $\rho(1)$, for instance, by looking for the column in which 1 is in the top row; the second row entry in that column is $\rho(1)$. In cycle notation, we calculate $\rho(1)$ by looking for the first number in the brackets with 1 directly to the right of 1. To find $\rho(4)$ using cycle notation, we see that nothing is to the right of 4 within that set of brackets, so we return to the first number, which is 1; thus $\rho(4) = 1$. The name is a reference to applying a permutation σ to an element of the set S repeatedly; it causes that element to "cycle" through the values in that cycle of the permutation. This permutation is exactly equivalent to the element $[1]$ of $\mathbb{Z}_4$, as we saw above.

Example 8.1: Find the cycle notation for the permutation
$$\begin{pmatrix} 1 & 2 & 3 & 4 & 5 \\ 3 & 5 & 1 & 2 & 4 \end{pmatrix}.$$

We begin by asking to which element 1 is mapped; this is 3. This means our permutation begins with $(13\ldots)$. Now, the third column of the two-line form shows that 3 is sent to 1. That "closes the cycle," so we get (13) for the first cycle. Now, we look at the next element that we haven't included; that is 2. The two-line form indicates that 2 goes to 5, and 5 goes to 4, and 4 goes back to 2. Thus the cycle notation is (13)(254). This permutation may also be written (542)(31), for instance, so the cycle representation is not unique in general. □

Example 8.2: Find the two-line notation for the permutation whose cycle form is (12)(3)(45). (It is more common to omit elements that are sent to themselves, or *fixed points*, as they are called; so the permutation might be written (12)(45).)

To find the two-line notation, we begin by listing the top line, so we get:

$$\begin{pmatrix} 1 & 2 & 3 & 4 & 5 \end{pmatrix}.$$

Now inspecting the cycle form, we see that 1 is mapped to 2, 2 is sent to 1, 3 stays at 3, and 4 and 5 exchange places. We get:

$$\begin{pmatrix} 1 & 2 & 3 & 4 & 5 \\ 2 & 1 & 3 & 5 & 4 \end{pmatrix}$$

and this is our two-line form. □

Given two groups G (permuting the elements of a set S) and H (permuting elements of a set T), we may form a new group $G \times H$, the *Cartesian product of G and H*, as follows. Given a permutation σ of G and a permutation τ of H, we define a permutation (σ, τ) on the set $S \times T$ in the obvious way: $(\sigma, \tau)(s, t) = (\sigma(s), \tau(t))$. The group operation will work within each coordinate of the ordered pair; so, $(\sigma_1, \tau_1) \cdot (\sigma_2, \tau_2) = (\sigma_1 \sigma_2, \tau_1 \tau_2)$. This is clearly well-defined as a function from $S \times T$ to itself, and it satisfies the group axioms; the proof is left to Problem 8.2.

Example 8.3: Find the Cartesian product of $C_2 \times C_2$ and give its multiplication table.

This group acts on the set $\{1, 2\} \times \{1, 2\} = \{(1, 1), (1, 2), (2, 1), (2, 2)\}$. If we denote the elements of C_2 as e and ρ, then $C_2 \times C_2$ consists of (e, e), (e, ρ), (ρ, e), and (ρ, ρ). For brevity, we will denote these as ee, $e\rho$, and so forth; and the set elements as 11, 12, and so on.

Now it is easy to see that the identity of the group $C_2 \times C_2$ is ee, and that each element is its own inverse. So, for instance, $\rho\rho \cdot \rho\rho = ee$. Our multiplication table is thus

	ee	$e\rho$	ρe	$\rho\rho$
ee	ee	$e\rho$	ρe	$\rho\rho$
$e\rho$	$e\rho$	ee	$\rho\rho$	ρe
ρe	ρe	$\rho\rho$	ee	$e\rho$
$\rho\rho$	$\rho\rho$	ρe	$e\rho$	ee

The reader should check the multiplication in each case. □

We will employ both cycle notation and two-line notation to represent permutations as we continue with our discussion of groups. In particular, the lengths of cycles and the numbers and kinds of fixed points will be important.

Example 8.4: Find the group of permutations that describe the symmetries of the square shown in Figure 8.1.

We have already mentioned the rotation ρ that describes rotation by 90°; the rotation ρ^2 that describes rotation by 180°, the rotation ρ^3 (270°) and e are the remaining planar motions that take the square into itself. The other motions are four reflections that take the square out of the plane. We may reflect the square about the line passing through the points 1 and 3, or 2 and 4, or through the vertical line bisecting the square, or the horizontal line that bisects the square.

$$\rho = (1234); \quad \rho^2 = (13)(24); \quad \rho^3 = (1432); \quad \rho^4 = e = (1)(2)(3)(4)$$

We will use α, β, γ, and δ to represent the listed reflections.

$$\alpha = (24); \quad \beta = (13); \quad \gamma = (12)(34); \quad \delta = (14)(23)$$

We have described eight symmetries, including the identity. Is this all? A careful study of the square would seem to indicate that no other motions can carry the square into itself. Because a group must be closed under composition of functions and inverses, we should check that all compositions and all inverses are accounted for. The multiplication table for the group is below.

$\circ$	e	ρ	ρ^2	ρ^3	α	β	γ	δ
e	e	ρ	ρ^2	ρ^3	α	β	γ	δ
ρ	ρ	ρ^2	ρ^3	e	γ	δ	β	α
ρ^2	ρ^2	ρ^3	e	ρ	β	α	δ	γ
ρ^3	ρ^3	e	ρ	ρ^2	δ	γ	α	β
α	α	δ	β	γ	e	ρ^2	ρ^3	ρ
β	β	γ	α	δ	ρ^2	e	ρ	ρ^3
γ	γ	α	δ	β	ρ	ρ^3	e	ρ^2
δ	δ	β	γ	α	ρ^3	ρ	ρ^2	e

Since we have found no new symmetries, we have a group (and in fact have found all the symmetries of the square). The reader will notice that each row and column contains the identity e as well as each other element of the group. This group is variously called the *dihedral group on four letters* or the *octic group*, and is denoted D_4. □

A number of points should be made here. First of all, notice that $\rho \circ \alpha = \gamma$, but $\alpha \circ \rho = \delta$. Unlike many common binary operations, then, function composition is not *commutative* in general. Groups in which the binary operation is commutative are called *abelian*. All the groups $\mathbb{Z}_n$ are abelian, because addition (and addition modulo n) is commutative. Also, the Cartesian product of two abelian groups is an abelian group; the proof of this assertion is left to Problem 8.2.

Next, if we ignore the reflections α, β, γ, and δ, we get a set of elements $\{e, \rho, \rho^2, \rho^3\}$. This set forms a group as well; it is closed under composition and each element has its inverse (the inverse of ρ is ρ^3, and ρ^2 is its own inverse). This is the group of rotations, and is referred to as the *cyclic group on four letters*, denoted C_4. In general, any regular n-sided polygon has a group D_n of $2n$ symmetries that contains the n rotations of the group C_n and n reflections about n axes of symmetry. For any n, C_n is abelian, but D_n is not, for $n \geq 3$. It is worth noticing that the group $\mathbb{Z}_n$ is essentially the same as the group C_n; that is, if we take the element $[1]$ of $\mathbb{Z}_n$ and write it as ρ, and write $[2]$ and ρ^2, and so forth, we get the same multiplication table. If we have two groups that, under a suitable renaming of the elements, have the same multiplication table, we say the groups are *isomorphic*, and the function that maps an element of one group to the corresponding element of the other is called an *isomorphism*.

We define a *subgroup* of a group G to be a subset of G that forms a group; that is, H is a subgroup of G if $H \subseteq G$, H includes the identity e, and H is closed under the group operation and under inverse. Thus, we have seen that C_n is a subgroup of D_n. When we define S_n to be the set of all permutations of a set of n objects, we call S_n the *(full) symmetric group on n letters*, and D_n is a subgroup of S_n. Another subgroup of D_4 is $\{e, \alpha, \beta, \rho^2\}$; we will denote this group by K. Confirmation that this is a subgroup is left to Problem 8.3.

We observe that the function α of D_4 sends 1 to 1, and 3 to 3; 1 and 3 are called *fixed points* of the permutation α. Clearly, every point is a fixed point of the identity e; also, non-identity elements of C_n have no fixed points. We will see that fixed points are a useful idea when we count distinct ways to color an object. The notation $FP(\sigma)$ will mean the set of fixed points of the permutation σ, so that $FP(\alpha) = \{1, 3\}$.

A group partitions the elements of the set S into *orbits*; the orbit of an element of S is the set of all elements of S to which it may be sent by one of the functions of the group. The group is said to be *transitive* on S if there is only one orbit. Equivalently, G is transitive if and only if, given any two elements a and b of S, there is an element σ_{ab} with the property that $\sigma_{ab}(a) = b$. Other characterizations of transitivity exist; see Problems 8.4 and 8.7.

Example 8.5: Find the orbits of the group K.

We observe that β and ρ^2 exchange 1 and 3; no element of K takes 1 to 2 or 4, so one orbit is the set $\{1, 3\}$. Now, α and ρ^2 both take 2 to 4, but no element of K takes either of 2 or 4 to either of 1 or 3. So there are two orbits of K, and $S = \{1, 3\} \cup \{2, 4\}$. $\qquad\square$

A symmetry of the square may be thought of as a permutation, not just of the corners of the square, but of the sides of the square as well. For example, if we label the sides of the square shown in Figure 8.1 by the numbers of the corners, we may look at ρ as the permutation that sends side 12 to side 23, side 23 to side 34, and so on. We sometimes say that this permutation of the sides is *induced* by ρ, and the group of permutations of the corners induces a group of permutations of the sides. It is not hard to see that the group of permutations of the corners is in fact isomorphic to the group of permutations of the sides. We also say that the group *acts on* the sides.

It is worth noticing that the permutations of the sides induced by the permutations of the corners will not have the same numbers of fixed points in general as the permutations that induce them. For instance, $\alpha = (24)$ fixes two corners, but the permutation of the sides induced by α exchanges side 12 with 14, and side 23 with 34. On the other hand, $\gamma = (12)(34)$ does not fix any corner, but the induced permutation fixes the sides 12 and 34.

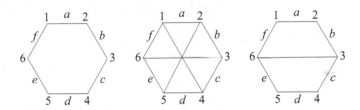

FIGURE 8.2: Three hexagons with labeled vertices and sides.

Example 8.6: Find the permutations in the dihedral group D_6 representing the symmetries of the first of the three hexagons in Figure 8.2. Find the actions of the permutations of this group on the sides of the hexagon. Do the permutations induce an action on the triangles within the second hexagon in this figure? Do these permutations induce an action on the quadrilaterals within the third hexagon? Explain why or why not.

We have the rotation $\rho = (123456)$ representing a 60° clockwise rotation of the figure, and the powers $\rho^2 = (135)(246)$, $\rho^3 = (14)(25)(36)$, $\rho^4 = (153)(264)$, $\rho^4 = (165432)$, and $\rho^6 = e$. Now we list the six reflections; they are $\alpha = (12)(36)(45)$, $\beta = (13)(46)$, $\gamma = (23)(14)(56)$, $\delta = (15)(24)$, $\epsilon = (16)(25)(34)$, and $\phi = (26)(35)$.

We can see that ρ will move the side a to the side b, which goes to side c, and so on; so the action of ρ on the sides is the cycle $(abcdef)$. In the same way, ρ^2 is $(ace)(bdf)$, $\rho^3 = (ad)(bc)(ef)$, $\rho^4 = (aec)(bfd)$, $\rho^5 = (afedcb)$, $\rho^6 = e$, $\alpha = (bf)(ce)$, $\beta = (ab)(cf)(de)$, $\gamma = (ac)(df)$, $\delta = (ad)(bc)(ef)$, $\epsilon = (ae)(bd)$, and $\phi = (af)(be)(cd)$. Notice that a permutation that fixes some vertices, such as β, may not fix any edge, and a permutation that leaves an edge unchanged (such as ϵ that sends edge c to itself) leaves no vertex unchanged.

Now, each permutation of D_6 acts on the interior triangles of the middle hexagon. Thus, ρ takes the triangle that includes the side a to the triangle that includes b, and so forth. Each triangle is sent to exactly one other triangle by any particular permutation in D_6, so this group acts on the triangles. If we were to label each triangle A (for the triangle that includes side a) and B (for that containing b) and so on, we could rewrite each permutation as a permutation on the letters A through F. Notice that the permutations we get on A through F are the same permutations as those on the edges, just with uppercase letters.

Finally, we look at the result of these permutations on the quadrilaterals of the rightmost hexagon. It is clear that, for instance, the permutation ρ does not send the quadrilateral with vertices 1, 2, 3, and 6 to either of the two quadrilaterals. Thus, ρ in particular does not induce an action on the quadrilaterals. However, some of the permutations of D_6 act on these two quadrilaterals; in particular, the set $\{e, \alpha, \delta, \rho^3\}$ is a subgroup of D_6 that acts on the two quadrilaterals. Both the elements e and α leave the quadrilaterals unchanged, and δ and ρ^3 exchange the two quadrilaterals.   □

8.3   Permutations and Colorings

In Section 8.2, we saw that a permutation on the corners of the square could be regarded as a permutation on the sides. In a similar way, we can look at the permutation of the colorings of the square induced by the permutation of the corners. If we color the corners of the square with k colors, there are clearly k^4 distinct colorings, many of which are equivalent to one another under symmetries of the square.

Consider for concreteness the coloring $\mathcal{C}_1$ in which corner 1 is given color c_1 and the other corners are given color c_2. If we apply ρ, we send corner 1 to corner 2, so that we have a new coloring in which corner 2 receives color c_1, and all the other corners receive color c_2. If we denote this coloring by $\mathcal{C}_2$, then we can say $\rho(\mathcal{C}_1) = \mathcal{C}_2$, where ρ is the permutation on the set of colorings induced by the permutation ρ of the set of corners. We will denote both permutations by ρ when no confusion will result.

Example 8.7: Color the corners of the square of Figure 8.1 with two colors (black and white) in such a way that the coloring is fixed by the permutation induced by ρ^2.

Without loss of generality, we will color corner 1 with black. Now, since ρ^2 exchanges corner 1 and corner 3, we must also color corner 3 with black. In the same way, if we color corner 2 white, we must also color corner 4 white. (We could find other ways to color the corners that would have the same result.) In general, all the corners that are in the same cycle as corner 1 must get the same color as corner 1, and the same for each other corner. □

Example 8.8: Find all colorings of the square with four colors that are fixed by the permutation ρ.

Since ρ has only one cycle, all four corners must get the same color. Since we have four colors to choose from, we have exactly four colorings; C_1, in which all four corners get color c_1, C_2 with all color c_2, C_3 with all color c_3, and C_4 with all color c_4. □

THEOREM 8.1 (Enumerating Equivalence Classes)
The number of colorings with k colors that are fixed by a permutation σ with n cycles is k^n.

Proof. We may choose one of k colors to use for each of the points in the first cycle; then all the points of that cycle must receive the same color. We choose one of the k colors independently for a point of the next cycle (if the permutation has more than one cycle). When we perform this process for each of the n cycles, the multiplication principle tells us that there are k^n ways to color the figure. □

8.4   An Important Counting Theorem

The main result of this section was known to Cauchy, proven by Frobenius, and published by Burnside, so that it is often called Burnside's lemma, or the Cauchy–Frobenius–Burnside theorem. The student is warned that the demonstration given here is necessarily somewhat incomplete. We have, however, given an idea of the proof that, while glossing over some group-theoretic detail, will give the student a good idea of why the result is true (and how a more complete proof could be made).

Given a group G (such as the group of symmetries of a square) acting on a set S (such as the set of ways to color the corners of this square), we wish to know how many distinct colorings (i.e., equivalence classes or orbits of colorings) there are. Let this number be N. Here $FP(\sigma)$ denotes the number of colorings fixed by σ, not the number of corners or sides; we view σ as a permutation of the colorings. Given a coloring C, we let $\text{Stab}(C)$, the *stabilizer* of C, be the set of permutations σ of G for which $\sigma(C) = C$. The stabilizer is a subgroup of G; see Problem 8.6.

How many colorings are equivalent to C under the action of permutations in G? We shall see that the answer is $|G| \, / \, |\text{Stab}(C)|$. For, suppose the colorings

C_1 and C_2 are equivalent (so that there is some σ in G for which $\sigma(C_1) = C_2$). Then for any τ in the stabilizer of C_1 (so $\tau(C_1) = C_1$), we have $\sigma \circ \tau(C_1) = C_2$. It follows that each equivalent coloring C_2 of C_1 is the image of C_1 for exactly $|\text{Stab}(C_1)|$ permutations in G. That means there are $|G| / |\text{Stab}(C_1)|$ such equivalent colorings. Let $|[C]|$ denote the number of colorings equivalent to the coloring C (we use square brackets to denote an equivalence class). Then

$$|\text{Stab}(C_1)| = |G| / |[C_1]| \text{ for any } C_1 \in S. \tag{8.1}$$

THEOREM 8.2 (Cauchy–Frobenius–Burnside Theorem)
For G, S, and N as above, we have

$$N = \frac{1}{|G|} \sum_{\sigma \in G} |FP(\sigma)|.$$

Proof. We shall count the set $\{(\sigma, C) : C \in S, \sigma \in \text{Stab}(C)\}$, in two different ways. Equating the two expressions will give us the result. The first thing we do is to observe that given a $\sigma \in G$, there are $|FP(\sigma)|$ colorings that it fixes; so the quantity in question is just $\sum_{\sigma \in G} |FP(\sigma)|$. Alternatively, we may sum over all colorings C in S the number of permutations that leave C fixed; this is $\sum_{C \in S} |\text{Stab}(C)|$. Equation (8.1) turns this into $\sum_{C \in S} |G| / |[C]|$. Factoring out $|G|$, we have

$$\sum_{\sigma \in G} |FP(\sigma)| = |G| \sum_{C \in S} 1/|[C]|.$$

Now, N is the number of equivalence classes that we are trying to enumerate. The sum on the right-hand side above may be grouped; the various colorings in a given equivalence class $[C]$ will each contribute $1/|[C]|$ to the sum. So all these colorings will contribute a 1 to the sum. It follows that this sum is just N, the number of equivalence classes. $\square$

Example 8.9: How many ways can we color a length of board with up to three colors if we place one color at each end and one color in the middle? Two colorings are equivalent if one may be obtained from the other by flipping the board end-for-end (corresponding to the group C_2).

We see that there are 27 ways to color the board in all (without considering equivalence classes); there are three colors for the left end, three for the middle, and three for the right end, giving 3^3. Now, the temptation to simply divide by two (since each coloring would appear to be equivalent to exactly one other, obtained by rotating the board by 180°) might be great, but it would lead us to conclude that there are 27/2 or $13\frac{1}{2}$ colorings, surely an absurd result. We will count the equivalence classes of colorings twice, once by combinatorial reasoning and once using Theorem 8.2. First, consider that the only colorings that are different but equivalent are the colorings that are not fixed by ρ, the permutation that exchanges the colors on the left and right ends of the board. We see that ρ has just two orbits; one orbit consists of the left and right ends of the board, and the other consists of the middle of the board. By Theorem 8.1, ρ fixes 9 colorings, so there are 18 it does not fix. These 18 colorings give us nine equivalence classes. The nine colorings fixed by ρ are equivalent only to themselves; each is its own equivalence class. Hence we have 18 different equivalence classes (ways to color the board).

Now, using Theorem 8.2, we use $|G| = 2$, and $|FP(e)| = 27$ because every possible coloring is fixed by the identity permutation. Since $|FP(\rho)| = 9$ by the foregoing reasoning, $N = (27 + 9)/2 = 18$. $\qquad\square$

Example 8.10: We wish to seat n people at a circular table; how many ways are there to do this? We let the group be C_n and observe that there are a total of $n!$ colorings (i.e., ways to order the people).

In Chapter 2, we observed that the number of ways to accomplish this is $(n-1)!$; we now use Theorem 8.2 to get the same result. Because we never repeat any of the "colors" (the n people occupy only one seat each), no permutation fixes any coloring except for the identity, which fixes all the colorings. Thus the number of equivalence classes is just $(1/|C_n|)(n!) = (n-1)!$. $\qquad\square$

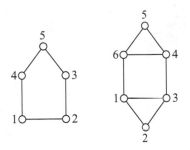

FIGURE 8.3: Two figures for coloring.

Example 8.11: In how many ways can we color the corners of the pentagon on the left of Figure 8.3 with k colors if we assume colorings are equivalent under reflection about the vertical axis? This reflection corresponds to the permutation $(12)(34)$; the figure's symmetry group thus has two elements (this permutation and the identity). Clearly, every coloring is fixed by the identity permutation; to compute the number of colorings fixed by the other permutation, we observe that corners 1 and 2 must receive the same color, and corners 3 and 4 must receive the same color.

There are k^5 total ways to color the corners with k colors (multiplication principle); all k^5 of them are fixed by the identity permutation. The other permutation fixes only k^3 colorings, because we may choose any of k colors for corner 1, any of k colors for corner 2, and any of k colors for corner 3; the colors for the other corners are determined. The Cauchy–Frobenius–Burnside theorem says that there are $\frac{1}{2}(k^5 + k^3)$ non-equivalent colorings. □

8.5 The Pólya and Redfield Theorem

What we have done so far is effective for counting the equivalence classes of colorings of structures with small groups. If $|G|$ is not too great, we can simply enumerate the colorings that are equivalent under the various permutations in G and find ourselves with the number of non-equivalent colorings after only a little work. However, what are we to do if we wish to color a complex object in which $|G|$ is very large? Just as in Chapter 6, we will find that generating functions can take much of the work and automate it.

In particular, each permutation is associated with a *cycle enumerator*, which is a product of variables that describe its cycles. First, we take a permutation and write it as a product of disjoint cycles, which may be done in essentially one way; for example, consider the permutation $(12)(456)$. It has one cycle of length 1, one of length 2, and one of length 3. We describe it by the monomial $z_1^1 z_2^1 z_3^1$. More generally, we put a factor of z_i^j to indicate j disjoint cycles, each of length i, in the permutation.

Example 8.12: Find the cycle enumerators of the eight permutations in D_4 as it acts on Figure 8.1.

The permutations, as outlined earlier, are e, $\rho = (1234)$, $\rho^2 = (13)(24)$, $\rho^3 = (1432)$, $\alpha = (24)$, $\beta = (13)$, $\gamma = (12)(34)$, and $\delta = (14)(23)$. Since e consists of four one-cycles, it has cycle enumerator z_1^4. Similarly, ρ and ρ^3 are both z_4^1. We find ρ^2, γ, and δ are all z_2^2, and α and β are both $z_1^2 z_2^1$. □

Now, to enumerate the non-equivalent ways to color a figure with k colors, we form a polynomial $C(z_1, z_2, \ldots z_n)$ by adding the cycle enumerators and dividing by $|G|$. This polynomial is called the *cycle index* of G (more properly, of the *action* of G on the objects being colored). For the previous example, this gives us $\frac{1}{8}(z_1^4 + 2z_4^1 + 3z_2^2 + 2z_1^2 z_2^1)$. What good does this polynomial do us? By substituting k for each of the z_is, we find the number of ways of coloring with k colors!

Example 8.13: How many ways are there to color the pentagon of Figure 8.3 with k colors?

The permutations consist of the identity e and the reflection $(12)(34)$. These have cycle enumerators of z_1^5 for e and $z_1^1 z_2^2$ for the reflection. Our cycle index is thus $C(z_1, z_2) = \frac{1}{2}(z_1^5 + z_1 z_2^2)$. Inserting k for each z, we get $(k^5 + k^3)/2$, just as with the example of the last section. □

THEOREM 8.3

Given a figure with a symmetry group G having cycle index $C(z_1, \ldots z_n)$, the number of colorings not equivalent under action of G using k colors is $C(k, \ldots k)$.

The detailed proof is omitted. Roughly speaking, however, each cycle enumerator evaluated at k will give the number of k-colorings that are fixed under the action of the corresponding permutation. Thus, the cycle index is the right-hand-side of the equation of Theorem 8.2 when evaluated at k, giving us the number of equivalence classes of colorings.

Example 8.14: Count the colorings of the hexagon of Figure 8.3.

For this hexagon, two colorings are equivalent under vertical reflection, horizontal reflection, or rotation by 180°. The symmetry group here thus (with the identity) has four elements; they are $\alpha = (13)(46)$, $\beta = (16)(25)(34)$, $\rho = (14)(25)(36)$, and e. The cycle enumerators are $z_1^2 z_2^2$ for α, z_2^3 for β and ρ, and z_1^6 for e. Thus, the total number of non-equivalent colorings with k colors is $C(k) = \frac{1}{4}(k^6 + k^4 + 2k^3)$. □

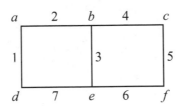

FIGURE 8.4: Coloring corners or edges.

It is important to make sure that the cycle index is appropriate to the objects being counted. Consider, for instance, the rectangle shown in Figure 8.4. We will consider coloring the corners, which are labeled with letters a through f, and the sides, labeled with the numbers 1 through 7. In each case, we consider two colorings equivalent if one is moved into the other by reflection across the horizontal axis, reflection across the vertical axis, or rotation by 180°. Notice that this is the group D_2 (two rotations and two reflections, if we count the identity as a 360° rotation), even though it is acting on six corners and seven sides.

Example 8.15: How many non-equivalent ways can we color the corners of Figure 8.4 with k colors?

The four permutations are: the identity, fixing each of the six corners; the horizontal reflection, corresponding to the permutation $(ad)(be)(cf)$; the vertical reflection, corresponding to $(ac)(df)$; and the rotation, $(af)(be)(cd)$. The cycle enumerators are, respectively, z_1^6, z_2^3, $z_1^2 z_2^2$, and z_2^3. This gives us a cycle index of $\frac{1}{4}(z_1^6 + 2z_2^3 + z_1^2 z_2^2)$. This gives us a total of $C(k) = (k^6 + k^4 + 2k^3)/4$ ways to color the corners with k colors. □

Example 8.16: How many non-equivalent ways can we color the sides with k colors?

The same four symmetries now have different cycle enumerators. The horizontal reflection is $(27)(46)$; the vertical reflection is $(15)(24)(67)$; and the rotation is $(15)(26)(47)$. Each of these fixes at least one edge. Our cycle index is $\frac{1}{4}(z_1^7 + z_1^3 z_2^2 + 2z_1^1 z_2^3)$. This gives us $C(k) = (k^7 + k^5 + 2k^4)/4$ ways to color the edges with k colors. Notice that this answer is just k^2 times the result for the previous example. □

We can now easily find the number of non-equivalent ways to color (using k colors, for any positive integral value of k) a figure with its symmetry group. This does not allow us to count the number of non-equivalent colorings subject to specific restrictions, however. In particular, we might ask how many ways there are if each of the k colors must be used at least once; how many ways there are if we must use more of one color than another; and so forth. These questions require a little more effort, but can still be addressed through a slight modification of the technique.

Suppose we are going to color with k colors; we introduce variables $d_1, d_2,$ $\ldots d_k$. If we take the cycle index $C(z_1, \ldots z_n)$ and replace each variable z_i with the expression $d_1^i + d_2^i + \cdots + d_k^i$, the result is a new polynomial in the variables $d_1, \ldots d_k$. We denote this function by $\hat{C}$.

$$\hat{C}(d_1, \ldots d_k) = C(d_1 + \cdots + d_k, d_1^2 + \cdots + d_k^2, \ldots, d_1^n + \cdots + d_k^n)$$

THEOREM 8.4 (The Pólya–Redfield Theorem)
Given a figure with symmetry group G having cycle index $C(z_1, \ldots z_n)$, the number of inequivalent colorings with k colors, in which color c_i is used m_i times, is the coefficient of $d_1^{m_1} d_2^{m_2} \ldots d_k^{m_k}$ in $\hat{C}(d_1, \ldots d_k)$ (where $m_1 + m_1 + \cdots + m_k = n$ is the number of points being colored).

Example 8.17: Color the corners of the square of Figure 8.1 with two colors, red and blue (using variables r and b).

We will use D_4 for the group; that is, two colorings are equivalent if there is a permutation in D_4 that maps one to the other. As we calculated earlier, our cycle index is $C(z_1, z_2, z_3, z_4) = (z_1^4 + 2z_4 + 3z_2^2 + 2z_1^2 z_2)/8$. Substituting z_i by $r^i + b^i$ gives us

$$\hat{C}(r, b) = \left((r + b)^4 + 2(r^4 + b^4) + 3(r^2 + b^2)^2 + 2(r + b)^2(r^2 + b^2)\right)/8.$$

This reduces, after sufficient algebra, to

$$\hat{C}(r, b) = r^4 + r^3 b + 2r^2 b^2 + rb^3 + b^4.$$

Each coefficient may be easily verified by combinatorial reasoning; for instance, to color the corners with two red and two blue, the two red corners may be next to each other or on opposite sides, giving the two equivalence classes of colorings. □

Example 8.18: How many non-equivalent colorings of the corners of the square of Figure 8.1 are there using four colors, if each color must be used precisely once?

Here we use variables d_1, d_2, d_3, and d_4. Rather than using the multinomial theorem of Chapter 2 to calculate $(d_1 + d_2 + d_3 + d_4)^4$, we will employ a computer algebra system to do the hard work for us. For instance, the *Mathematica* sequence that follows will work.

```
chr[z1_, z2_, z3_, z4_] := (z1^4 + 2*z4 + 3*z2^2 + 2*z1^2*z2)/8

chr[d1 + d2 + d3 + d4, d1^2 + d2^2 + d3^2 + d4^2,
    d1^3 + d2^3 + d3^3 + d4^3, d1^4 + d2^4 + d3^4 + d4^4]

Together[Expand[%]]
```

The first line defines the cycle index as `chr`; at the end of the line, using Shift+Enter causes *Mathematica* to evaluate the expression and it will recognize the function `chr[z1, z2, z3, z4]` as defined. So, for instance, entering `chr[3,3,3,3]` and Shift+Enter will give the number of non-equivalent ways to color the corners with three colors.

The second line outputs the expression for $\hat{C}$ as a function of d_1 through d_4, but not in an easily readable form. The third line causes *Mathematica* to expand the powers and collect like terms so that we may read the result. The output (again, we end each line with Shift+Enter) is:

```
(d1^4 + d1^3d2 + 2d1^2d2^2 + d1d2^3 + d2^4 + d1^3d3 +
    2d1^2d2d3 + 2d1d2^2d3 + d2^3d3 + 2d1^2d3^2 +
    2d1d2d3^2 + 2d2^2d3^2 + d1d3^3 + d2d3^3 + d3^4 +
    d1^3d4 + 2d1^2d2d4 + 2d1d2^2d4 + d2^3d4 +
    2d1^2d3d4 + 3d1d2d3d4 + 2d2^2d3d4 +
    2d1d3^2d4 + 2d2d3^2d4 + d3^3d4 + 2d1^2d4^2 +
    2d1d2d4^2 + 2d2^2d4^2 + 2d1d3d4^2 +
    2d2d3d4^2 + 2d3^2d4^2 + d1d4^3 + d2d4^3 + d3d4^3 +
    d4^4)
```

The coefficient of d1 d2 d3 d4 is 3, as we see in the middle of the fifth line of output; there are therefore exactly three non-equivalent colorings of the corners with four colors in which each color is used exactly once. If we had not used the reflections (used the group C_4 instead of D_4) we would have gotten 6, corresponding to the six cyclic permutations of four distinct objects.

A similar approach will work for *Maple*. In this case, we use the character sequence - > to produce the right arrow, and we use the right arrow key $\rightarrow$ to delimit some operations. For instance, if we wished to type x^2 + y^2, we would need to use the right arrow key to get out of the exponent. That is, we would type z^2→+y^2→ because all keystrokes after the ^ are interpreted as part of the exponent otherwise. Here is the *Maple* code with right-arrow key noted:

```
chr := (z1,z2,z3,z4)->(z1^4→+2z4+3z^2→+2z1^2→z2)/8→

chr(d1+d2+d3+d4, d1^2+d2^2+d3^2+d4^2, d1^3+d2^3+d3^3+d4^3,
    d1^4+d2^4+d3^4+d4^4)/8

expand(%)
```

The result will be the same as for the *Mathematica* code.

Exercises 8A

1. Find the multiplication table of C_5, given $\rho = (12345)$.

2. Find the two-row form of the permutation on five letters whose cycle form is:
   (i) (15)(234)
   (ii) (1245)

3. Find the cycle form of the permutation whose two-row form is:
   (i) $\begin{pmatrix} 1 & 2 & 3 & 4 & 5 \\ 3 & 5 & 1 & 2 & 4 \end{pmatrix}$
   (ii) $\begin{pmatrix} 1 & 2 & 3 & 4 & 5 \\ 2 & 5 & 4 & 1 & 3 \end{pmatrix}$

4. Find the group of all symmetries of a 2×1 rectangle.

5. Find the Cartesian product $C_2 \times C_3$ and its multiplication table.

6. Use the permutations of D_3, the symmetries of an equilateral triangle, to find the cycle index of the triangle. Now replace each variable z_i by $(r^i + b^i)$ and expand (by hand or using a computer algebra system such as *Maple* or *Mathematica*) to find all the non-equivalent ways of two-coloring the corners of the triangle with the various distributions of colors.

7. A four-foot pole is to be colored in bands of one foot in length. Two colorings are equivalent under flipping the pole end-for-end. Label each one-foot band, find the cycle index, and replace each z_i by $(r^i + b^i)$. Use the resulting expression to find the number of ways to two-color the pole so that each color is used at least once.

8. Consider the problem of coloring the corners of a regular pentagon, where colorings are equivalent under the cyclic group C_5. Find an expression for the number of distinct ways to k-color this figure; then find an expression for two-coloring with colors r and b using the Pólya–Redfield theorem.

9. Consider the problem of coloring the corners of a regular heptagon (seven-sided polygon), where colorings are equivalent under the dihedral group D_7. Find an expression for the number of distinct ways to k-color this figure; then find an expression for two-coloring with colors r and b using the Pólya–Redfield theorem.

Exercises 8B

1. Find the multiplication table of D_3, the symmetries of an equilateral triangle with vertices 1, 2, and 3.

2. Find the two-row form of the permutation on six letters whose cycle form is:
    (i) $(123)(456)$;
    (ii) $(23)(45)$.

3. Find the cycle form of the permutation on six letters whose two-row form is:

(i) $\begin{pmatrix} 1 & 2 & 3 & 4 & 5 & 6 \\ 4 & 5 & 6 & 3 & 2 & 1 \end{pmatrix}$;

(ii) $\begin{pmatrix} 1 & 2 & 3 & 4 & 5 & 6 \\ 6 & 5 & 4 & 3 & 2 & 1 \end{pmatrix}$

4. Draw a diagonal from corner 1 to corner 3 of the square of Figure 8.1. Find the subgroup of D_4 that leaves this diagonal fixed. *Hint.* : This group has four elements.

5. Find the Cartesian product $C_2 \times C_4$ and its multiplication table.

6. Use the permutations of D_2, the symmetries of a non-square rectangle, to find the cycle index of the rectangle (assume we are coloring the corners). Now replace each variable z_i by $(r^i + b^i)$ and expand (by hand or using a computer algebra system such as *Maple* or *Mathematica*) to find all the non-equivalent ways of two-coloring the corners of the rectangle with the various distributions of colors.

7. A five-foot pole is to be colored in bands of one foot in length. Two colorings are equivalent under flipping the pole end-for-end. Label each one-foot band, find the cycle index, and replace each z_i by $(r^i + b^i)$. Use the resulting expression to find the number of ways to two-color the pole so that each color is used at least once.

8. Consider the problem of coloring the corners of a regular pentagon, where colorings are equivalent under the dihedral group D_5. Find an expression for the number of distinct ways to k-color this figure; then find an expression for two-coloring with colors r and b using the Pólya–Redfield theorem.

9. Consider the problem of coloring the corners of a regular heptagon (seven-sided polygon), where colorings are equivalent under the cyclic group C_7. Find an expression for the number of distinct ways to k-color this figure; then find an expression for two-coloring with colors r and b using the Pólya–Redfield theorem.

Problems

Problem 1: Show that if G is a subgroup of H, and G is transitive on S, then H is transitive on S. Find a counterexample to show that the converse is false.

Problem 2: Prove that if G and H are abelian groups, then so is $G \times H$.

Problem 3: Show that the set K contained in D_4 is a subgroup.

Problem 4: Show that a group G is transitive on a set $S = \{1, 2, \ldots n\}$ if and only if for any $i \in S$ there is a permutation $\sigma_i \in G$ where $\sigma_i(1) = i$.

Problem 5: Find an isomorphism between the group $C_2 \times C_2$ and the group K.

Problem 6: Show that the stabilizer $\text{Stab}_G(s)$ of any element $s \in S$ is a subgroup of G.

Problem 7: A student wrote $|G/[c_1]|$ instead of $|G|/|[c_1]|$, based on the identity $|a|/|b| = |a/b|$ taught in algebra. The algebra book notwithstanding, the student's teacher took off many points! Explain.

Problem 8: We are to color a square with a color in each corner and a color in the center; two colorings are equivalent if one may be moved into another by rotations of 90° (corresponding to the group C_4). How many distinct colorings are there with k colors? Notice that even though this group looks like C_4, it acts on five points. How many colorings are there if we only color the corners? Can you find a relationship between the solutions to these two problems?

Problem 9: We wish to color a 3×1 rectangle so that each of its 1×1 squares is given a color. Colorings are equivalent under a rotation of the rectangle by 180°. Determine the number c_k of ways to do this with k colors. Now use the methods of Chapter 6 to find the generating function of c_k. Use the denominator of this generating function to find a recurrence relation for c_k.

Problem 10: Find all 16 permutations in the symmetry group D_8 of symmetries of the regular octagon. Now group these into sets according to how many cycles they have, and find the cycle index of the regular octagon. In how many ways can the corners be colored with k colors?

Problem 11: A regular tetrahedron is a solid with four faces, each an equilateral triangle. It has four corners which we label 1, 2, 3, and 4, and six edges. The group of symmetries of the tetrahedron has 12 elements. These include $\alpha = (234)$, $\beta = (134)$, $\gamma = (124)$, and $\delta = (123)$; the squares of these elements; three other elements produced by products of these; and of course the identity. Find all twelve symmetries as permutations of the corners; find the cycle index of the tetrahedron. Now find an expression for the number of ways to color the corners of the tetrahedron with k colors.

Problem 12: As in the previous problem, we consider a regular tetrahedron. Find the twelve symmetries as permutations of the edges of the tetrahedron, and find the cycle index. How many ways can the edges of the tetrahedron be colored with k colors?

Problem 13: We wish to color the circles on the square shown in Figure 8.5. Find the cycle index using the group C_4 (rotations but no reflections). If you have access to a computer algebra program, find $\hat{C}(r, b)$ for the two-colorings of the figure.

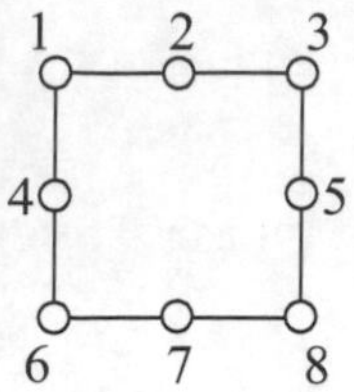

Introduction to Combinatorics

FIGURE 8.5: Square with eight points to color.

Problem 14: Find the cycle index for Figure 8.5 using the group D_4 (all rotations and reflections). If you have access to a computer algebra program, find $\hat{C}(r, b)$.

Problem 15: Generalize the exercises to find the number of ways to k-color the corners of the regular polygon with p sides (where p is an odd prime) where colorings are equivalent under the group C_p. Repeat using the group D_p to allow reflections.

Problem 16: Find the number of ways to k-color the corners of a regular polygon with $2p$ sides, where p is an odd prime, and colorings are equivalent under the cyclic group C_{2p}. Repeat using the group D_{2p} to allow reflections.

Chapter 9

Partially Ordered Sets

9.1 Introduction

The first objects that we encounter in mathematics are ordered; that is, given (for instance) two different integers, we know that one of them must be larger than the other. Later in mathematics, we find objects that are not ordered in the same way; for instance, there are two different groups of order four, and it makes no sense to say that one of them is "bigger than" or "comes before" the other. Between these two extremes, there are collections for which it is possible to make a sensible statement that some members of the collection are bigger or smaller than others, but for other pairs of members, no comparison is possible. Such a collection of objects, together with the appropriate definition of "bigger than," is called a *partially ordered set*, or *poset* for short.

More rigorously, a partially ordered set is a set S together with a relation $\prec$ that is irreflexive, antisymmetric, and transitive (see Appendix A to review these terms). We may also use the relation $\preceq$, which is reflexive, antisymmetric, and transitive; it is understood that that $a \preceq b$ means "Either $a \prec b$ or $a = b$." Either way, the relation is called a *partial order* on S. If the poset has the property that any two elements x and y are related or *comparable* (that is, either $x \prec y$ or $y \prec x$), we say it is a *total order*, and refer to it as "an ordered set." A poset that is not a total order has at least one pair of elements that are *incomparable*, i.e., there are x and y in the poset for which neither $x \prec y$ nor $y \prec x$ is true. We write $x\|y$ to indicate that x and y are incomparable. Since the usual sets of numbers $\mathbb{N}, \mathbb{Q}, \mathbb{R}$ and $\mathbb{Z}$ are ordered sets, we see that a partially ordered set is merely a generalization of these familiar objects.

A poset is said to be *locally finite* if, for any comparable pair $x \prec y$ there are only finitely many z satisfying $x \prec z \prec y$. Thus, $\mathbb{N}$ and $\mathbb{Z}$ are locally finite; $\mathbb{Q}$ and $\mathbb{R}$ are not. If $x \prec y$ but there is no z for which $x \prec z \prec y$, we say that y *covers* x. If the poset contains one or more elements x with the property that $x \not\prec z$ for any z, this element x is a *maximum* element of the poset; similarly, if there is no z with $z \prec x$, then x is a *minimum* element of the poset.

9.2   Examples and Definitions

Consider a set X of n elements $x_1, x_2, \ldots x_n$, and let S be the collection of all subsets of X. The partial order will be the relation $\subset$, so that $S_1 \subset S_2$ means that S_1 is a subset of S_2. We sometimes say this is *the power set of X ordered by containment.* If $n \geq 3$, we might have $S_1 = \{x_1, x_2\}$ and $S_2 = \{x_2, x_3\}$. In this case neither of the statements $S_1 \subset S_2$ nor $S_2 \subset S_1$ is true so that S_1 and S_2 are incomparable. If $S_3 = \{x_1, x_2, x_3\}$ then clearly $S_1 \subset S_3$ and $S_2 \subset S_3$; we sometimes might write $S_3 \supset S_1$ to indicate $S_1 \subset S_3$. This poset is sufficiently well known that it has a number of names; we shall call it $\mathcal{B}_n$, referring to Boolean algebra. It is also possible to look at the poset of multisubsets of a multiset; thus, X might consist of more than one "copy" of some elements. So, for example, if $X = \{0, 0, 1\}$, its multisubsets include $\{0\}$, $\{0, 0\}, \{1\}$, $\{0, 1\}$, and the empty set $\emptyset$, as well as X. This concept will be more thoroughly explored in the exercises and problems.

If $x \prec y$ in a poset P, and there is no z in P with $x \prec z \prec y$, we say y *covers* x. Thus, the set $\{1, 2\}$ covers both $\{1\}$ and $\{2\}$ in the poset $\mathcal{B}_3$. However, even though $\{1\} \prec \{1, 2, 3\}$ in the partial order, we do not say that $\{1, 2, 3\}$ covers $\{1\}$ because there are other sets (in particular, $\{1, 2\}$ and $\{1, 3\}$) that lie between them.

It is convenient to diagram a finite poset in such a way as to show the partial ordering; we form a graph whose vertices are the elements of the poset, and an edge goes down from x to y if x covers y in the partial order. Such a graph is called the *Hasse diagram* of the poset. We illustrate with the poset $\mathcal{B}_3$ in Figure 9.1.

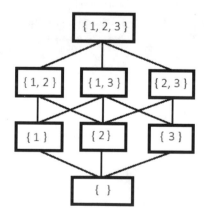

FIGURE 9.1:   Poset of subsets of a set of three elements.

Another common family of posets comes from group theory. Given a group G, we can consider its collection of subgroups ordered by containment. That is, if G_1 and G_2 are subgroups of G, we say $G_2 \prec G_1$ if G_2 is a subgroup of G_1. We illustrate with the Hasse diagram of the subgroups of $\mathbb{Z}_{12}$ in Figure 9.2.

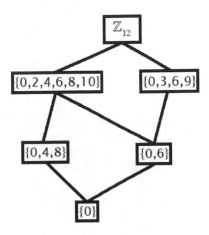

FIGURE 9.2: Poset of subgroups of $\mathbb{Z}_{12}$, ordered by containment.

Another family of posets is found in number theory; consider an integer n and all of its divisors. We will say $k \prec n$ provided that k is a divisor of n. This is really an infinite poset on the set of all integers, but we may restrict ourselves to the positive integer divisors of a specified integer. In Figure 9.3, we see the divisors of 30. We denote the poset of the divisors of n by D_n.

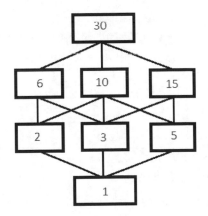

FIGURE 9.3: Positive divisors of 30 ordered by divisibility.

Yet another family of posets consists of the posets of all partitions of a set of n elements ordered by refinement. Recall that a *partition* of a set is a collection of disjoint subsets whose union is that set. Suppose Π_1 and Π_2 are partitions of the same set of n elements; we say that Π_2 is a *refinement* of Π_1 provided that each subset in Π_2 is entirely contained in one of the subsets of Π_1. (We discussed partitions of finite sets in Section 7.2.) Consider, for example, the set $\{1, 2, 3, 4\}$. There are $B_4 = 15$ partitions of this set. The partition $\{1, 2\}\{3\}\{4\}$ is a refinement of the partition $\{1, 2\}\{3, 4\}$, but it is not a refinement of $\{1\}\{2, 3\}\{4\}$.

An element x of a poset is said to be *maximal* provided that there is no y with $x < y$, and *minimal* if there is no y with $y < x$. It is important not to confuse these two definitions with two similar-looking ones; we say x is a *maximum* element provided that $y \leq x$ for all y in the poset. In this case, every element is comparable to the maximum element. In the same way, a *minimum* element x is one for which every other element is comparable, and greater. Not every poset has a maximum or a minimum, although every poset we have seen so far has both. Clearly, if a poset has a maximum element, that is the only maximal element, and similarly a minimum element is the unique minimal element.

THEOREM 9.1 (Extreme Elements)
A finite poset possesses at least one maximal and at least one minimal element.

Proof. We will prove there exists a maximal element, and leave the proof of a minimal element to Problem 1. Choose an element x_1; if there is no element greater than it, it is maximal and we are done. Otherwise, find x_2 with $x_1 \prec x_2$; if there is no element greater than x_2, we are done. Continue this process; it must terminate, as there are only finitely many elements in the poset. When it terminates, we will have found a maximal element. □

The theorem is easily seen to be false for infinite posets; consider, for instance, the set of all integers with the usual order. A poset without a maximum or a minimum element is shown in Figure 9.4.

9.3 Bounds and Lattices

If a poset containing elements a and b also contains an element x with $x \preceq a$ and $x \preceq b$, we say that x is a *lower bound of a and b*; similarly an *upper bound* of a and b is an element x with $a \preceq x$ and $b \preceq x$. There may be elements of a poset that have no upper or lower bound; consider, for example, the elements c and e of Figure 9.4, which have neither. We say that x is the *least upper*

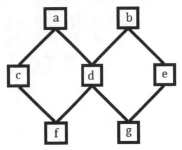

FIGURE 9.4: Poset with no unique minimum or maximum element.

bound of a and b if x is an upper bound of a and b, and for any y that is also an upper bound of a and b we must have $x \preceq y$. There is a corresponding definition for *greatest lower bound*; x is the greatest lower bound of a and b provided that x is a lower bound of a and b, and whenever y is a lower bound of a and b then $y \preceq x$. (Notice that this implies that the greatest lower bound and least upper bound of two elements is unique when it is defined.) When they exist, the greatest lower bound and least upper bound of two elements a and b are denoted by $a \wedge b$ (also called the *meet* of a and b) for the greatest lower bound, and by $a \vee b$ (the *join*) for the least upper bound.

Not every pair of elements in a poset has a meet or a join; again, consider the poset of Figure 9.4. The elements c and e have no upper or lower bounds at all, so certainly no meet or join. However, in some posets every pair of elements has a meet and a join; such a poset is called a *lattice*. Many of the posets we have discussed are lattices.

The meet and join operations take familiar forms in some kinds of posets; for instance, in the poset $\mathcal{B}_n$ the meet of two sets $S \wedge T$ is the intersection $S \cap T$, and the join $S \vee T$ is the union $S \cup T$. Similarly, in D_n it is not hard to see that $a \wedge b$ is the greatest common divisor of a and b, whereas $a \vee b$ is the least common multiple. The terms *meet* and *join*, as well as the shapes of the operators, help us remember which is which; *meet* is akin to set intersection, and is smaller. *Join* is akin to union, and is the larger.

THEOREM 9.2 (Properties of Meet and Join)
In any lattice, for any three elements a, b, and c, the meet and join operations satisfy:

1. *Associative law:* $a \wedge (b \wedge c) = (a \wedge b) \wedge c$ *and* $a \vee (b \vee c) = (a \vee b) \vee c$.

2. *Commutative law:* $a \wedge b = b \wedge a$ *and* $a \vee b = b \vee a$.

3. *Idempotent law:* $a \wedge a = a$ *and* $a \vee a = a$.

4. *Absorption law:* $a \wedge (a \vee b) = a$ *and* $a \vee (a \wedge b) = a$.

5. $a \wedge b = a$ *if and only if* $a \vee b = b$ *if and only if* $a \preceq b$.

Proof. We will prove the associative law for the join and leave the others to exercises. Because we are dealing with a lattice, we know that any two elements have a meet and a join, so there is no question that both sides of the equation are defined; we merely need show that they are equal. Let $x = a \wedge (b \wedge c)$ and $y = (a \wedge b) \wedge c$. By definition, we see $x \succeq a$ and $x \succeq (b \wedge c)$, so $x \succeq b$ and $x \succeq c$. Now $a \wedge b$ is the least element w for which $w \succeq a$ and $w \succeq b$, so $x \succeq (a \wedge b)$. (It is not possible for x and w to be incomparable, for if this were so, a and b would not have a unique least upper bound.) It follows that $x \succeq (a \wedge b) \wedge c = y$. An exactly similar proof shows that $y \succeq x$ as well; because the partial order relation is irreflexive, we must have $x = y$. $\square$

THEOREM 9.3 (Unique Maximum and Minimum)
Every finite lattice has a unique maximum element $\bar{1}$ and a unique minimum element $\bar{0}$.

Proof. Given the finite lattice, we know from Theorem 9.1 that there is at least one minimal element and one maximal element. Take such a minimal element, which we will call $\bar{0}$. If x is any other element of the lattice, find $\bar{0} \wedge x = z$. By definition, $z \preceq \bar{0}$, which means (since $\bar{0}$ is minimal) that $z = \bar{0}$. Since $\bar{0} \wedge x = \bar{0}$, by Theorem 9.2(5), we have $0 \preceq x$. Thus $\bar{0}$ is the unique minimum element. The existence of the unique maximum element is similar. $\square$

A lattice is said to be *distributive* provided that meet and join each distribute over the other; that is, that

$$a \wedge b \vee c = (a \wedge b) \vee (a \wedge c) \text{ and } a \vee (b \wedge c) = (a \vee b) \wedge (a \vee c).$$

We know from set theory (see Appendix A) that union and intersection of sets distribute in this way, so that every subset lattice is a distributive lattice. Not every lattice is a distributive lattice; see, for instance, the lattice in Figure 9.5.

9.4   Isomorphism and Cartesian Products

The reader has probably noticed that the Hasse diagrams of Figure 9.1 and of Figure 9.3 look very much alike; the posets have the same number of elements, and their Hasse diagrams are arranged in the same pattern. Accordingly, we define two posets to be *isomorphic* if there is a bijection ϕ between their elements that is order-preserving; that is, $x \preceq y$ if and only if $\phi(x) \preceq \phi(y)$. Thus we can see that D_{30} is isomorphic to $\mathcal{B}_3$; for instance, $\phi(\{1\}) = 2$, $\phi(\{2\} = 3$, and $\phi(\{3\}) = 5$ gives rise to an isomorphism between

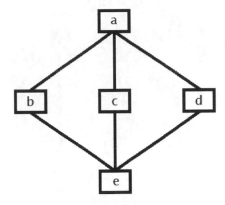

FIGURE 9.5:   A lattice that is not distributive.

the two. Clearly, any two total orders on n elements are isomorphic; other isomorphisms are explored in the exercises.

Example 9.1: Find an isomorphism between the poset D_{12} and the poset of submultisets of the multiset $\{a, a, b\}$.

We see from the Hasse diagrams that these appear to be isomorphic. Create an isomorphism by mapping each occurrence of a in a submultiset to a factor of 2 in a divisor of 12, and each b to a factor of 3. Thus $\{a, b\}$ corresponds to 6, $\{a, a\}$ to 4, and so forth. We can easily confirm that this is an isomorphism.
□

Clearly, if two posets have the same Hasse diagram (apart from the labels on the vertices), they are isomorphic; we must be very careful, however, with the converse. For instance, consider the two posets of the diagram in Figure 9.6. The two graphs are isomorphic as graphs, but considered as directed graphs with edges oriented downwards, they are different, and represent different posets.

Given two posets X and Y, there are a few ways of constructing other posets from them. The simplest way is perhaps the *direct sum*. This is denoted $X + Y$ and is defined on the set $X \cup Y$; we say $a \prec b$ in $X \cup Y$ if $a \prec b$ in X, or $a \prec b$ in Y, and any element of X is incomparable to any element in Y. The Hasse diagram of $X + Y$ is the disjoint union of the Hasse diagrams of X and of Y, respectively.

The *Cartesian product* of X and Y is the poset on the set of ordered pairs $\{(x, y) : x \in X, y \in Y\}$ in which $(x_1, y_1) \prec (x_2, y_2)$ whenever $x_1 \prec x_2$ in X and $y_1 \prec y_2$ in Y.

Example 9.2: Find the Cartesian product of the poset $\mathcal{B}_1 \times \mathcal{B}_1$.

Recall that $\mathcal{B}_1$ is the poset whose Hasse diagram consists of two vertices and a single edge; we may think of the elements of $\mathcal{B}_1$ as $\{1\}$ and $\{\} = \emptyset$.

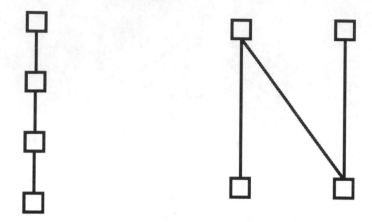

FIGURE 9.6:   Two posets with almost the same Hasse diagram.

Then the product has four elements; they are $(\{1\}, \{1\})$, $(\{1\}, \emptyset)$, $(\emptyset, \{1\})$, and $(\emptyset, \emptyset)$. We show the Hasse diagram of this poset below, next to the poset of $\mathcal{B}_2$ for comparison purposes.   □

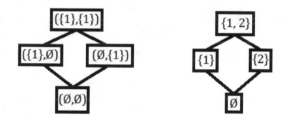

FIGURE 9.7:   The Hasse diagram for $\mathcal{B}_1 \times \mathcal{B}_1$ and for $\mathcal{B}_2$.

The similarity is not a coincidence. For a more general statement, see Problem 4.

It turns out that every finite poset is isomorphic to some subset of $\mathcal{B}_n$, for a suitable n. We shall set out to prove this result. Define a *down-set* of the poset X to be a set $Y \subseteq X$ with the property that, for any y in Y, if $z \prec y$ then $z \in Y$; that is, if Y contains any elements smaller than that element. (Note that Y may be empty; thus every poset has at least two down-sets, namely itself and the empty set.)

Example 9.3: Determine whether the set $\{10, 6, 3, 2, 1\}$ is a down-set of the poset D_{30} whose Hasse diagram is shown in Figure 9.3.

We see that the element 5 precedes 10 in the partial order, but 5 is not in the given set, so it is not a down-set. When we omit 10, the resulting set $\{6, 3, 2, 1\}$ is a down-set.   □

THEOREM 9.4 (Representation of Posets)

Any finite partially ordered set is isomorphic to a poset of sets ordered by inclusion.

Proof. Given a finite poset X, for each element x of X denote by L_x the down-set $\{y \in X : y \preceq x\}$. (It is not in general true that every down-set of a poset is of the form L_x for some x; for instance, in the poset D_{30} of Figure 9.3, the set $\{5, 2, 1\}$ is a down-set but is not of the form L_x.) We create a new poset Y whose elements are $\{L_x : x \in X\}$ and $L_x \prec L_y$ whenever $L_x \subset L_y$. Define $\phi : X \to Y$ by $\phi(x) = L_x$. We demonstrate that this is an isomorphism. First suppose $\phi(x) = \phi(y)$; by definition, $x \in L_x$, and $L_x = L_y$. Now since $x \in L_y$ we must have $x \preceq y$. Similarly, we conclude $y \preceq x$; thus $x = y$. Also, ϕ is clearly onto.

Now suppose $x \prec y$ in X; we demonstrate that $L_x \subset L_y$. But this is just a restatement of the transitive property; any element z of L_x satisfies $z \preceq x$, and $x \prec y$; so $z \prec y$, and $z \in L_y$. $\qquad\square$

In fact, more is true; for instance, if X is any finite distributive lattice, then there is some poset for which X is isomorphic to its poset of down-sets. (This refers to *all* down-sets, not just those of the form L_x.) For a deeper discussion of this and related ideas, we refer the reader to a source such as [107].

9.5 Extremal Set Theory: Sperner's and Dilworth's Theorems

A *chain* is a totally ordered subset of a poset; that is, $C \subseteq P$ is a chain of P if for any two distinct x, y in C we have either $x < y$ or $y < x$. The *height* of a poset is the length of any maximum-length chain of that poset. An *antichain* is a subset of a poset in which no two elements are comparable; the *width* of the poset is the number of elements in a maximum-sized antichain. In the case of the poset $\mathcal{B}_3$ of the figure above, we see that its height is 4, because the chain consisting of $\emptyset$, $\{1\}$, $\{1, 2\}$, and $\{1, 2, 3\}$ is a maximum-length chain. Its width is 3, as illustrated by the antichain $\{1\}$, $\{2\}$, and $\{3\}$. A question of combinatorial interest is finding the size of the largest antichain in a poset; this question was entirely settled for $\mathcal{B}_n$ in 1928.

A collection of sets is called a *Sperner family* if none of them is a proper subset of another. It is clear that if we look at subsets of a set of cardinality n, the subsets that all have k elements for a given k form a Sperner family, as the proper subsets of each must have fewer than k elements. Recall that $\lfloor n/2 \rfloor$ is the greatest integer less than or equal to $n/2$, and $\lceil n/2 \rceil$ is the least integer greater than or equal to $n/2$. Clearly, then, we can find an antichain in $\mathcal{B}_n$ of $C(n, \lfloor n/2 \rfloor)$ elements by taking the subsets of cardinality $\lfloor n/2 \rfloor$. Can we do better? Sperner's theorem says that we cannot.

THEOREM 9.5 (Sperner's Theorem)
Let S be a Sperner family of subsets of an n-set. Then $|S| \leq C(n, \lfloor n/2 \rfloor)$, *and if equality holds then either S is the collection of all subsets of cardinality* $\lfloor n/2 \rfloor$ *or the collection of subsets of cardinality* $\lceil n/2 \rceil$.

Proof. Given $X = \{x_1, x_2, \ldots, x_n\}$ take any permutation σ of these elements; we find a chain $\emptyset \subset \{\sigma(x_1)\} \subset \{\sigma(x_1), \sigma(x_2)\} \subset \cdots \subset X$. As there are $n!$ such permutations, there are clearly $n!$ chains of length n (maximal chains) in $\mathcal{B}_n$. What proportion of those chains contain a particular set A of k elements (say $x_1, x_2, \ldots x_k$)? There are $C(n, k)$ such sets A; by symmetry, each is contained in an equal number of maximal chains, and that number is $n!/C(n, k) = k!(n - k)!$. Now take our Sperner family S. By definition, no chain contains more than one member of S. If any maximal chain contains no member of S, we can add any set from this chain to S to make a larger Sperner family. Assume that we have done this and now have a maximal Sperner family so that S contains exactly one set from every maximal chain. We now partition the set of maximal chains according to which set of S they contain; if S_k is a member of S of cardinality k, it is in $n!/C(n, k)$ maximal chains. Thus we have

$$\sum_{S \in \mathcal{S}} n!/C(n, |S|) = n! \sum_{S \in \mathcal{S}} \frac{1}{\binom{n}{|S|}} = n!$$

This is equivalent to $\sum_{S \in \mathcal{S}} 1/C(n, |S|) \leq 1$; to make $|S|$ as large as possible, we want to make $1/C(n, |S|)$ as small as possible, so we want the largest possible value of $C(n, |S|)$. This clearly occurs when $|S| = \lfloor n/2 \rfloor$ or $|S| = \lceil n/2 \rceil$.

Now if n is even, we are done; each set in S must have exactly $n/2$ elements to make the sum as small as possible. If n is odd, set $n = 2r + 1$. We could make a maximum antichain with sets of cardinality r or of cardinality $r + 1$; we need to show that we cannot make a maximum antichain with a mixture. Suppose that S contains an r-set S_0. If T_0 is any $r + 1$-set containing S, clearly $T_0 \notin S$. Now, remove an element of T_0 to get a new set $S_1 \neq S_0$ of cardinality r. Notice that T_0 and S_1 are part of a chain; exactly one element of each chain must be in S, and since T_0 is not, S_1 must be in S. We now add a new element to S_1 to obtain an $r + 1$-set T_2, different from T_0 or T_1; as before $T_2 \notin S$ since $S_1 \subset T_2$. We now remove some element of T_2 to obtain a new set S_2; do so in such a way as to guarantee that S_2 is different from S_0 or S_1, and the same reasoning suggests that S_2 must be in S. It seems clear that we can continue this process until all of the sets of cardinality $r + 1$ have been shown not to be elements of S, and all of the sets of cardinality r have been shown to be in S. It is intuitively clear that we can do this; indeed, if S and S' are any two sets of cardinality R, we may travel between the two by alternately adding to S an element of S' that it does not contain to get a new set T, and deleting from T any element of S that is not in S'. It follows that if S contains any r-set, it contains every r-set and no $r+1$-sets. If S contains a set of cardinality $r+1$, on the other hand, the same process shows that it contains all of those sets, and none of cardinality r. □

The last result from extremal set theory that we will prove is Dilworth's theorem. This result deals with partitioning a poset into chains; what is the smallest number of chains that will suffice for each element of the poset to be in exactly one chain? Clearly, if the poset has width a, we need at least a chains, since no chain can contain two or more elements of an antichain. The surprising result of Dilworth is that we can always cover the poset in exactly a chains.

THEOREM 9.6 (Dilworth's Theorem)
If the poset X has width a, then there is a set of a disjoint chains whose union is X.

There are several ways to prove this theorem, some more involved than others. Our proof below is adapted from that of Marshall Hall ([50]).

Proof. We use induction on the number n of elements of X; if $n = 0$ or $n = 1$, the result is trivial. We therefore take n to be greater than 1 and assume the truth of the Theorem for any poset with fewer points. Suppose we find a maximal chain C of our poset (where by *maximal* we mean that the addition of any element to C makes it no longer a chain). Consider the poset X' obtained from X by removing the elements of C; if X' has width no more than $a-1$, we apply the induction hypothesis to X' to cover it in $a-1$ chains, replace the chain C, and we are done. Therefore, we must assume that C does not include any point from some antichain $Y = \{y_1, y_2, \dots y_a\}$. Define the *down-set generated by Y* to be $Y^- = \{x \in X : x \preceq y_i$ for some $1 \le i \le a\}$; the set of all elements of X that are "smaller" than some element of Y. Similarly, $Y^+ = \{x \in X : y_i \preceq x$ for some $1 \le i \le a\}$ is the *up-set generated by Y*. Notice that if there is any element of X not contained in $Y^- \cup Y^+$, we could add this element to Y to make a larger antichain; as Y was a maximal antichain, this is impossible, so $Y^- \cup Y^+ = X$, and $Y^- \cap Y^+ = Y$.

At this point, the idea is to say that Y^- is a poset that is smaller than X, and has width a, so by the induction hypothesis it can be partitioned into a chains; and Y^+ is a poset that is smaller than X, so it may be partitioned similarly. Now, we "glue" together the chains in Y^+ to those in Y^- by the element of Y that they have in common; so the chain in Y^+ for which y_i is the least element will join to the chain of Y^- for which y_i is its greatest element, for every $1 \le i \le a$. How can this fail?

It can fail rather easily, in fact. For instance, Y^+ might be the entire poset, because Y could be the set of all minimal elements, or of all maximal elements. In this case the induction hypothesis would not apply. (The largest antichain consisting of minimal elements must consist of *all* the minimal elements, and similarly for the maximal elements; see the Exercises.) However, if Y consists neither of all maximal elements nor all minimal elements, the proof is finished. So assume that Y consists entirely of maximal elements (if Y consists of minimal elements, we will turn the poset upside-down, partition it into chains, and then turn it right side up again). Find a minimal element x_1 with $y_1 \prec x_1$; that is, follow any chain from y_1 downward as far as possible and give the minimal element of that chain (which must be minimal in the poset) the label x_1. Now we have a chain C_1; look at the poset $X \setminus C_1$. What is the width of this poset?

We claim now that $X \setminus C_1$ has width at most $a - 1$. We have already established that every largest antichain is either the set of all maximal or of all minimal elements; C_1 contains one of each, so we have removed one element from each antichain of size a. Now the induction hypothesis applies, and we may partition $X \setminus C_1$ into $a - 1$ chains. Throw the chain C_1 back in, to find a partition of X into a chains, and we are done. $\qquad\square$

Dilworth's theorem is useful for a number of purposes, not least of which is that it provides a fairly short proof of Philip Hall's theorem (Theorem 1.2) which we give here, along with the statement of the theorem:

A collection D of sets has a system of distinct representative if and only if it never occurs that some n sets contain between them fewer than n elements.

Proof. Let $D = \{B_1, B_2, \ldots, B_k\}$ containing among them v elements $\{1, 2, \ldots, v\}$ and suppose that any union of n of these sets contains at least n elements. We form our poset P on the elements $B_1, B_2, \ldots, B_k, 1, 2, \ldots, v$ and we rule that $i \prec B_j$ if and only if $i \in B_j$; D is an antichain, as is the set $\{1, 2, \ldots v\}$. However, these may not be the maximum antichains. An antichain of largest cardinality might consist of some of the elements and some of the sets. By renumbering if necessary, let the largest antichain be $1, 2, \ldots \ell, B_1, B_2, \ldots B_f$; here we note that $\ell + f \geq v \geq k$ because this largest antichain must be at least as big as the antichain D and the antichain $\{1, 2, \ldots, v\}$. Because this is an antichain, no element below $\ell + 1$ may belong to any B_i for $i \leq k$, so we have $B_1 \cup B_2 \cup \ldots \cup B_f \subseteq \{\ell + 1, \ell + 2, \ldots, v\}$. By our hypothesis, this union contains at least f elements, so we have $f \leq v - \ell$.

Use Dilworth's theorem to partition our poset into $\ell + f$ chains. We can again renumber our sets and elements so that the chains look like

$$\{B_1, 1\}, \{B_2, 2\}, \ldots \{B_r, r\}, r + 1, \ldots, v, B_{r+1}, \ldots B_k.$$

Now we see that $v + k - r = \ell + f$, because we have $v + k$ total objects in the poset; there are $\ell + f$ chains, of which r are length-2 chains, $v - r$ are single elements, and $k - r$ are sets from D. Thus $r = k + (v - \ell - f)$. Since $f \leq v - \ell$, we get $v - \ell + f \geq 0$, and so $r \geq k$. But r (the number of sets B_i used in length-two chains) cannot be greater than k (the total number of sets

B_i); thus $v = k$. That means that the set $\{1, 2, \ldots, k\}$ is a system of distinct representatives for D. $\qquad\qquad\qquad\qquad\qquad\qquad\qquad\qquad\qquad\qquad\qquad\square$

Exercises 9A

1. Find the Hasse diagram for the poset D_{36}.

2. Find the Hasse diagram for the poset of the multisubsets of the multiset $\{a, a, b, b\}$.

3. Prove that meet and join commute (Theorem 9.2(2)).

4. Prove the idempotent laws for meet and join (Theorem 9.2(3)).

5. Prove that $a \wedge b = a$ if and only if $a \preceq b$.

6. Illustrate Theorem 9.4 by assigning sets to each element of the poset of Figure 9.4 so that it is ordered by inclusion.

7. Show that the largest antichain consisting only of minimal elements must consist of all the minimal elements.

8. Find all minimal elements of the poset of Figure 9.8.

9. Find the width of the poset of Figure 9.8.

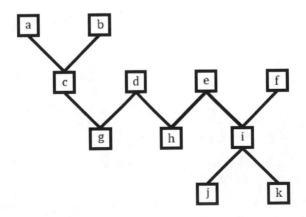

FIGURE 9.8: Poset for exercises and problems.

Exercises 9B

1. Find the Hasse diagram for the poset D_{60}.

2. Find the Hasse diagram for the poset of the multisubsets of the multiset $\{a, a, b, c\}$.

3. Prove the associative law for the join (Theorem 9.2(1)).

4. Prove the absorption laws for meet and join (Theorem 9.2(4)).

5. Prove that $a \vee b = b$ if and only if $a \preceq b$.

6. Illustrate Theorem 9.4 by assigning sets to each element of the poset of Figure 9.5 so that it is ordered by inclusion.

7. Show that the largest antichain consisting only of maximal elements must consist of all the maximal elements.

8. Find all maximal elements of the poset of Figure 9.8.

9. Find the height of the poset of Figure 9.8.

Problems

Problem 1: Show that a finite poset must have a minimal element.

Problem 2: Show that no poset has a Hasse diagram that contains a K_3.

Problem 3: Let n be the product of four distinct prime numbers; find an explicit isomorphism between D_n and $\mathcal{B}_4$.

Problem 4: Find an isomorphism between $\mathcal{B}_n \times \mathcal{B}_m$ and $\mathcal{B}_{n+m}$; derive as a corollary the result that each subset lattice $\mathcal{B}_n$ is isomorphic to a product of copies of $\mathcal{B}_1$.

Problem 5: Prove that the union of two down-sets is a down-set; prove that the intersection of down-sets is a down-set.

Problem 6: Find all down-sets of the poset of Figure 9.5 and find the poset of these sets ordered by inclusion.

Problem 7: Find all largest antichains of the poset of Figure 9.8.

Chapter 10

Introduction to Graph Theory

10.1 Degrees

Recall the definitions of graphs from Section 1.1.4. The *degree* of the vertex x was defined to be the number of edges that have x as an endpoint. We write $deg_G(x)$ for the degree of x in graph G, or simply $deg(x)$ if only one graph is being discussed. A graph is *regular* if all vertices have the same degree, and in particular a regular graph of degree 3 is called *cubic*. We denote the smallest of all degrees of vertices of G by $\delta(G)$, and the largest by $\Delta(G)$.

The *degree sequence* of a graph consists of all the degrees of the vertices. For example, degree sequence $(4, 2, 2, 1, 1, 0)$ corresponds to the graph

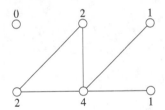

(Degrees are shown next to the vertices.) We normally write such a sequence in non-ascending order.

It is interesting to ask whether a given sequence is the degree sequence of any graph. If so, the sequence is called *graphical*; a graphical sequence is called *valid* if there is a *connected* graph with those degrees. Some sequences are easily seen not to be graphical. Theorem 1.1 says that the sum of degrees in any graph must be even, so the sum of the elements of any graphical sequence must be even. Moreover, if a graph has $d + 1$ vertices with degree greater than zero (non-isolated vertices), no vertex can have degree greater than d: $(3, 1, 0, 0)$ is obviously not graphical. (This last argument does not apply to multigraphs.)

Example 10.1: Which of the following sequences are graphical: $(3, 3, 1, 1, 0)$, $(3, 3, 2, 1, 1)$, $(3, 3, 2, 1, 0)$?

Suppose a graph has five vertices and two of them have degree 3. If those two vertices are not adjacent, each must be joined to all three of the remaining vertices (See the left-hand graph in Figure 10.1) so these all have degree at

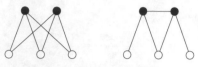

FIGURE 10.1:   Testing degree sequences.

least 2. Therefore the two vertices of degree 3 are adjacent. There are only two possibilities: either both are adjacent to the same two vertices out of the remaining three, or not. In the former case, both of those vertices have degree at least 2 (it may be greater, if they are adjacent to each other or to the final vertex), and this is not possible for any of the degree sequences given. So the graph must look like the right-hand graph in the figure, possibly with other edges between the lower vertices. This rules out the first and third sequences, so they are not graphical; however, the middle sequence is graphical, and is realized by the right-hand graph.   □

Of course, we know that the third sequence is not graphical, because the sum of its elements is odd.

One way to construct a new graph from an old one is to add one vertex and join it to some of the other vertices. The following theorem of Havel [53] and Hakimi [49] essentially shows that this process can be reversed.

THEOREM 10.1 (The Havel–Hakimi Theorem)
If $v \geq 2$ and $S = (d_0, d_1, \ldots, d_{v-1})$ is a sequence of integers satisfying

$$d_0 \geq d_1 \geq \ldots \geq d_k \geq d_{k+1} \geq \ldots \geq d_v \geq 0,$$

then S is graphical if and only if T is graphical,

$$T = (d_1 - 1, d_2 - 1, \ldots, d_{d_0} - 1, d_{d_0+1}, \ldots, d_{v-1}).$$

Proof.   Suppose you have a graph whose degree sequence is T. You can construct a graph with degree sequence S by simply adding a new vertex and joining it to the d_0 vertices with degrees $d_1 - 1$ to $d_{d_0} - 1$. So the "if" part is easy.

So let us assume S is graphical. Suppose there is a graph G whose vertices $x_0, x_2, \ldots, x_{v-1}$ satisfy $deg(x_i) = d_i$ for every i. For convenience we will call $x_1, x_2, \ldots, x_{d_0}$ the *earlier* vertices and $x_{d_0+1}, \ldots$ the *later* vertices.

Suppose the vertices of G can be ordered so that x_0 is adjacent to each of the earlier vertices. (Ordering would be relevant if there is more than one vertex with degree d_0 or more than one vertex with degree d_{d_0}.) In that case we shall say G is *type I*; otherwise it is *type II*. If G is type I, simply delete x_0 and all the edges joining it, and you have a graph with degree-sequence T.

Now we assume G is type II. Then there must be some earlier vertices x_j that are *not* adjacent to x_0. Suppose there are t such vertices ($t > 0$),

$x_{j_1}, x_{j_2}, \ldots, x_{j_t}$. Let's write

$$A = \{x_{j_1}, x_{j_2}, \ldots, x_{j_t}\}.$$

So there must also be exactly t later vertices x_k that *are* adjacent to x_0. We write

$$B = \{x_{k_1}, x_{k_2}, \ldots, x_{k_t}\}$$

for this set of vertices. Finally, we define $n(G)$ to be the sum of degrees of vertices adjacent to x_0.

If all the vertices in A and B had the same degree, we could simply reorder the vertices by exchanging x_{j_i} and x_{k_i} for every i, and obtain an ordering where x_0's neighbors are precisely the earlier vertices; but that would mean G is type I. So there exist x_j in A and x_k in B such that $d_j > d_k$. There must be a vertex that is adjacent to x_j but not to x_k; say x_m is one such. Construct a new graph, G_1 say, by deleting edges $x_m x_j$ and $x_0 x_k$ and replacing them by edges $x_0 x_j$ and $x_m x_k$. G_1 has the same degree sequence as G. Moreover, $n(G_1) = n(G) + d_j - d_k$, which is greater than $n(G)$.

If G_1 is type I we are finished. But suppose G_1 is type II. Then we could go through the whole process again, and produce a graph G_2, and so on.

If none of the graphs $G_1, G_2, \ldots$ is type I, then the sequence $n(G), n(G_1), n(G_2), \ldots$ is a strictly increasing infinite sequence of integers, so eventually it will be greater than the sum of all degrees of vertices of G. But this is obviously impossible. So eventually we must construct a type I graph with the same degree sequence as G, and we can use it to construct the desired graph. $\qquad\square$

Given a non-decreasing sequence $S = (d_0, d_1, \ldots, d_{v-1})$ of non-negative integers, Theorem 10.1 enables us either to construct a graph with S as its degree sequence or to show that S is not graphical. Construct the degree sequence T as given in the theorem. (We'll call this *reducing S*.) If T is obviously graphical, then form a graph with degree sequence T and adjoin a new vertex to the first d_0 vertices. If T is obviously not graphical, then S is not graphical. Otherwise, reduce T, and repeat until you find a sequence for which the answer is obvious. Some obviously graphical sequences include an all-zero sequence, corresponding to the *null* or empty graph with no edges and a sequence of $2n$ 1s, for a graph consisting of n disjoint edges. At any stage, 0s correspond to isolated vertices, and do not change the answer.

Example 10.2: Is the sequence $(4, 4, 2, 2, 1, 1)$ graphical?

When you reduce $(4, 4, 2, 2, 1, 1)$ you get $(3, 1, 1, 0, 1)$, or in standard form $(3, 1, 1, 1, 0)$. This in turn reduces to $(0, 0, 0, 0)$, the degree sequence of a null graph with 4 vertices. So the original is graphical. The calculation could be represented as

$$(4, 4, 2, 2, 1, 1) \Rightarrow (3, 1, 1, 0, 1) = (3, 1, 1, 1, 0) \Rightarrow (0, 0, 0, 0).$$

The graph is constructed as follows:

Example 10.3: Is the sequence $(6, 5, 4, 2, 2, 2, 1)$ graphical?

$$(6, 5, 4, 2, 2, 2, 1) \Rightarrow (4, 3, 1, 1, 1, 0) \Rightarrow (2, 0, 0, 0, 0)$$

which is not graphical (you cannot have one vertex of degree 2 and all the rest isolates). □

To illustrate the remark about added 0s, you can immediately conclude from the example that $(4, 4, 2, 2, 1, 1, 0, 0)$ and $(4, 4, 0, 2, 2, 1, 0, 1, 0)$ are graphical but $(6, 5, 4, 2, 2, 2, 1, 0)$ is not.

10.2 Paths and Cycles in Graphs

A lot of the terminology of graph theory comes from the application of networks to modeling road systems. For example, the multigraph

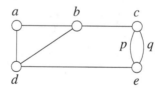

could represent five camps denoted a, b, c, d, e and the walking tracks between them; there are two tracks joining c to e, and they are labeled p and q, and other tracks are denoted by their endpoints. (These representations are ordered.) A sequence of vertices and edges of the form $(\ldots, x, xy, y, yz, z, \ldots)$ will be called a *walk*, because on the diagram it represents a trail that one could walk among the camps.

Consider the following walks:

$$
\begin{aligned}
W_1 &= (a, ab, b, bd, d, da, a, ad, d, de, e); \\
W_2 &= (a, ad, d, de, e, p, c, cb, b); \\
W_3 &= (c, p, e, p, c, cb, b, bd, d, de, e, q, c).
\end{aligned}
$$

Each edge of W_1 is uniquely determined by its endpoints—it does not contain p or q—so we could simply write it as (a, b, d, a, d, e), omitting the edges. This will always be true of walks in graphs. The walk W_2 has no repeated edges. A walk with this property is called *simple*. W_3 returns to its starting point; a walk that does this is called *closed*. We shall make the following formal definitions:

Definition. A *walk of order* n (or of *length* $n - 1$) in a network is a finite sequence

$$(x_1, y_1, x_1, y_2, x_2, y_3, \ldots, y_{n-1}, x_n)$$

of vertices $x_1, x_2, \ldots$ and edges $y_1, y_2, \ldots$ where the endpoints of y_i are x_i and x_{i+1}. x_1 and x_n are the *start* and *finish*, respectively, of the walk, and we refer to a *walk from x_1 to x_n*. The following are special types of walk:

Simple walk: A walk in which no edge is repeated;

Path: A simple walk in which no vertex is repeated;

Circuit or *closed walk*: A simple walk in which the start and finish are the same;

Cycle: A simple closed walk in which $x_1, x_2, \ldots, x_{n-1}$ are all different. No vertex is repeated, except that the start and finish are equal.

A cycle might also be called a *simple circuit*. Many mathematicians also use the phrase "closed path," although strictly speaking this is not possible, as the start and finish make up a repetition.

Given a path in a graph or multigraph G, you could define a subgraph made up of the vertices and edges in the path. This subgraph is called a *path in G*. If we think of it as a graph by itself, an n-vertex path is denoted P_n and is a graph with n vertices $x_1, x_2, \ldots, x_n$ and $n - 1$ edges $x_1x_2, x_2x_3, \ldots, x_{n-1}x_n$. Similarly an n-vertex *cycle* or n-cycle C_n is a graph with n vertices $x_1, x_2, \ldots, x_n$ and n edges $x_1x_2, x_2x_3, \ldots, x_{n-1}x_n, x_nx_1$, and we can think about cycles as subgraphs in other graphs. A 3-cycle is often called a *triangle*.

Suppose a walk W from x_1 to x_n, where $x_1 \neq x_n$, passes through a point more than once; say x_i is the first occurrence of the point and x_j the last. Then the walk $x_1, x_2, \ldots, x_i, x_{j+1}, \ldots, x_n$ is part of W—we could call it a *subwalk*—and contains fewer repeats. If we continue this process, we eventually finish with no repeats, a path. That is, any x_1-x_n walk contains an x_1-x_n path. So:

THEOREM 10.2 (Walks Contain Paths)
If a graph contains a walk from x_1 to x_n, $x_1 \neq x_n$, it contains a path from x_1 to x_n.

In the same way, one could argue that any closed walk contains a cycle, but there is one exception to be made: it is possible to walk from x_1 to x_n and then walk back by retracing your steps. The graph of such a walk will not contain a cycle, unless the x_1-x_n walk contained one. But any *circuit* must contain a cycle.

We say two vertices are *connected* if there is a walk from one to the other. (In view of Theorem 10.2, we could replace "walk" by "path.") This gives us a formal definition of connectivity: a graph is connected if and only if all its pairs of vertices are connected. We can now define a *component* as the subgraph formed by taking a maximal set of vertices with every pair connected, plus all the edges touching them.

The *distance* between two vertices is the minimum length among the paths joining them (or the *length of the shortest path*). The distance from x to y is written $d(x,y)$. For consistency we write $d(x,x) = 0$.

Example 10.4: Find the distance from x to each other vertex in the following graph.

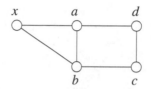

There is a path of length 1 (a single edge) from x to a, and also to b. The distance between two distinct vertices cannot be smaller than 1, so $d(x,a) = d(x,b) = 1$. To find the shortest path from x to d we look at all the possible paths; they are

$$\begin{array}{lll}
(x,a,d) & \text{of length} & 2 \\
(x,a,b,c,d) & \text{of length} & 4 \\
(x,b,a,d) & \text{of length} & 3 \\
(x,b,c,d) & \text{of length} & 3
\end{array}$$

and the shortest length is $d(x,d) = 2$. Similarly, $d(x,c) = 2$.   □

Listing all paths is fine in small cases, but very wasteful of time in a large graph. In 1959, the Dutch computer scientist Edsger Dijkstra [27] devised an efficient algorithm to find the distance from a given vertex to every other vertex in a graph.

Dijkstra's algorithm works as follows. Suppose you want to find the distance from vertex x to every other vertex in your graph. Trivially $d(x,x) = 0$, and $d(x,y) = 1$ for all neighbors of x.

We now use the following argument. Suppose you want to find the distance from vertex x to every other vertex in your graph. Say you know $d(x,z) = n$. Then there is a path $P = (x, x_1, x_2, \ldots, x_{n-1}, z)$ from x to z. If t is any neighbor of z, then there will be a walk of length $n + 1$ from x to t—simply append t to P. This might not be a path, because t might be x_{n-1} (which is one of z's neighbors). But in that case we know a shorter path to t. So $d(x,y)$ is at most $n + 1$.

So we proceed as follows. Trivially $d(x,x) = 0$; label x as x_0. We know $d(x,y) = 1$ for all neighbors of x. Label these neighbors $x_1, x_2, \ldots, x_p$ in some order (where $p = deg(x)$). We say x has been *processed*. Now we *process* x_1. We consider all the neighbors of x_1 and discard all that are already labeled. Label the remaining vertices $x_{p+1}, x_{p+2}, \ldots$. Each is distance 2 from x.

Continue in this way. When you finish processing x_i, go on to x_{i+1}. Discard any neighbor of x_{i+1} that is already labeled, and label the remainder with the smallest unused labels. Each of them is distance $d(x, x_{i+1})$ from x. Go on until all vertices are labeled.

Example 10.5: Apply the above algorithm to

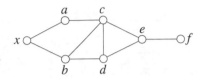

We start by labeling x as x_0 and note that $d(x, x_0) = 0$. The neighbors of x_0 are a and b. Neither is yet labeled. Label a as x_1 and b as x_2. Set $d(x, a) = d(x, x_1) = d(x, b) = d(x, x_2) = 1$. Processing x_0 is completed.

Now we process x_1. Its neighbors are x and c. We skip x as it is already labeled. We label c as x_3 and set $d(x, c) = d(x, x_3) = 1 + d(x, x_1) = 2$. Next we process x_2. The only unlabeled neighbor is d. So d is labeled x_4, and $d(x, d) = d(x, x_4) = 1 + d(x, x_2) = 2$.

The only unlabeled neighbor of x_3 is e. So e is x_5 and $d(x, e) = d(x, x_5) = 1 + d(x, x_3) = 3$. x_4 has no unlabeled neighbors. Finally, x_5 has one unlabeled neighbor, f. So f is x_6 and $d(x, f) = d(x, x_6) = 1 + d(x, x_5) = 4$. All vertices are now labeled.

This could be written in a table, as shown in Figure 10.2. $\qquad \square$

Of course, this example is more easily handled by finding all paths. But the algorithm can be applied to much larger graphs. Writing the table is tedious, but it illustrates how all the data could be stored on a computer. Several implementations of this algorithm are available on the Internet. One example is [111], but there are many others; just type the phrase "Dijkstra's algorithm" into your favorite search engine.

10.3 Maps and Graph Coloring

By a *map* we mean the kind of map you would usually find in an atlas, a map of a continent (showing countries) or of a country (showing states). However, we need to make one restriction. We require that you can get from one point in a country (or state) to another without crossing a border; that is, each country or state is connected. We'll call such a map *simple*. For example, a map of North America is not simple, because the United States

Vertex	Neighbors		
$x = x_0$ $d(x,x) = 0$	a, b	$a = x_1$ $b = x_2$	$d(x,a) = 1$ $d(x,b) = 1$
$a = x_1$ $d(x,a) = 1$	x, c x	x already labeled $c = x_3$	$d(x,c) = 2$
$b = x_2$ $d(x,b) = 1$	x, c, d	x already labeled c already labeled $d = x_4$	$d(x,d) = 2$
$c = x_3$ $d(x,c) = 2$	a, b, e	a already labeled b already labeled $e = x_5$	$d(x,e) = 3$
$d = x_4$ $d(x,d) = 2$	b, e	b already labeled e already labeled	
$e = x_5$ $d(x,c) = 2$	c, d, f	c already labeled d already labeled $f = x_6$	$d(x,f) = 4$

FIGURE 10.2: Tabular form of Dijkstra's algorithm.

is not connected – to get from Alaska to the other states you need to cross either Canada or the ocean. We say two countries or states are *neighbors* if they have a border—a single common point does not count.

Given a map, we construct a graph as follows: the vertices correspond to the countries (or states), and two vertices are adjacent when the two countries share a border. It is always possible to draw this graph in such a way that no two of the edges cross—the easiest way is to draw the graph on top of the map, with each vertex inside the corresponding country. A graph that can be drawn without edge crossings is called *planar*. So every map corresponds to a planar graph. It is easy to draw a map from any given planar graph, so the correspondence goes both ways.

Not all graphs are planar: try as you might, you cannot draw K_5 without crossings. On the other hand, some graphs are planar even though they do not look that way. Figure 10.3 shows two representations of the same graph, K_4. One has crossings, the other does not. Remember, we say the graph is planar if it *can* be drawn in the plane without crossings, even though the representation we are using has some crossings.

Suppose you want to print a map in an atlas. It is usual to use different colors for the various countries. Some colors may be repeated, but the map is easier to read if you avoid using the same color on two countries with a common border. The corresponding problem in a graph is to label the vertices in such a way that no two adjacent vertices receive the same label. The labels are called *colors*, and a labeling is called a *coloring* of the graph, even though the ideas can be applied to non-planar graphs, where the relationship to maps does not exist. A coloring in which no two adjacent vertices receive the same

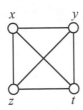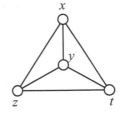

FIGURE 10.3:   Two representations of K_4.

color is called *proper*.

Map and graph coloring has been studied since 1852 when Francis Guthrie asked whether it was true that any map can be colored using four colors. The first printed reference is due to Cayley [18]. A year later an incorrect 'proof' by Kempe appeared; its incorrectness was pointed out by Heawood 11 years later. Another failed proof is due to Tait in 1880; a gap in the argument was pointed out by Petersen in 1891. It was not until 1976 that Appel and Haken [4] proved the four color theorem:

THEOREM 10.3 (Four Color Theorem)
Any planar graph can be colored with four colors.

Their proof was long, complicated and required extensive use of computers. Later proofs include [88] and [46].

In a graph coloring, the set of all vertices receiving the same color is called a *color class*. There are a number of applications of color classes outside map printing (in fact, there is not much practical application of the four color theorem to map printing.) For example, coloring is used in conflict resolution.

Example 10.6: A *habitat zoo* consists of several large habitats, each containing several species. In such a zoo, you cannot put natural enemies together. Use graph coloring to design the zoo.

We form a graph whose vertices are the species to be housed in the zoo. An edge means the two species are "enemies," and cannot be housed in the same habitat. Now color this graph. Two species can be put into the same habitat if they receive the same color.   □

In this example we would like the number of habitats to be as small as possible. So we want the number of colors to be as small as possible. We define the *chromatic number* $\chi(G)$ of a graph G to be the smallest number of colors possible for a proper coloring.

Another application is school timetabling. In this case the classes are represented by vertices, and an edge means that some student will be taking both classes. Vertices in the same color class can be scheduled simultaneously.

The complete graph K_n has n vertices, and every pair is joined, so $\chi(K_n) = n$. The cycle C_n has chromatic number 2 when n is even and chromatic number 3 when n is odd. The *complete bipartite graph* $K_{m,n}$, which contains no odd cycles, has chromatic number 2.

Obviously, if a graph is properly colored, every subgraph will also be properly colored. So, if G contains a subgraph K_n, its chromatic number will be at least n. This gives us a tool to find chromatic numbers of small graphs: find the largest complete graph that is contained in your graph. Color all its vertices differently. Then try to color the remaining vertices using those colors. Very often this is easy to do. If there is no complete graph, is there an odd-length cycle? If so, at least three colors will be needed.

Another tool is:

THEOREM 10.4 (Brooks' Theorem) [12]
Suppose the graph G is not a complete graph, nor a cycle of odd length. Then $\chi(G) \leq \Delta(G)$.

The proof is a little difficult, and can be found in graph theory texts (for example, [122, p. 980]).

We now present an easy algorithm that will always color a graph G in no more than $\Delta(G)+1$ colors—not the minimum, but close. Suppose the vertices of G are $x_1, x_2, \ldots$ and the colors are $c_1, c_2, \ldots, c_{\Delta(G)+1}$. Apply color c_1 to x_1. Then go through the vertices in order; when x_i is to be colored, use c_j, where j is the smallest positive integer such that c_j has not yet been applied to any vertex adjacent to x_i.

Example 10.7: Apply the above algorithm to

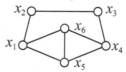

The maximum degree is 3. We apply c_1 to x_1. As $x_2 \sim x_1$ we cannot use c_1, and x_2 receives c_2. The neighbors of x_3 are x_2 and x_4, so color c_2 is the only one barred (vertex x_4 has not yet received *any* color); c_1 is still free, so x_3 gets c_1. Similarly x_4 gets c_2. Now x_5 is adjacent to x_1 (color c_1) and x_4 (color c_2) so it receives c_3. Finally x_6 is adjacent to all three colors, so it gets c_4. □

We could represent this coloring by

$$\{x_1, x_3\} \leftarrow c_1, \quad \{x_2, x_4\} \leftarrow c_2, \quad \{x_5\} \leftarrow c_3, \quad \{x_6\} \leftarrow c_4.$$

As we said, this method does not necessarily return a minimal coloring; in fact, the given graph has chromatic number 3, realized by

$$\{x_1, x_4\} \leftarrow c_1, \quad \{x_2, x_5\} \leftarrow c_2, \quad \{x_3, x_6\} \leftarrow c_3.$$

Finally, there is an interesting example in the April 2009 issue of *Math Horizons*. The cover shows a four-colored map of all United States counties, illustrating an article by Mark McClure and Stan Wagon [73].

Exercises 10A

1. Prove that in any graph there must be two vertices with the same degree. Show that this is not true for multigraphs, by exhibiting a multigraph with degree sequence $(3, 2, 1, 0)$.

2. Which of the following sequences are graphical? If the sequence is graphical, draw a graph that realizes it.

   (i) $(4, 2, 2, 2, 1, 1)$ (ii) $(5, 4, 3, 2, 2, 2)$

   (iii) $(3, 3, 3, 3, 2, 2)$ (iv) $(5, 3, 3, 3, 1, 1, 1, 0)$

   (v) $(6, 5, 2, 2, 1, 1, 1, 0)$ (vi) $(6, 5, 4, 3, 3, 3, 1, 1, 0)$

3. Find all paths from x to f in the graph

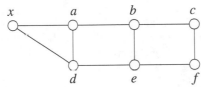

4. Find the distance from x to each other vertex in the graph of the preceding exercise.

5. Find the distance from b to each other vertex in the following graphs.

(i) 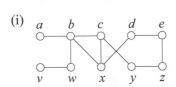   (ii)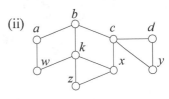

6. The graph G contains two paths from vertex x to vertex y, namely $(x, a, \ldots, y)$ and $(x, b, \ldots, y)$. Assuming vertices a and b are distinct, prove that G contains a cycle that passes through x. Show that the assumption "a and b are distinct" is necessary.

7. G consists of two subgraphs, G_1 and G_2, that share one common vertex. What is the relation between $\chi(G)$, $\chi(G_1)$ and $\chi(G_2)$?

8. Find the chromatic numbers of the following graphs.

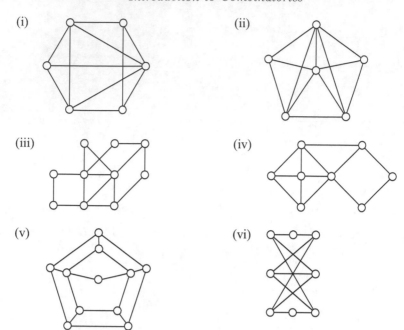

9. By subdividing an edge in a graph we mean taking an edge xy and replacing it with a new vertex u and new edges xu and uy. Suppose G is a bipartite graph, and suppose H is produced from G by subdividing one or more edges.

   (i) Show that H must have chromatic number no greater than 3.

   (ii) Either prove that H *must* have degree 3, or produce a counterexample.

10. Suppose x is a vertex of odd degree in a graph G. Show that G must contain a walk that starts at x and ends at some other vertex of odd degree.

Exercises 10B

1. Show that no cubic graph has an odd number of vertices.

2. A bipartite graph has $2n$ vertices. What is the maximum number of edges it can have?

3. Which of the following sequences are graphical? If the sequence is graphical, draw a graph that realizes it.

   (i) $(4, 4, 3, 3, 3, 1, 1, 0)$ (iv) $(4, 3, 3, 2, 2, 2, 2)$

   (ii) $(3, 3, 2, 2, 1, 1)$ (v) $(7, 4, 2, 2, 2, 2, 1, 1, 1, 0)$

   (iii) $(6, 5, 4, 2, 2, 2, 1, 0)$ (vi) $(8, 5, 4, 3, 3, 3, 1, 1, 1, 1, 0, 0)$

4. Show that there is no graph with degree sequence $(4, 4, 4, 2, 2)$, but there is a multigraph with this degree sequence.

5. Find all paths from a to d in the graph

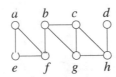

6. Find the distance from c to each other vertex in the graph of the preceding exercise.

7. Find the distance from b to each other vertex in the following graphs.

   (i)

   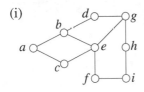

   (ii)

   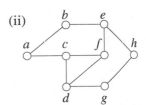

8. The *square* G^2 of a graph G has the same vertex-set as G; x and y are adjacent if and only if $d(x, y) = 1$ or 2 in G. Find the squares of:

   (i) $K_{1,n}$;   (ii) P_5;   (iii) C_5.

9. Suppose x, y and z are three vertices in a connected graph. Show that

$$d(x, z) \leq d(x, y) + d(y, z).$$

10. The graph G contains at least two vertices. One vertex has degree 1, and every other vertex has degree greater than 1. Show that G contains a cycle.

11. Prove that any graph with v vertices and at least $1 + \frac{1}{2}(v-1)(v-2)$ edges must be connected.

12. Find the chromatic numbers of the following graphs.

(i)

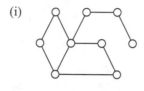

(ii)

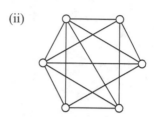

(iii)

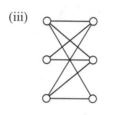

(iv)

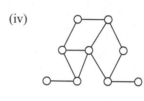

(v)

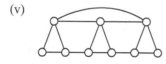

(vi)

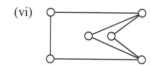

Problems

Problem 1: Show that it is possible to find a graph with degree sequence $(5, 5, 5, 3, 2, 2, 1, 1)$, but it is not possible to find such a graph in which the two vertices of degree 1 are adjacent.

Problem 2: Graph G has v vertices. What is the smallest number of edges that G can have, as a function of v and $\delta(G)$?

Problem 3: A graph that is isomorphic to its complement is called *self-complementary*.

(i) Prove that the path P_4 and the cycle C_5 are self-complementary.

(ii) Show that a self-complementary graph can contain no isolated vertices.

(iii) Find all self-complementary graphs on five vertices.

(iv) Prove that the number of vertices of a self-complementary graph must be congruent to 0 or 1 modulo 4.

Problem 4: Show how to construct a cubic graph with v vertices whenever v is even, $v \geq 4$.

Problem 5: Is Theorem 10.1 true for multigraphs?

Problem 6: [31] Graph G has v vertices and degree sequence $(d_1, d_2, \ldots, d_v)$, written in non-ascending order. Prove that, if $0 < r < v$,

$$\sum_{i=1}^{r} d_i \leq r(r-1) + \sum_{i=r+1}^{v} \min\{r, d_i\}.$$

Problem 7: The square of a graph was defined in Exercise 10B.8. A graph is called a *perfect square* if it is the square of some graph. Find all perfect squares on four vertices.

Problem 8: If x is a vertex in a connected graph G, the *eccentricity* of x, $\epsilon(x)$, is defined to be the largest value of $d(x, y)$, where y ranges over all the vertices in G. The *radius* R of G is the smallest value of $\epsilon(x)$, and the diameter D is the largest value of $\epsilon(x)$, for all vertices x. Prove that

$$R \leq D \leq 2R.$$

Problem 9: For what values of n is K_n a subgraph of $K_{n,n}$?

Chapter 11

Further Graph Theory

11.1 Euler Walks and Circuits

It is reasonable to say that graph theory was founded by the Swiss mathematician and physicist Leonhard Euler (see Appendix C). As we mentioned in Section 1.1.1, Euler addressed the St. Petersburg Academy in 1735 on the problem of the Königsberg bridges. The Prussian city of Königsberg was set on both sides of the Pregel River. It included a large island called The Kneiphof, and the river branched to the east of it. So there were four main land masses—we shall call them the four *parts* of the city, and label them A, B, C, D—connected to each other by seven bridges. For convenience we repeat our rough map of the city, on the left in Figure 11.1:

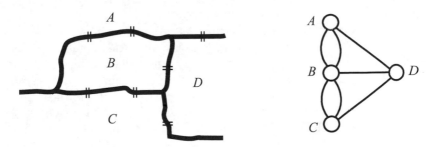

FIGURE 11.1: The original Königsberg bridges.

The problem was to find a walk through the city that would cross each bridge once and only once, and visit all four parts. The only way to go between the different pieces of land was to cross the bridges. In modern terminology (terminology that would not have existed without Euler's work!) he observed that the city could be represented by a multigraph, with the parts represented by vertices and the bridges represented by edges, as shown in the second part of Figure 11.1. A solution to the problem would be a simple walk in the multigraph that included every edge; even better would be a closed simple walk, so that a person could start and finish from home.

Euler proved that no such walk is possible. But he did much more. He essentially invented graph theory and showed how it could be used to represent any set of islands and bridges, or indeed any set of towns and roads joining them, and concocted an algorithm for traversability problems.

In graph terminology, we define an *Euler walk* to be a simple walk that contains every edge of the network, and an *Euler circuit* to be a closed Euler walk. A network that has an Euler circuit is called *eulerian*.

Suppose G has an Euler walk, and suppose X is a vertex of G. (So there are $d(X)$ bridges leading to the area X in the corresponding map.) Every time the walk passes through X, you traverse two of these bridges. By the end of the walk, this accounts for an even number of edges. The only way X can have odd degree is if it is either the start or the finish of the walk. A walk can have at most one start and one finish, so a network cannot have an Euler walk if it has more than two odd vertices. Similarly, an eulerian graph must have all its vertices even.

Euler showed [32] that these conditions are sufficient, if we add the obvious proviso that the graph must be connected:

THEOREM 11.1 (Euler Circuit Characterization)
If a connected network has no odd vertices, then it has an Euler circuit. If a connected network has two odd vertices, then it has an Euler walk whose start and finish are those odd vertices.

Proof. We prove the theorem by describing an algorithm to construct an Euler walk, and then show that it always works.

Suppose a network has no odd vertex. Select any vertex x, and select any edge incident with x. Go along this edge to its other endpoint, say y. Delete the edge from the network. Then choose any other edge incident with y and go along this edge to its other endpoint; delete the edge you just travelled. Continue in this way: in general, on arriving at a vertex, delete the edge just used, select one of the remaining edges incident with it that has not yet been used, and go along the edge to its other endpoint. At the moment when this walk has led into the vertex z, where z is not x, an odd number of edges touching z have been used up and then deleted (the last edge to be followed, and an even number previously—two each time z was traversed). Since z is an even vertex (as are all the vertices), there is at least one edge incident with it that is still available. Continue the walk—eventually you will run out of edges (remember, we assume that all networks are finite)—until you can go no further. The only possible way this can happen is that the last vertex you reach must be x—that is, you have constructed a closed walk. It will necessarily be a simple walk and it must contain every edge incident with x.

If all the edges have been used, we have a closed circuit. Otherwise, the remaining edges—those you haven't yet deleted—make up a multigraph with every vertex even. Moreover, it must have at least one vertex in common with the walk, or else the original network would not have been connected, and that case was not allowed. Select such a vertex b, and find a closed simple walk starting from b in the "remainder," just as you did before. Then unite the two walks as follows: at one place where the original walk contained c, insert the new walk. For example, if the two walks are

$$x, y, \ldots, a, b, c, \ldots, x$$

and

$$b, s, \ldots, t, b,$$

then the resulting walk will be

$$x, y, \ldots, a, b, s, \ldots, t, b, c, \ldots, x.$$

(There may be more than one possible answer, if b occurred more than once in the first walk. Any of the possibilities may be chosen.) The new walk is a closed simple walk in the original network. Repeat the process of deletion, this time deleting the newly formed walk. Continue in this way. Each walk contains more edges than the preceding one, so the process cannot go on indefinitely. It must stop: this will only happen when one of the walks contains all edges of the original network, and that walk is an Euler walk.

The case where there are two odd vertices, say p and q, and every other vertex is even, now follows very easily from the "all vertices even" case. Form a new network by adding another edge pq to the original. The new network has every vertex even, so you can find a closed Euler walk in it. Moreover, it follows from the proof that you can choose p as the first vertex and the new edge pq as the first edge. Then delete this first edge; the result is an Euler walk from q to p. □

Example 11.1: Find an Euler circuit in the graph

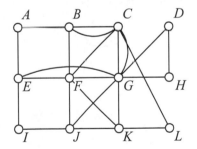

First we check that the graph is connected and every degree is even. Then select any starting vertex–A say–and trace a circuit at random, with the rules:
- You cannot use an edge twice;
- If you reach a vertex, you must continue if there is any edge touching it that has not been used.

In the example, say we choose to go

$$ABFCGKFEA.$$

(We omit the parentheses, and commas between the vertices, when no confusion will arise.) We then have to stop because both edges touching A have been used. Let's call this the *first circuit*. Deleting its edges, we get

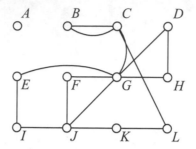

We now start again from any one of the vertices that were touched in the first circuit but still have edges in the above diagram. A good choice is F. One possible circuit is

$$FJIEGDHGF.$$

Again you can go no further, E's edges are used up. This is the *second circuit*.

Now combine the first and second circuits as follows: go along the first circuit until you come to the first occurrence of F; then go round the whole of the second circuit; then finish the first circuit. In other words, replace the first F in $ABFCGKFEA$ by $FJIEGDHGF$: the result is

$$AB\ FJIEGDHGF\ CGKFEA$$

(where the space shows where we changed circuits; there isn't really a gap there).

All the edges in the second circuit are deleted, leaving

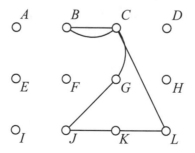

and you look for another circuit in the remainder. In this case, starting at B, you find the circuit (actually the whole thing)

$$BCGJKLCB.$$

Feed this third circuit into the big circuit as before:

$$A\ BCGJKLCB\ FJIEGDHGFCGKFEA.$$

We have finished; we have an Euler circuit. □

By the way, Königsberg is now called Kaliningrad, and is in Russia. Two of the seven original bridges were destroyed by bombs during the Second World War. Two others were later demolished but were replaced by a modern highway incorporating two new bridges. One other was rebuilt, but two of the original bridges remain from Euler's time. The new current configuration and its network are shown in Figure 11.2. So the town now has an Euler walk! (But not a closed one.)

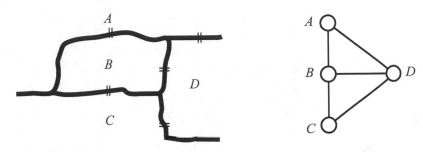

FIGURE 11.2: The Königsberg bridges today.

Euler walks and their generalizations are used in what are called *route inspection problems*. For example, suppose a highway inspector needs to drive along and check all the main roads in an area. The road map can be translated into a network G by interpreting all the intersections as vertices and the roads as edges. Since the inspector does not want to cover the roads more than once on each trip, and since he must start and finish from the city's vehicle facility, the most efficient route will correspond to an Euler circuit in G.

What if there is no such circuit? Then some of the roads must be traversed twice. This is not allowed in an Euler circuit, so we *add roads*: for example, if you need to travel the road from A to B twice, you represent that road by two edges in the corresponding network. This process of adding duplicate edges to make the network eulerian is called *eulerizing* the network. The minimum number of edges needed to eulerize G is called the *eulerization number* of G, and an *efficient* eulerization is one that uses that number.

Example 11.2: Find an efficient eulerization of the graph in Figure 11.3.

There are two odd vertices, namely u and y. You would like to add an edge from u to y, but this is not allowed: you must duplicate an edge that is already present. So you try to find a solution that duplicates two edges. Obviously one edge would need to touch u, and the other to touch y. There are three possible solutions; one is to duplicate uv and vy. □

As we said, you would prefer to add an edge uy. There are some situations

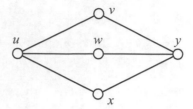

FIGURE 11.3:   Find an efficient eulerization.

where this would be possible. But for many examples—like snow-plowing—you have to use the existing roads, so duplication of edges is required.

In many cases there is a cost associated with an edge. For example, the cost of plowing a street might be proportional to the length of the street or to the time it takes to traverse it. In that case you could assign a cost to each edge, and define the cost of an eulerization to be the sum of the costs of the new edges added. The most desirable eulerization would then be the cheapest one possible. This is the form in which the problem was first studied, by Kwan Mei-Ko [74]. Several algorithms were described in [29]; see also [68]. Ko wrote in terms of mail delivery. The author was Chinese, the original paper was written in Chinese, it appeared in a Chinese journal and the translation appeared in the journal *Chinese Mathematics*, so Alan J. Goldman of the U.S. National Bureau of Standards suggested the name "Chinese Postman problem," and the name stuck.

As a final note, we observe that the whole study of Euler circuits can also be carried out for directed networks. This will make sense when divided roads or one-way streets need to be taken into account.

11.2   Application of Euler Circuits to Mazes

Apart from such obvious applications as mail delivery and highway inspection, there are several other applications of Euler circuits and eulerization. As an example we shall explore one of them.

A maze can be modeled by a network: each intersection of paths, where you have a choice of ways to go, corresponds to a vertex, as does the end of every "blind alley," and edges represent the direct paths between these points. Figure 11.4 shows a small section of a maze and the corresponding network.

The network of a maze contains one or two special vertices, representing the entrance and the exit (which may be in the same place or not). If the maze can be solved, there must be a walk from start to finish. Finding such a walk is easy enough if you have a map of the maze, but these are not always available. In fact, your aim may be to construct such a map—for example,

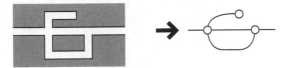

FIGURE 11.4: Graphical representation of a maze.

you may be exploring a newly discovered system of caves. So we shall present an algorithm for solving an unknown maze. We assume that you have some way of marking parts of the maze, such as leaving markers or drawing chalk arrows. Every edge will be marked immediately after it is traversed, and every vertex will be marked just after you first leave it; we'll refer to the vertices and edges as *marked* after this is done and *unmarked* otherwise.

Proceed as follows. Start at the entrance. Select any available edge, call it the *ingress*, and take it. Do not take the ingress again while any other edge is available. Apart from that, when you arrive at a vertex, select the next edge using the following rules:

(i) From a new vertex, any edge may be taken;

(ii) From an old vertex, if the edge just traversed was new, then turn around and retake it;

(iii) Never use any edge three times.

This walk can be explained as follows: it is an Euler circuit in the network formed by doubling each edge in the original. It will visit every edge. If your aim is to get through the maze, you leave the first time you encounter the exit. If you want to construct a map, complete the whole circuit.

Example 11.3: In the following diagram, is it possible to draw an unbroken line that goes through each line segment exactly once (you are not allowed to pass through the corners)?

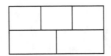

If we treat this as a map with five countries, we get the map shown on the left in Figure 11.5. However, we need to look at the outside boundaries, corresponding to the ocean coast. This is handled by adding a sixth "country" (or ocean) that encloses the other five. So we put in another vertex, connected to all five, and look at the corresponding graph, shown on the right in the figure. The question now asks, is there an Euler walk in this new graph? The graph has two odd vertices, so the answer is "yes." □

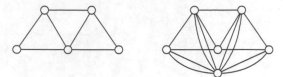

FIGURE 11.5: Two graphs from the map.

11.3 Hamilton Cycles

In 1857 Thomas Kirkman posed the following problem in a paper that he submitted to the Royal Society: Given a graph, does there exist a cycle passing through every vertex? Two years later Sir William Rowan Hamilton, who was Astronomer Royal of Ireland, invented a puzzle called the *icosian game*. The board is a regular graph with 20 vertices and 30 edges. (The graph is a representation of a dodecahedron, with the vertices representing the corners.) The object of the game was to visit every vertex using each edge at most once. This was essentially an example of Kirkman's problem, because the path generated in visiting the vertices would be the sort of cycle Kirkman was discussing, but historically a cycle that contains every vertex is called a *Hamilton cycle*. (Kirkman's name is better known in the history of block designs.) The name *Hamilton path* is used for the case of a path that contains every vertex but is not closed.

A graph is called *hamiltonian* if it has a Hamilton cycle. The first question is whether or not a given graph is hamiltonian. Some cases are easy: any complete graph larger than K_2 has a Hamilton cycle, and in fact any permutation of the vertices gives rise to one of them. In any cycle in $K_{m,n}$, alternate vertices must be in different parts of the vertex-set, so $K_{m,n}$ is hamiltonian if and only if $m = n$.

Example 11.4: Find Hamilton cycles in the following graphs, if possible.

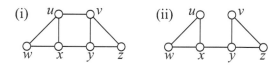

In (i), the cycle $wxyzvuw$ is obviously Hamilton. In (ii), any Hamilton path would necessarily include edge xy. If it is traversed from x to y, then the first vertex in the path must be one of $\{u, w, x\}$ and the last must be one of $\{v, y, z\}$—you cannot get back across xy. The edge xy is called a *bridge* (we'll define this idea formally in Section 11.4, below), and its existence shows that no Hamilton cycle exists.   □

The following necessary condition was found by Ore in 1960:

THEOREM 11.2 (Ore's Condition) [83]
Suppose G is a graph with v vertices, $v \geq 3$, and for every pair of nonadjacent vertices x and y,

$$deg(x) + deg(y) \geq v$$

then G is hamiltonian.

Proof. Suppose the theorem is not true. We can assume that all pairs of nonadjacent vertices satisfy the given degree condition, and that if p and q are nonadjacent vertices then the graph formed by adding edge pq, denoted $G+pq$, will be hamiltonian (if not, then join pq and use the new graph instead of G). We would say G is *maximal* for the condition.

As $v \geq 3$, K_v is hamiltonian, so G is not complete. Select two vertices, x and y, that are not adjacent. Then G is not hamiltonian, but $G^* = G+xy$ is, so xy must lie in every Hamilton cycle of G^*. Trace such a cycle of the form $(yx \ldots y)$ and delete the first edge; the result is a Hamilton path $(x = x_1 x_2 \ldots x_v = y)$ in G. If $x \sim x_i$ for some i with $2 < i < p$, then $y \not\sim x_{i-1}$, or else

$$(xx_i x_{i+1} \ldots y x_{i-1} x_{i-2} \ldots x_2 x)$$

would be a Hamilton cycle in G, which is impossible.

Say the neighbors of x are $x_2, x_{j_2}, x_{j_3}, \ldots, x_{j_d}$, where $2 < j_2 < j_3 < \ldots < j_d$ and $d = deg(x)$. Then y is *not* adjacent to any of $x_{j_2-1}, x_{j_3-1}, \ldots, x_{j_d-1}$, or to x. So there are d vertices not adjacent to y, and therefore $deg(y) \leq (v-1)-d = v - d - 1$. Therefore

$$deg(x) + deg(y)d + deg(y) \leq d + v - d - 1 < v,$$

a contradiction. $\qquad\square$

This gives a sufficient condition, but it is not necessary; many hamiltonian graphs do not satisfy the theorem. In fact, we know of no simple way to tell whether a graph is hamiltonian. At first sight, the Euler circuit and Hamilton cycle problems look similar, but there is a simple test to see whether a graph is eulerian, and a straightforward algorithm to construct an Euler cycle when it exists.

Often there are costs associated with the edges, such as airfares or travel times. In those cases, we define the cost of a Hamilton cycle to be the sum of the costs of the edges. These ideas were first studied in the 1930s by various mathematicians. Whitney pointed out that the problem of finding the *cheapest* Hamilton circuit could be motivated as follows: A traveling salesman needs to stop at several cities (ending back at home); find the shortest/cheapest route that can be used. He referred to it as the *Traveling Salesman Problem* or TSP. (For more history, see [98].) It is often sufficient to talk about the TSP for complete graphs.

The obvious approach to the TSP is to list all possible Hamilton cycles and work out the cost of each. However, this can be an extremely long computation. If there are n cities, there are $n!$ different ways to order them. Clearly

(a, b, c, d) costs the same as (b, c, d, a), so you only need to check one in n cycles, and the cycle gives the same result if you travel it in reverse order, but this leaves $(n-1)!/2$ essentially different cycles. Computer time will be prohibitive, even for small cases. For example, if you process one hundred routes per second, it will take more than 55 hours to examine every Hamilton cycle in K_{12}. If you consider a salesman who needs to visit 30 or 40 cities, the problem is far too large.

So we consider some algorithms that give an approximate solution. We'll look at two, the *nearest neighbor algorithm* and the *sorted edges algorithm*. We usually assume our graph is a copy of K_n and there is a distance or cost $c(x, y)$ associated with each edge xy (the airfare, or distance, or whatever). The algorithms can be applied to incomplete graphs, but there is no guarantee that any cycle will be produced.

To explain the ideas behind the algorithms, suppose an explorer wants to visit all the villages in a newly discovered area, and wants to do it as efficiently as possible. The "costs" are the distances between settlements.

The sorted edges algorithm models the way she might proceed if she has a map of the whole area, possibly produced by an aerial survey, with distances marked. She might hope to find an efficient route by taking the shortest feasible connections.

Sorted edges algorithm: Initially sort the edges from shortest (cheapest) to longest and list them in order. Select the shortest edge. At each stage delete from the list the edge you just used, and any edges that would form a circuit or if added to those already chosen, or would be the third edge chosen that was adjacent to some vertex (we'll call this a *threeway*), then select the shortest remaining edge. In this case the result does not depend on the starting point.

In the nearest neighbor algorithm, the explorer starts at one of the villages. She asks the locals, "what is the nearest village to here?" If possible she goes to that village next.

Nearest neighbor algorithm: First choose any vertex x_1 as a starting point. Find the vertex x_2 for which $c(x_1, x_2)$ is smallest. Continue as follows: at each stage go to the nearest neighbor, except that you never close off a cycle (until the end) or backtrack. In other words, after x_i is chosen $(i < n)$, x_{i+1} will be the vertex *not* in $\{x_1, x_2, \ldots, x_i\}$ for which $c(x_i, x_{i+1})$ is minimum. Finally, x_n is joined to x_1. Notice that the result of this algorithm depends on the starting point.

In either example, it may be that two candidate edges have the same cost, and either may be chosen. However, if the algorithms are to be implemented on a computer, some sort of choice must be made. For your convenience in working problems, let's assume we always take the edge that occurs earlier in alphabetical order (if edges bc and bd have the same length, choose bc).

Example 11.5: Figure 11.6 shows a copy of K_4 with costs on the edges. Apply the two algorithms to this graph.

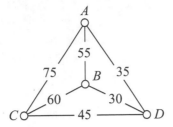

FIGURE 11.6:   A labeling of K_4.

In the nearest neighbor algorithm starting at A, the first stop is D, because $c(AD) < c(AB) < c(AC)$ (that is, $35 < 55 < 75$). Next is B ($c(DB) < c(DC)$, AD is not allowed). Finally, we go to C. The cycle is $ADBCA$, or equivalently $ACBDA$, cost 200. Nearest neighbor from B yields $BDACB$ (the same cycle), from C, $CDBAC$ (cost 205), and from D, $DBACD$ (same cycle as from C).

In sorted edges we first make the list

edge	cost
BD	30
AD	35
CD	45
AB	55
BC	60
AC	75

Then we proceed: select BD; AD is usable, so select it; CD is *not* acceptable (threeway at D); AB is not acceptable (triangle ABD); BC and AC are acceptable. The cycle is $ACBDA$, cost 200. □

Figure 11.7 shows all three Hamilton cycles in the copy of K_4, with their costs. Note that the cheapest route, (A, B, C, D), cost 195, was not found by either method. These algorithms are not guaranteed to give a perfect result. But they give a *good* result in a reasonable time. They are more effective for larger graphs. The examples in this book are all relatively small, so that you can follow them, but in the real world you would not apply the nearest neighbor or sorted edges algorithms to graphs like these.

As we said, either algorithm can be applied to graphs that are not complete, but neither is guaranteed to give a result in that case. If the graph is close to complete, the algorithms usually work, but not always.

Example 11.6: Find the routes generated by the sorted edges algorithm and by the the nearest neighbor algorithm starting at each vertex, if possible, in

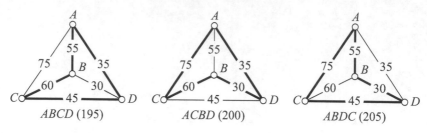

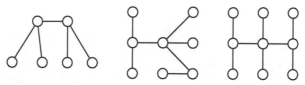

FIGURE 11.7:   All Hamilton cycles of the K_4.

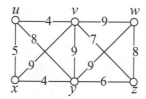

FIGURE 11.8:   Some trees.

The sorted edges algorithm gives the edges (in order) uv, xy, ux, yz, wz, vw, forming a cycle $uvwzyx$ of total cost 36 (which happens to be the cheapest cycle possible). Now for nearest neighbor: Starting from u, we choose uv (the cheapest edge from u), then vz (vu would be a repeat), then zy, then yx. No further edge is available: xu makes a 5-cycle and xv makes a 4-cycle. Similarly, starting from v we get the path $vuxyz$, and from x we get $xyzvu$, neither of which can be completed. If we start at w, we get $wzyxuvw$, which is the same cycle as we found by sorted edges, and from initial point z we get the same answer. If we start from y, we get $yxuvzw$, of total cost 37.   □

On the other hand, if there are a number of pairs of nonadjacent vertices (we say the graph is not very *dense*), a complete search may be feasible. We give a technique for doing this at the end of the next section.

11.4   Trees

A *tree* is a connected graph that contains no cycles. Figure 11.8 shows three examples of trees. Paths P_n and stars $K_{1,n}$ are other examples.

A tree is a minimal connected graph in the following sense. If any edge is deleted, the resulting graph is not connected. An edge is called a *bridge* if its deletion disconnects the graph (or, in the case of a disconnected graph, disconnects the component in which the edge lies). So we have just said that every edge in a tree is a bridge. In fact we can prove a stronger result. First we show:

LEMMA 11.3

An edge xy in a graph is a bridge if and only if it belongs to no cycle in the graph.

Proof. Write G for the component of the graph that contains x and y. Delete edge xy from G; call the resulting graph H.

(i) First assume xy is a bridge. Then H is not connected. So there are two vertices in G, w and z say, that are not joined by a walk in H. But they *are* joined by at least one walk in G. The only possibility is that xy was an edge in every one of those walks. In particular, there is at least one walk of the form $wWxyZz$ in H, where W and Z are some strings of vertices. (If x appears before y in every walk, we can swap labels w and z.)

Suppose xy is in a cycle in G, say $xx_1x_2\ldots yx$. Even after the edge xy is deleted, the path $xx_1x_2\ldots y$ is still in H, and there is a w-z walk in H, namely

$$wWxx_1x_2\ldots yZz.$$

But we said there was no such walk. This is a contradiction, so the cycle could not have existed.

This proof is rather wordy and pedantic, but it becomes much clearer if you look at Figure 11.9. In the left-hand diagram the heavy lines indicate the walk that includes the edge xy, with the rest of the cycle shown in light dots.

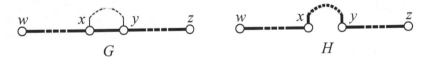

FIGURE 11.9: Illustration of Lemma 11.3.

(ii) Suppose xy is a not bridge, so H is connected. So there is an x-y path in H, and it does not contain edge xy. Add in that edge, and you get a cycle in G that contains xy. □

From this we deduce a simple characterization of trees:

THEOREM 11.4 (Bridge Characterization of Trees)

A connected graph is a tree if and only if every edge is a bridge.

(Writing out a formal proof of this is left as Exercise 9A.15.)

There is a similar characterization involving vertices. A vertex is called a *cutpoint* if its deletion disconnects the graph (or, in the case of a disconnected graph, disconnects the component in which it lies). If a connected graph is a tree then every vertex of degree greater than 1 is a cutpoint.

Another important characterization involves the numbers of vertices and edges.

THEOREM 11.5 (Order Characterization of Trees)

A connected graph G with v vertices is a tree if and only if it has exactly $v - 1$ edges.

Proof. (i) Suppose G is a tree with v vertices. We prove by induction on v that G has $v - 1$ edges.

The result is true for $v = 1$–the only graph on 1 vertex, K_1, is a tree. Suppose it is true for $v < n$, and suppose G is a tree with n vertices. Select an edge xy and delete it. The result is a union of two components, say G_x (containing x) and G_y. G_x is a tree, because it contains no bridge, and it has fewer than n vertices (at least one, y, is gone from G). Say it has m vertices. By the induction hypothesis it has $m - 1$ edges. Then G_y is also a tree, with $n - m$ vertices, and $(n - m) - 1$ edges. G is the union of three edge-disjoint parts, namely G_x, G_y and xy, so it has $(m - 1) + (n - m - 1) + 1 = n - 1$ edges.

(ii) Suppose G has v vertices but is not a tree. It has an edge that is not a bridge. Delete that edge. If the resulting edge is not a tree, repeat the process. Continue until only bridges remain. The result is a tree, and has $v - 1$ edges. Since at least one edge has been deleted, G had more than $v - 1$ edges. □

From part (ii) of the above proof we see that any graph with v vertices and fewer than $v - 1$ edges cannot be connected.

COROLLARY 11.5.1

Any tree has at least two vertices of degree 1.

Proof. Say a tree has v vertices, of which u have degree 1. Every other vertex has degree at least 2, so the sum of the degrees is at least $u + 2(v - u) = 2v - u$. There are $v - 1$ edges, so the sum of the degrees is $2v - 2$. So $2v - u \le 2v - 2$, and $u \ge 2$. □

Trees can be used to enumerate things. One example is the use of trees to find all Hamilton cycles in a graph.

Suppose we want to find all Hamilton cycles in a graph G. We use G to construct a new graph, a tree, which we shall call $T(G)$ or simply T.

First we select any vertex of G as a starting point; call it x. We construct the first vertex of the tree and label it with x. This vertex of the tree is called the *root*. For every edge of G that contains x, draw an edge of the tree. The set of these new vertices is called the *first level* of vertices. Each vertex in the first level is labeled with the corresponding vertex of G.

At level $i+1$, we process every vertex y as follows. Look at the string of labels on the vertices (called a *branch* leading from the root to the new vertex). If this string is $x, x_1, \ldots, x_i, y$, then $P = xx_1 \ldots x_iy$ is a path in G, and we add an edge of the tree corresponding to every edge of G that could be an edge of the Hamilton cycle starting with P (that is, it would not duplicate an edge of G or complete a small cycle in G), and label the new endpoint with the name of the corresponding vertex.

As we go on, some of the branches cannot be continued. At level v, the only possible label is x, and each branch that reaches level v corresponds to a Hamilton cycle in G. Each Hamilton cycle will be represented precisely twice, because the cycle is the same if traversed forwards or backwards.

Example 11.7: Find all Hamilton cycles in the graph

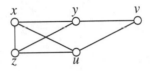

First we select x as the starting point. There are three vertices joined to x in G, namely y, z and u. So we start three branches, and label them as shown.

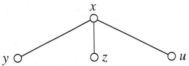

A branch will go from y to vertices that we label z and v, but not to one labeled x (because edge yx would be a duplication) or u (there is no edge yu in G). Similarly there are two branches from each of the first level vertices labeled z and u.

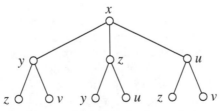

We continue in this way. For example, the branch x, y, v can extend only to u, because y would correspond to a repeated edge and v is not connected to x or z. The branch x, z, u can extend only to v, because z would be a repeated edge and x would form a small cycle.

After x, y, z, u, v there is no possible continuation, because neither vx nor vz is an edge of G, vy would form a triangle and vu would be a duplication. Others, like x, z, y, v, u, x can be completed: the branch contains every vertex of G and finishes back at x—not a *small* cycle, because all the vertices are there. The final tree is

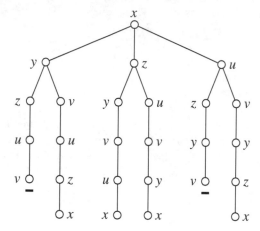

So the example has two Hamilton cycles: $xyvuzx$ and $xzyvux$. Each is represented twice.   □

11.5   Spanning Trees

Suppose you have to connect a number of computers in an office building. For speed and reliability, the management wishes to have a hard-wired connection between every pair of computers. The connection need not be direct; it is possible to relay a message through an intermediate computer. To model this, the computers are represented as vertices and the possible direct connections are represented as edges. The set of edges that are chosen must form a connected subgraph. Moreover this subgraph must contain all the vertices—this is called a *spanning* subgraph.

There will be a positive cost associated with making a direct link between a pair of computers. So, instead of making direct connections between *every* pair of computers for which it is feasible, management might plan to avoid any cycles. The resulting graph would still need to span the system, so they want to find a connected subgraph that contains all the vertices, a *spanning tree* in the original graph.

Every connected graph contains a spanning tree. To see this, consider any connected graph. If there is no cycle, it is a tree. So suppose the graph contains a cycle, such as $abcdea$. The pairs of vertices ab, bc, cd, de and *ea* are all joined by edges. If you need a path from *a* to *e* (for example, to transmit a message between computers), you could use the single edge, or the longer path $a-b-c-d-e$. So remove one edge from the cycle. The resulting spanning subgraph is still connected. If it still contains a cycle, delete an edge from it. Continue in this way: graphs are finite, so the process must stop eventually, and you have a spanning tree.

Spanning trees are important for a number of reasons. A spanning tree is a subgraph with a relatively small number of edges—and therefore easy to examine—but it contains a significant amount of information about the original graph. They have proven very useful in designing routing algorithms and in a number of areas of network design. We shall touch on some of this here, but for a more complete discussion please consult books on trees such as [141].

Sometimes it is useful to know how many different spanning trees there are in a given graph or multigraph. We'll write $\tau(G)$ for the number of spanning trees in G. Some cases are very easy: if T is a tree, then $\tau(T) = 1$; if G is derived from a tree T by replacing one edge of T by a multiple edge of multiplicity k, then $\tau(G) = k$; the cycle C_n has $\tau(C_n) = n$.

Example 11.8: Count the number of trees in the following (multi)graphs:

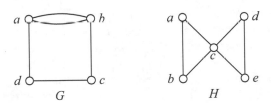

$$G \qquad\qquad H$$

A 4-cycle *abcd* has four spanning trees, and three of them include edge *ab*. Each of those will give rise to two trees in G, one for each of the edges from *a* to *b*. There is one other (the path (*bcda*)). So $\tau(g) = 7$. In H, each spanning tree will consist of a spanning tree in cycle (*abc*) and a spanning tree in cycle (*cde*). Each has three, so the combined graph has $\tau(H) = 3 \times 3 = 9$. □

It is easy to work out $\tau(G)$ in small cases, but an algorithm is needed for large G. If G is a graph or multigraph and $A = xy$ is any edge of G, G_{xy} (or G_A) is formed by deleting both x and y, inserting a new vertex A, and replacing every edge zx and every edge zy by an edge from z to A. G_{xy} is called the *contraction* of G by xy. We also define a (multi)graph $G - xy$ (or $G - A$), formed from G by deleting xy. Examples of these graphs are shown in Figure 11.10.

THEOREM 11.6 (Deletion-Contraction Counting)

$$\tau(G) = \tau(G - A) + \tau(G_A).$$

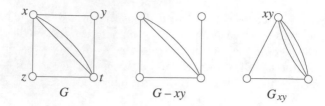

FIGURE 11.10: Graphs for counting trees.

Proof. Suppose A is an edge of G. We count the spanning trees of G that contain A and those that do not. A spanning tree that does not contain A is a spanning tree of $G - A$; conversely, the spanning trees of $G - A$ are spanning trees of G and do not contain A. So there are $\tau(G - A)$ spanning trees of G that do not contain A. Similarly, it is easy to see that each spanning tree of G that contains A can be formed by taking a spanning tree of $G - A$ and adding a to it, and the spanning tree of $G - A$ is uniquely determined. So the number of spanning trees of G that contain A is $\tau(G_A)$.   □

Example 11.9: Calculate $\tau(G - xy)$ and $\tau(G_{xy})$ for the graph G of Figure 11.10, and use this to find $\tau(G)$.

Every spanning tree of $G - xy$ must contain edge yt, or else y will be omitted. If neither edge xt is included, the other two edges must both be used, and there is one such tree. Otherwise, there are two choices for the edge from xt, and two choices for the edge to contain z (either xz or zt, but not both). So there are four spanning trees with an xt edge, and five in all.

Similarly, for each edge from xy to t in G_{xy} there will be two spanning trees, and there is one other, a total of seven.

So $\tau(G - xy) = 5$, and $\tau(G_{xy}) = 7$, and $\tau(G) = 5 + 7 = 12$.   □

Here are two multigraphs that are useful in counting spanning trees. An *n-fold path* is formed from a path by replacing each edge with a multiple edge of multiplicity n. An *n-fold cycle* is defined similarly. The number of spanning trees in an n-fold path is

$$\tau(nP_v) = n^{v-1}$$

and the number of spanning trees in an n-fold cycle is

$$\tau(nC_v) = vn^{v-1}.$$

Verification is left as an exercise (Exercise 9A.14).

Another important example is the complete graph. Cayley discovered the following result in 1889:

THEOREM 11.7 (Cayley's Theorem) [19]
The complete graph K_n has n^{n-2} spanning trees.

The proof is left to Problems 11.10 and 11.11. Note that we are counting *all* spanning trees, not just the number of isomorphism classes. For example, K_3 has three spanning trees, as the theorem says, but they are all isomorphic.

Example 11.10: Calculate $\tau(G)$, where G is the following graph. Check your calculation by deleting and contracting a different edge.

The method of decomposing the relevant graphs is indicated in Figure 11.11. From (i) we see that $\tau(G) = \tau(G_1) + \tau(G_2) = \tau(G_1) + \tau(G_3) + \tau(G_4)$. As G_1 is a 5-cycle, $\tau(G_1) = 5$; $\tau(G_3) = 2$ (it is a tree with one double edge); and G_4 is a 2-fold P_3, so $\tau(G_3) = 2^{3-1} = 4$. So $\tau(G) = 5 + 2 + 4 = 11$. From (ii) we have $\tau(G) = \tau(G_5) + \tau(G_6) = \tau(G_7) + \tau(G_3) + \tau(G_8) + \tau(G_4)$. Now $\tau(G_7) = 1$, $\tau(G_8) = 4$; so $\tau(G) = (1 + 2) + (4 + 4) = 11$. □

Let's return to the idea of connecting computers in an office building, and consider the graph whose vertices are the computers and whose edges represent hard-wired direct links. As we said, it is cheapest if the connections form a tree. But which tree should be chosen?

We shall assume that the costs are additive. So the costs can be interpreted as weights on the edges, and the cost of installing the communication system will equal the sum of the costs of the installed edges. We want to find a spanning tree of minimal cost, or *minimal spanning tree.*

A finite graph can contain only finitely many spanning trees, so in theory you could list all spanning trees and their weights, and find a minimal spanning tree by choosing one with minimum weight. But this process could take a very long time, since $\tau(G)$ can be very large. So efficient algorithms that find a minimal spanning tree are useful.

We shall outline an algorithm due to Prim [85] for finding a minimal spanning tree in a connected graph G with v vertices.. It finds a sequence of trees, $T_1, T_2, \ldots$. To start, choose a vertex x. Trivially, the cheapest tree with vertex-set $\{x\}$—the *only* tree with vertex-set $\{x\}$—is the K_1 with vertex x_0. Call this T_1. Then find the cheapest tree with two vertices, one of which is x; that is, find the cheapest tree that can be formed by adding one edge to T_1. Call this tree T_2. Continue in this way: if you have constructed T_i, then T_{i+1} is the cheapest tree that can be formed by adding one edge to it. The process can always be carried out provided $i < v$—the graph is connected, so it must always be possible to find a new vertex that is joined to one of the vertices of T_i, so there is at least *one* candidate tree for the choice of T_{i+1}, and finally T_v is the desired minimal spanning tree.

THEOREM 11.8 (Minimality of Prim's Algorithm)
The spanning tree constructed by Prim's algorithm is minimal.

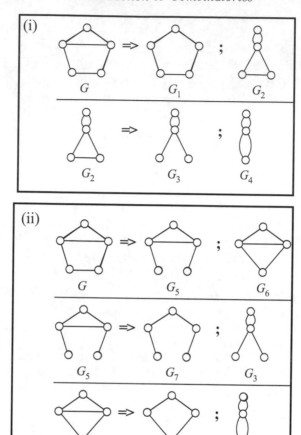

FIGURE 11.11:   Counting trees.

Proof. We continue the notation we used in describing the algorithm. For convenience, write y_i for the edge that was added in going from T_{i-1} to T_i, and V_i for the vertex-set of T_i. The cost of edge y is denoted $c(y)$.

Suppose T^1 is a minimal spanning tree in G. If $T^1 = T_v$ then we are done. Otherwise, find the smallest i such that y_i is not an edge of T^1. We could say that y_i is the first edge used in the algorithm that is not in T^1. One endpoint of y_i is in V_{i-1} and the other is not.

Since T^1 is a spanning tree of G, there is a path in T^1 joining the two endpoints of y_i. As one travels along the path, one must encounter an edge z joining a vertex in V_{i-1} to one that is not in V_{i-1}. Now, at the iteration when y_i was added to T, z could also have been added, and it would have been added instead of y_i if its cost was less than that of y_i. So $c(z) \geq c(y_i)$. Let

T^2 be the graph obtained by removing z from T^1 and adding y_i. It is easy to show that T^2 is connected, has the same number of edges as T^1, and the total cost of its edges is not larger than that of T^1, therefore it is also a minimum spanning tree of G and it contains y_1 and all the edges added before it during the construction of T_v. Repeat this process; each T^j differs from T_v in fewer and fewer edges, and eventually you achieve T_v. So T_v is a minimal spanning tree. □

The new edge used to construct T_{i+1} may not be uniquely defined in this algorithm, and indeed T_v may not be uniquely defined. This is to be expected: after all, there may be more than one minimal spanning tree in G. You need to decide on a "tiebreaking" strategy before you begin. Let us assume, unless we state otherwise, always to take the edge that comes earlier in alphabetical order.

Example 11.11: Apply Prim's algorithm to the graph

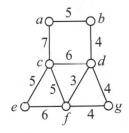

(1) We start at a. The two possible edges are ab and ac. As ab is cheaper, select $y_1 = ab$.
(2) The only possible choices for the next edge are ac and bd, so the cheaper one, bd, is chosen. T_3 is the path abd.
(3) There are now four possible edges: ac (cost 7), cd (cost 6), df (cost 3) and dg (cost 4), so the choice is df.
(4) Now there is a choice. The possible edges are ac (cost 7), cd (cost 6), cf (cost 5), dg (cost 4), ef (cost 6) and fg (cost 4). Either dg or fg could be used. Using alphabetical ordering, we choose dg.
(5) We have ac (cost 7), cd (cost 6), cf (cost 5) and ef (cost 6) (fg is not allowed). The choice is cf.
(6) Finally, from ce (cost 5) and ef (cost 6) we choose ce. □

The sequence of trees is

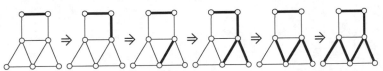

Prim's algorithm was a refinement of an earlier algorithm due to Kruskal [67]. In his algorithm, you start by listing all edges in order of increasing cost.

The first approximation is the K_2 consisting of the edge of smallest cost. The second approximation is formed by appending the next edge in the ordering. At each stage the next approximation is formed by adding on the smallest edge that has not been used, provided only that it does not form a cycle with the edges already chosen. In this case the successive approximations are not necessarily connected, until the last one.

In large graphs, the initial sorting stage of Kruskal's algorithm can be very time-consuming.

Exercises 11A

1. Are the following networks eulerian? If so, find an Euler circuit. If not, what is the eulerization number?

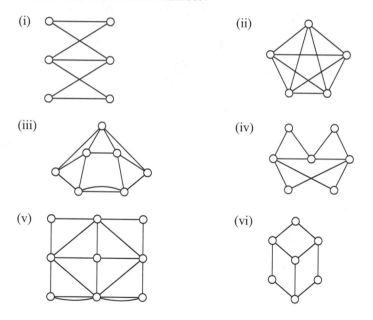

2. Are the graphs in Chapter 10, Exercise A.8 eulerian? If so, find an Euler circuit. If not, what is the eulerization number?

3. Draw the network of the following maze, and use it to solve the maze.

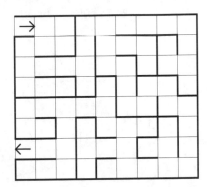

4. Find Hamilton cycles in the following graphs.

(i)

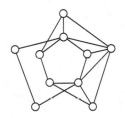

(ii)

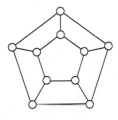

(iii)

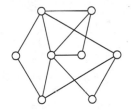

(iv)

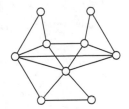

5. Explain why the following graph has no Hamilton path.

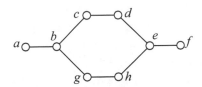

6. The following diagram is a plan of the maze at the Hampton Court palace, near London. You enter at the bottom; the object is to reach the center. Produce a network corresponding to the maze and solve it.

7. Find a connected cubic graph with 10 vertices that is not hamiltonian.

8. The vertices of K_5 are labeled 1, 2, 3, 4, 5, and edge xy has associated cost $x + y$.

   (i) Find the Hamilton cycle obtained by applying the sorted edges algorithm to this graph, and find the cycles obtained using the nearest neighbor algorithm starting from each vertex.

   (ii) Show that every Hamilton cycle on this graph has the same length.

9. The complete graph K_5 has vertices a, b, c, d, e. In each part we list the set of costs associated with the edges. Find the costs of the routes generated by the nearest neighbor algorithm starting at a, by the nearest neighbor algorithm starting at c, and by the sorted edges algorithm.

   (i) $ab = 22$ $ac = 19$ $ad = 20$ $ae = 21$ $bc = 19$
        $bd = 24$ $be = 22$ $cd = 19$ $ce = 18$ $de = 24.$

   (ii) $ab = 53$ $ac = 70$ $ad = 55$ $ae = 58$ $bc = 54$
        $bd = 54$ $be = 46$ $cd = 72$ $ce = 63$ $de = 77.$

   (iii) $ab = 24$ $ac = 23$ $ad = 25$ $ae = 24$ $bc = 29$
         $bd = 25$ $be = 27$ $cd = 22$ $ce = 26$ $de = 26.$

   (iv) $ab = 26$ $ac = 27$ $ad = 32$ $ae = 25$ $bc = 21$
        $bd = 24$ $be = 28$ $cd = 29$ $ce = 23$ $de = 24.$

10. In each part we show a graph labeled with the costs associated with the edges. Find the costs of the routes generated by the sorted edges algorithm and by the nearest neighbor algorithm starting at each vertex, if possible.

(i) 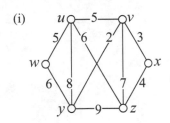   (ii)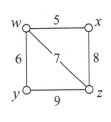

11. Show that the following graph has a Hamilton circuit, but that the sorted edges algorithm fails and the nearest neighbor algorithm fails for every possible starting vertex.

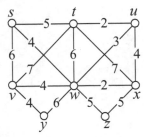

12. Show that there are exactly six non-isomorphic trees on six vertices.

13. Suppose a tree T has a vertex of degree n. Show that T has at least n vertices of degree 1.

14. Verify the formulas for:
   (i) the number of spanning trees in an n-fold path;
   (ii) the number of spanning trees in an n-fold cycle.

15. Find minimal spanning trees in the following graphs.

(i) 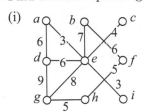   (ii)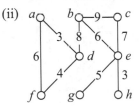

(iii) 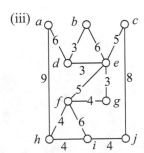   (iv)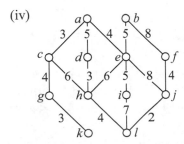

Exercises 11B

1. Are the following networks eulerian? If so, find an Euler circuit. If not, what is the eulerization number?

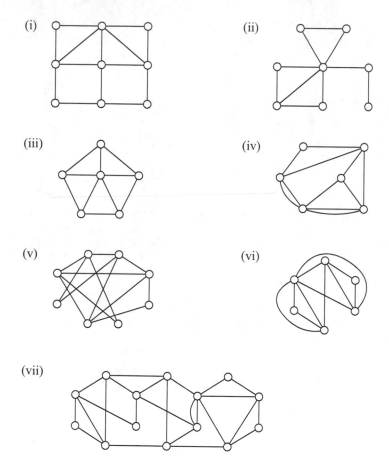

(i)

(ii)

(iii)

(iv)

(v)

(vi)

(vii)

2. Are the graphs in Chapter 10, Exercise 10B.12 eulerian? If so, find an Euler circuit. If not, what is the eulerization number?

3. The graph G has precisely two vertices x and y of odd degree. A new graph or multigraph H is formed from G by adding an edge xy. If H is connected, prove that G is connected.

4. Draw the network of the following maze, and use it to solve the maze.

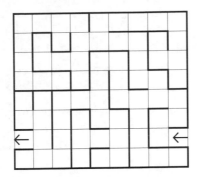

5. For what values of m and n does $K_{m,n}$ have a Hamilton path?

6. Find Hamilton cycles in the following graphs.

(i)

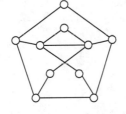

(ii)

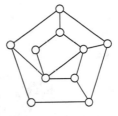

(iii)

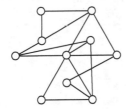

(iv)

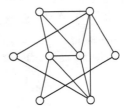

7. Show that the following graph contains a Hamilton path but no Hamilton cycle.

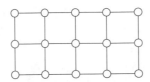

8. The complete graph K_5 has vertices a, b, c, d, e. In each part we list the set of costs associated with the edges. Find the costs of the routes generated by the nearest neighbor algorithm starting at a, by the nearest neighbor algorithm starting at c, and by the sorted edges algorithm.

  (i) $ab = 44 \quad ac = 39 \quad ad = 30 \quad ae = 51 \quad bc = 49$
  $\quad bd = 46 \quad be = 42 \quad cd = 37 \quad ce = 38 \quad de = 44.$

(ii) $ab = 23 \quad ac = 30 \quad ad = 27 \quad ae = 28 \quad bc = 34$
$\quad\quad bd = 29 \quad be = 41 \quad cd = 32 \quad ce = 42 \quad de = 37.$

(iii) $ab = \ 8 \quad ac = \ 5 \quad ad = \ 7 \quad ae = 11 \quad bc = \ 9$
$\quad\quad bd = \ 7 \quad be = \ 8 \quad cd = 11 \quad ce = \ 9 \quad de = 11.$

(iv) $ab = 34 \quad ac = 37 \quad ad = 32 \quad ae = 45 \quad bc = 29$
$\quad\quad bd = 34 \quad be = 29 \quad cd = 42 \quad ce = 43 \quad de = 39.$

9. In each part we show a graph labeled with the costs associated with the edges. Find the costs of the routes generated by the sorted edges algorithm and by the nearest neighbor algorithm starting at each vertex, if possible.

(i)
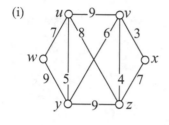

(ii)
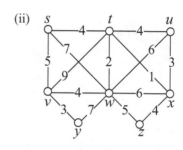

10. Verify the results of Example 11.3.

11. Show that it is possible to decompose K_9 into four Hamilton cycles, no two of which share an edge.

12. Show that a graph on v vertices that has no cycle is connected if and only if it has precisely $v - 1$ edges.

13. Prove that a graph is a tree if and only if: it has no cycles, but if you join any two nonadjacent vertices the new graph has exactly one cycle.

14. Find minimal spanning trees in the following graphs.

(i)
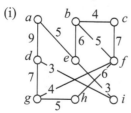

(ii)
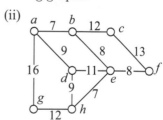

(iii)
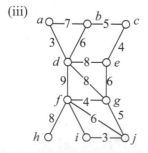

(iv)

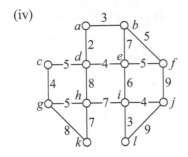

15. Prove that if a connected graph is a tree then every vertex of degree greater than 1 is a cutpoint, but the converse does not hold.

Problems

Problem 1: We did not mention loops in our discussion of Euler walks. How would you treat them? What sort of road would translate as a loop?

Problem 2: The *girth* of a graph G is defined to be the length of the shortest cycle in G. (The girth of a tree is either defined to be ∞ or else not defined.)

   (i) Prove that if a regular graph of degree r has girth 4, then it has at least $2r$ vertices.

   (ii) Prove that if a regular graph of degree r has girth 5, then it has at least $r^2 + 1$ vertices.

Problem 3: Draw a graph that has:

   (i) An Euler circuit but no Hamilton cycle;

   (ii) A Hamilton cycle but no Euler circuit;

   (iii) Both an Euler circuit and a Hamilton cycle;

   (iv) Neither an Euler circuit nor a Hamilton cycle.

Problem 4: The vertices of K_n are labeled $1, 2, 3, \ldots, n$, and edge xy has associated cost $x + y$. Prove that the cost of every Hamilton cycle in this graph is the same.

Problem 5: The graph G has $2n$ vertices, and every vertex has degree at least n. Show that G is hamiltonian.

Problem 6: The graph $K_{a,b,c}$ is defined to have three sets of vertices, of sizes a, b and c, and two vertices are adjacent if and only if they are in different sets.

(i) Prove that $K_{n,2n,3n}$ is hamiltonian for every positive integer n.

(ii) Prove that $K_{n,2n,3n+1}$ is not hamiltonian for any positive integer n.

Problem 7: Perfect squares were defined in Problem 10.10.3. Show that no tree with three or more vertices is a perfect square.

Problem 8: See Problem 10.10.3. The *center* of a graph consists of all the vertices x such that $\epsilon(x) = R$. Prove that the center of a tree consists of either one vertex or two adjacent vertices.

Problem 9: T is a tree with precisely four vertices of degree 1 and precisely one vertex of degree 4. Find the degrees of the remaining vertices of T, and show that T is the union of two edge-disjoint paths.

Problem 10: Suppose $d_1, d_2, \ldots, d_v$ are positive integers whose sum is $2v-2$. Write $\mathcal{T}$ for the set of all trees on v vertices $x_1, x_2, \ldots, x_v$ such that x_i has degree d_i for every i. Show that the number of elements of $\mathcal{T}$ is

$$\frac{(v-2)!}{(d_1-1)!(d_2-1)!\ldots(d_v-1)!}.$$

Problem 11: Show that there are precisely v^{v-2} different spanning trees in K_v.

Chapter 12

Coding Theory

Suppose you want to send an e-mail message to a friend. Your message may be an ordinary string of English words, but your computer can only send electronic messages. So a program is used to convert your words into a string of instances of two states, "power on" and "power off." For our convenience let's write 1 for "power on" and 0 for "power off"; your message no longer looks like "Let's meet for lunch," but something like

1001101100110000110111000101100110111000101100110110010010110110000.

Such a sequence of 0s and 1s is called a *binary string*.

The substitution process is called *encoding*, and the corresponding process of going back from the string to the original message is called *decoding*. Coding theory has been widely studied in mathematics, and there are a number of good books on the subject. One example is [91].

Coding has also been studied from the viewpoint of secrecy and security; this area is usually called *cryptography*. But we shall restrict our attention to avoiding errors in the encoding and decoding process.

12.1 Errors and Noise

Sometimes there will be errors in a message. In electronic transmissions they are often due to *noise*. If your message was sent a long distance, it may be bounced off a satellite, and photographs of the Moon and Venus are transmitted as radio messages in space; random electrical discharges, sunspots, etc. can introduce errors. The medium through which the message is transmitted is called a *channel*, and some channels are noisier than others.

To see a way around the noise problem, consider what happens in everyday English, when there is a misprint. If you are reading a book and you come across the word " bive", you are fairly sure this is not a real word. Assuming one letter is wrong, there are many possibilities: *dive, five, live, bile, bite* are only a few. If this were all the information you had, you could not tell which was intended. But the context—the other words nearby—can often help. For example, if the sentence is "...the next bive years ..." then the author probably meant "five." If it says "...a review of a bive performance ..." then "live" is most probable. The phrase "...teeth used in the bive ..." suggests "bite", and "...let me bive you an example ..." probably should be "give." The context gives you *extra information*.

If the encoding results in a binary string, there is no information from context. You don't even have a context until after the string is decoded. To get around this, we add some more symbols. One very simple method is to break the message up into substrings or *words*, say of length 4, and after each subsequence add another symbol (a *check digit*), 0 or 1, chosen so that the subsequence has an even number of 1s. For example, suppose you want to send

$$10011011001100001101.$$

First break it up into the form

$$1001 \quad 1011 \quad 0011 \quad 0000 \quad 1101$$

and then insert check digits:

$$10010 \quad 10111 \quad 00110 \quad 00000 \quad 11011.$$

This method is useful when there is a way of checking with the sender: for example, if the message was received as

$$10110 \quad 10111 \quad 00110 \quad 00000 \quad 11011$$

you could say "I got the second through fifth words OK; would you repeat the first?" But this is not useful for electronic communications, where a large number of symbols are sent every second. In the next section we'll look at a way around this.

Another problem is that the method is useless if more than one error occurs in a word. The same occurs in everyday English: if "bive" contains two errors, there are many more possibilities: "bone," "bane,", "brie," and so on. There is no solution to this problem. There are encoding methods that can detect (and correct!) many errors, but it is always possible that more errors occur. For example, in the photos sent back from space, there will be the occasional small area where the image cannot be decoded.

12.2 The Venn Diagram Code

We assume you are sending a message composed of 0s and 1s. If it is an ordinary English sentence, maybe you represent a as 000001, b as 000010, ..., A as 011011, and so on. These 64 ($= 2^6$) strings will enable you to represent the 26 capitals and 26 lowercase numbers, together with 12 punctuation marks. This type of representation is widely used in computers—the ASCII (American Standard Code for Information Interchange) code, which was developed for computers, uses seven binary symbols for each letter, giving a wide range of punctuation and special symbols. *Extended ASCII* uses eight binary symbols for each.

Suppose you are happy to assume that no more than one error will occur in every seven transmitted symbols: the cases where more errors occur will be sufficiently few that you can recover the message. Proceed as follows. First, break the message up into substrings, or *message words*, of four symbols. Then replace each message word by a *codeword* as follows. If the message is $ABCD$ (each of A, B, C and D represents a 1 or a 0), define three more binary digits, E, F and G, chosen as follows so that each of the sums $A+B+C+E$, $A+C+D+F$ and $B+C+D+G$ is even.

This whole process is called *encoding* or *encryption*.

Example 12.1: How do you encrypt 110011010110?

First break the message up as

$$1100 \quad 1101 \quad 0110.$$

If $ABCD$ is 1100, then $A+B+C+E = 1+1+0+E$; for this to be even, $E = 0$. Similarly $A+C+D+F = 1+0+0+F$, $F = 1$, and $B+C+D+G = 1+0+0+G, G = 1$. The first codeword is 1100011. The other words are 1101000 and 0110010. □

The easiest way to represent this process is as follows. Draw a diagram (essentially a three-set *Venn diagram*) of three overlapping circles, and write the numbers A, B, C and D in the positions shown in the left-hand part of Figure 12.1. Then the other numbers are chosen so that the sum in each circle is even. These are E, F and G, as we defined them, and shown in the right-hand part.

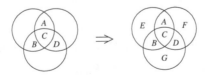

FIGURE 12.1: Venn diagram encoding.

After the codeword is sent, the receiver, (*decoder* or *decrypter*) must reconstruct the original message. Suppose a string of seven binary symbols is received. The decoder assumes that at most one error has been made in the encryption. If more than one error occurs in seven symbols, the wrong message will be found. We have to assume that this will occur very infrequently. (If errors are more common, other methods can be used.)

Example 12.2: Suppose you receive 1111010. How is it decoded?

$A + B + C + E = 3$, so one of these symbols must be wrong. (Since there is assumed to be at most one error, there cannot be three wrong symbols.)

$A + C + D + F = 3$, so one of these symbols must be wrong.

$B + C + D + G = 2$, so these four symbols are all correct.

The first and third conditions tell us that exactly one of A and E is incorrect; the second and third conditions tell us that exactly one of A and F is incorrect. So A is wrong; it should have been 0, not 1, the correct codeword is 0110010 and the original message was 0110.

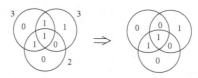

FIGURE 12.2:   Venn diagram decoding.

This decoding process is very simple when the Venn diagram is used. The decoder fills in the diagram and puts the number of 1s next to each circle, as shown on the left-hand side of Figure 12.2. The two top circles are odd; the wrong symbol must be the one in the unique cell common to both of those circles.   □

This process will always work. Provided there is at most one error, the Venn diagram method will yield the correct codeword. Moreover, every binary string of length 7 can be decoded by this method. We'll prove this in the next section.

12.3   Binary Codes, Weight, and Distance

There are 16 possible message words in Venn diagram encoding, so there are 16 codewords, as shown in Figure 12.3.

The complete set of 16 words is called a *binary code*, or simply *code*. In general, a binary code means any set of binary codewords. Obviously you could encode messages using more than two symbols, but we will concentrate on the binary case because of its widespread applications in electronic communication.

In most cases, we require all words in a code should be same length; such a code is called a *block code*, and the common length is called the *length* of the code. (There are a few important exceptions; see Section 12.7.)

Message	Codeword	Message	Codeword
0000	0000000	1000	1000110
0001	0001011	1001	1001101
0010	0010111	1010	1010001
0011	0011100	1011	1011010
0100	0100101	1100	1100011
0101	0101110	1101	1101000
0110	0110010	1110	1110100
0111	0111001	1111	1111111

FIGURE 12.3: The codewords in Venn diagram encoding.

Given two binary strings, we define the *distance* between them to be the number of positions in which they differ. The *minimum distance* between any two codewords in a code (that is, the smallest of all differences between two codewords) is called the *distance* (or *minimum distance*) of the code. The *weight* of a codeword is the number of 1s it contains. That is, it is the distance from $000\ldots00$. The weight of a *code* is the minimum weight of all its non-zero words (if $000\ldots$ is in the code, you don't include it in calculating the minimum weight). We use the notations $d(x, y)$ for the distance between binary strings x and y and $w(x)$ for the weight of x. The weight and distance of a code $\mathcal{C}$ are denoted $w(\mathcal{C})$ and $d(\mathcal{C})$. (The use of the same notation for distance between strings and between vertices of a graph is quite intentional; see Problem 12.7.)

Example 12.3: The $(n+1, n)$ *parity-check code* was introduced informally in Section 12.1. It has 2^n codewords of length $n + 1$. The messages are the possible binary strings of length n, with another symbol added at the end to make the number of 1s even. What are the words of the $(3, 2)$ parity-check code? What is the minimum distance of this code?

The possible messages are 11, 10, 01 and 00, so the codewords are 110, 101, 011 and 000. By inspection, the code has minimum distance 2. $\quad\square$

We'll look further at parity-check codes in Example 12.3 and Exercise 12A.5.

Example 12.4: Here is a code with five words:

$$
\begin{aligned}
A &= 1\,0\,1\,0\,1\,0\,1 \\
B &= 0\,1\,0\,1\,0\,1\,0 \\
C &= 1\,1\,1\,0\,0\,0\,0 \\
D &= 0\,0\,0\,1\,1\,1\,1 \\
E &= 0\,0\,0\,0\,0\,0\,0
\end{aligned}
$$

What are the length, distance and weight of this code?

The code is length 7. The distances between the words are

A to B :	7	B to D :	3
A to C :	3	B to E :	3
A to D :	4	C to D :	7
A to E :	4	C to E :	3
B to C :	4	D to E :	4.

So the minimum distance—the distance of the code—is 3.

The weights of A, B, C and D are 4, 3, 3 and 4. As only non-zero words are to be checked, the code has weight 3.   □

The distance between binary strings obeys the usual rule for metrics:

THEOREM 12.1 (Triangle Inequality)
If x, y and z are binary strings, then $d(x, z) \leq d(x, y) + d(y, z)$.   □

The proof is left as an exercise.

The 16 words that can arise from Venn diagram encoding form a code of length 7. If you check the table, you will see that the code has weight 3, and in fact the minimum distance between words is 3. So this is a weight 3 code. (At the moment, the easiest way to see this is to check all the pairs of words. We'll see an easier method in Section 12.4.)

Suppose a codeword x from this code is to be transmitted, and exactly one error occurs. Let's say the actual string received is y. Then the triangle inequality tells us that the distance from y to any other codeword (other than x) must be at least 2. So, if it is assumed that the number of errors is at most 1, then the correct message will be reconstructed. We express this by saying that the Venn diagram code will correct any single error. It is called a *single error-correcting code*.

In nearly all cases, decoding follows the rule that, if the received string is not a word of the code, it is decoded on the assumption that the number of errors was smallest possible. This is called *nearest neighbor decoding*.

Suppose a code has minimum distance d. If d or more errors are made when codeword x is transmitted, it is possible that a different codeword could be received, and the receiver would mistakenly assume that no error occurred. However, this is impossible when the number of errors is smaller than d. In that case the error would certainly be detected. If the number of errors is smaller than $d/2$, nearest neighbor decoding will lead to the correct message. These two facts are usually expressed as

THEOREM 12.2 (Distance Criteria)
A binary code of distance d will detect any $d - 1$ errors and correct any $(d - 1)/2$ errors.   □

(Of course, when d is even, this last number is rounded down to $d/2 - 1$, because a fractional number of errors is meaningless.)

It follows from Theorem 12.2 that the Venn diagram method will correct any single error in a word.

Example 12.5: The $(n + 1, n)$ parity-check code was defined in Example 12.3. How many errors can it detect or correct?

The code has distance 2. Any single error can be detected, but no errors can be corrected. (For example, if you receive 11100, you cannot tell whether the original was 01100, 10100, 11100, 11110 or 11101.)

In constructing a block code with distance d, we must select a number of codewords (binary strings) of length n such that every two words differ in at least d places. We would like to find as many codewords as possible. What is the maximum number? This problem is not completely solved, but there are some useful results.

Example 12.6: What is the maximum number of words in a block code with length 3 and distance 2?

We know that the $(3, 2)$ parity-check code has length 3 and distance 2, and this code has four words. Are five words possible?

Suppose we have such a set S of five words. Write $S = S_0 \cup S_1$, where S_i consists of the words starting with i. One of these must have at least three elements; say it is S_1, and suppose S_1 contains the distinct words $1\,a\,b$, $1\,c\,d$ and $1\,e\,f$. Then $a\,b$, $c\,d$ and $e\,f$ are distinct binary words each of distance 2. But this would mean, for example, that a, c and e are all different, and this is impossible since we only have two symbols to choose from. $\qquad\square$

12.4 Linear Codes

The set of integers modulo 2 is denoted $\mathbb{Z}_2$, and consists of the two elements 0 and 1 whose sum is defined by

$$0 + 0 = 0 \quad 0 + 1 = 1$$
$$1 + 0 = 1 \quad 1 + 1 = 0$$

and whose product is

$$0 \times 0 = 0 \quad 0 \times 1 = 0$$
$$1 \times 0 = 0 \quad 1 \times 1 = 1.$$

The *join* or *bit sum* or *sum* of two binary sequences is defined as follows: the sum of individual entries is the sum in $\mathbb{Z}_2$, as defined above, and to add *sequences*, you add terms separately, with no carry; for example,

$$
\begin{array}{r}
1\ 0\ 1\ 0\ 1\ 0\ 1\ 1 \\
+\ 1\ 1\ 1\ 0\ 0\ 0\ 1\ 0 \ . \\
\hline
0\ 1\ 0\ 0\ 1\ 0\ 0\ 1
\end{array}
$$

In other words, sequences are added as vectors whose elements, instead of real numbers, are members of $\mathbb{Z}_2$.

Given this definition the distance between two codewords is the weight of their sum.

The set of all binary sequences of length n form an abelian group, under the sum (or join) operation. The identity element is $000\ldots0$, and each element is its own inverse. We shall denote this set as $V_n(2)$. It can also be interpreted as a vector space over the field $\mathbb{Z}_2$.

A *linear code* is one in which the binary sum of any two of the codewords is also a codeword. In the binary case, this is the same as saying that the code is a group under the sum operation. Two important properties of linear codes are that the zero element $000\ldots0$ is always one of the codewords, and the minimum distance of code equals the weight of code.

Example 12.7: Show that the Venn diagram code is linear.

We have described Venn diagram encoding as follows: "Given a message word $A_1\,A_2\,A_3\,A_4$, with elements from $\{0,1\}$, the corresponding codeword is $A_1\,A_2\,A_3\,A_4\,A_5\,A_6\,A_7$, where $A_1 + A_2 + A_3 + A_5$, $A_1 + A_3 + A_4 + A_6$ and $A_2 + A_3 + A_4 + A_7$ are even." An alternative formulation is, "Given a message word $A_1\,A_2\,A_3\,A_4$, with elements from $\mathbb{Z}_2$, the corresponding codeword is $A_1\,A_2\,A_3\,A_4\,(A_1 + A_2 + A_3)\,(A_1 + A_3 + A_4)\,(A_2 + A_3 + A_4)$." Then the binary sum of $A_1\,A_2\,A_3\,A_4\,A_5\,A_6\,A_7$ and $B_1\,A_4\,B_3\,B_4\,B_5\,B_6\,B_7$ is

$$
\begin{aligned}
&(A_1 + B_1)(A_2 + B_2)\,(A_3 + B_3)\,(A_4 + B_4)\,(A_5 + B_5)\,(A_6 + B_6)\,(A_7 + B_7) \\
={}&(A_1 + B_1)(A_2 + B_2)\,(A_3 + B_3)\,(A_4 + B_4)\,([A_1 + A_2 + A_3] + [B_1 + B_2 + B_3]) \\
&([A_1 + A_3 + A_4] + [B_1 + B_3 + B_4])\,([A_2 + A_3 + A_4] + [B_2 + B_3 + B_4]) \\
={}&(A_1 + B_1)\,(A_2 + B_2)\,(A_3 + B_3)\,(A_4 + B_4)\,([A_1 + B_1] + [A_2 + B_2] + [A_3 + B_3]) \\
&([A_1 + B_1] + [A_3 + B_3] + [A_4 + B_4])\,([A_2 + B_2] + [A_3 + B_3] + [A_4 + B_4])
\end{aligned}
$$

which is a codeword—the codeword corresponding to the message

$$
(A_1 + B_1)\,(A_2 + B_2)\,(A_3 + B_3)\,(A_4 + B_4).
$$

So the binary sum of two codewords is a codeword: the code is linear. $\qquad\square$

We can generalize this as follows. Take all binary strings of length n as messages. Add check digits in a linear fashion: if the message is $A_1 A_2 \ldots A_n$ and the check digits are $A_{n+1}, \ldots, A_{n+j}, \ldots$, then each A_{n+j} equals a linear form in $A_1 A_2 \ldots A_n$ (over $\mathbb{Z}_2$). (The alternative formulation of the Venn diagram code is an example of this.) Then possibly apply a fixed permutation to every codeword.

Suppose a code can correct e errors. It is quite possible that a binary string could differ from every possible codeword in more than e places. A very simple example is the code $C = \{0000, 1111\}$. If two errors are made, for example if 1010 is received, there is no reason to decode as 0000 or as 1111. We would prefer a code that gives a unique solution under nearest neighbor decoding.

THEOREM 12.3 (Hamming Bound)
Suppose C is an e-error-correcting code with n codewords of length t. Then

$$n \leq \frac{2^n}{\sum_{i=0}^{e} \binom{t}{i}}.$$

Proof. Write S for the set of all binary vectors of length n. For each codeword a in in C, define $S(a)$ to be the set of all members of S that are distance e or less from a. Suppose x is a common member of $S(a)$ and $S(b)$. If message x is received, we cannot tell whether the intended message was a or b. So in this case the code cannot correct a transmission with e (or fewer) errors—a contradiction. Therefore the sets $C(a)$, for $a in C$, are all disjoint. There are $\binom{t}{i}$ vectors distance precisely i from a, so $|C(a)| = \sum_{i=0}^{e} \binom{t}{i}$. There are n words in C, so there are at least $n \sum_{i=0}^{e} \binom{t}{i}$ binary vectors of length n. But there are precisely 2^n such vectors. So $2^n \geq n \sum_{i=0}^{e} \binom{t}{i}$, and $n \leq \frac{2^n}{\sum_{i=0}^{e} \binom{t}{i}}$. $\square$

The inequality in Theorem 12.3 is called the *Hamming bound* after Richard Hamming (See Appendix C).

A code C of N binary codewords of length t, of width $d(C) = d - 2e + 1$, is called *perfect* if it meets the Hamming bound, that is

$$n \leq \frac{2^n}{\sum_{i=0}^{e} \binom{t}{i}}. \tag{12.1}$$

Then for every possible binary word w_0 of length n, there is a codeword w in C whose distance from w_0 is at most e. This word w is unique—C is e-error correcting, so there cannot be another codeword w' with $d(w_0, w') \leq e$. So a perfect code can transmit the maximum possible number of different messages.

The Venn diagram code is a perfect code: it has $e = 1$, $n = 16$ and $t = 7$. Then

$$\sum_{i=0}^{e} \binom{t}{i} = \binom{7}{0} + \binom{7}{1} = 1 + 7 = 8 = 128/16 = (2^7)/16 = (2^t)/n,$$

and (12.1) is satisfied.

12.5   Hamming Codes

We could describe Venn diagram encoding as follows: "Given a message word $A_1\,A_2\,A_3\,A_4$, with elements from $\mathbb{Z}_2$, the corresponding codeword is $A_1\,A_2\,A_3\,A_4\,(A_1 + A_2 + A_3)\,(A_1 + A_3 + A_4)\,(A_2 + A_3 + A_4)$." We can use matrices to prescribe this encoding: if the message is $A_1\,A_2\,A_3\,A_4$, the three check digits are the elements of the vector

$$\begin{bmatrix} 1 & 1 & 1 & 0 \\ 1 & 0 & 1 & 1 \\ 0 & 1 & 1 & 1 \end{bmatrix} \begin{bmatrix} A_1 \\ A_2 \\ A_3 \\ A_4 \end{bmatrix}.$$

There is also a matrix method of testing whether a seven-symbol string $A_1\,A_2\,A_3\,A_4\,A_5\,A_6\,A_7$ is a codeword. Write H for the 3×7 matrix

$$H = \begin{bmatrix} 1 & 1 & 1 & 0 & 1 & 0 & 0 \\ 1 & 0 & 1 & 1 & 0 & 1 & 0 \\ 0 & 1 & 1 & 1 & 0 & 0 & 1 \end{bmatrix}$$

and U for the vector

$$U = (A_1\,A_2\,A_3\,A_4\,A_5\,A_6\,A_7).$$

That is, V is the string $A_1\,A_2\,A_3\,A_4\,A_5\,A_6\,A_7$ written as a vector over $\mathbb{Z}_2$. Then the string is a codeword if and only if

$$HU^T = (0\,0\,0)^T.$$

Now suppose an error has been made, and the string W has been sent instead of the codeword U. We call the difference, $V = W - U$ (and because we are working over zz_2, we could just as easily say $V = W + U$), the *error pattern*. When we test W we get

$$HW^T = HU^T + HV^T = (0\,0\,0)^T + HV^T = HV^T.$$

The useful property of the Venn diagram code is that, if we assume there is at most one error, HV^T tells us exactly which symbol was wrong.

In order to generalize this, we look at the columns of H. If we translate the columns into numbers by making column $(a\,b\,c)^T$ correspond to the integer $4a + 2b + c$, each of 1, 2, 3, 4, 5, 6, 7 occurs precisely once. We say that $(a\,b\,c)^T$ is the *binary column representation* of $a + 2b + c$, the representation in base 2 written as a column (with leading zeroes when necessary). Moreover, the last three columns—the columns that correspond to the check digits—represent 1, 2 and 4, the first three powers of 2.

Instead of three rows, let's take n rows. We construct an $n \times (2^n - 1)$ matrix H whose jth column is the binary column representation of j. The code we construct has as its codewords binary vectors of length $2^n - 1$. Each message is a string of $2^n - 1 - n$ binary symbols. We shall refer to the code as a $(2^n - 1, 2^n - 1 - n)$ *code*. The message is converted into a codeword by adding n check digits; the check digits will be in positions $1, 2, 4, \ldots, 2^{n-1}$ (that is, the corresponding columns represent the n powers of 2 starting with 2^0), and the message symbols are placed in order in the remaining positions. For example, in the case $n = 3$ (a $(7, 4)$ code), we have $2^n - 1 - n = 4$; the message $A\,B\,C\,D$ (the letters representing some string of 0s and 1s) would give rise to a codeword $* * A * B\,C\,D$ where the asterisks are the check digits.

How would the check digits be selected? If $U = (u_1, u_2, \ldots)$ is to be a codeword, we require $HU^T = 0$ (the zero vector of appropriate length). This is a set of n linear equations. The columns of H corresponding to the check digits will each contain only one 1 and every other symbol 0, so each check digit occurs in only one equation; and, in fact, the first non-zero element is the check digit. So we can read off the values very easily. For this reason, H is called the *encoding matrix*.

Example 12.8: Show how to find the check digits in the case $n = 3$.

In this case the matrix H is

$$H = \begin{bmatrix} 0 & 0 & 0 & 1 & 1 & 1 & 1 \\ 0 & 1 & 1 & 0 & 0 & 1 & 1 \\ 1 & 0 & 1 & 0 & 1 & 0 & 1 \end{bmatrix}$$

and $HU^T = 0$ is

$$\begin{aligned} u_4 + u_5 + u_6 + u_7 &= 0, \\ u_2 + u_3 \qquad\quad + u_6 + u_7 &= 0, \\ u_1 \quad + u_3 \quad + u_5 \quad + u_7 &= 0. \end{aligned}$$

So we see immediately that the check digits are

$$\begin{aligned} u_4 &= u_5 + u_6 + u_7, \\ u_2 &= u_3 + u_6 + u_7, \\ u_1 &= u_3 + u_5 + u_7. \end{aligned}$$

(Remember that in $\mathbb{Z}_2$, addition and subtraction are essentially the same.) $\square$

Suppose it is intended to transmit the codeword U, but an error occurs and the string $U + V$ is sent—V is the pattern. Assume that at most one error has occurred; V contains at most one 1. Now calculate $H(U + V)^T$. If this is zero, we assume the received word was correct, and we reconstruct the message by deleting the check digits. Otherwise, $H(U + V)^T = HV^T$ will equal a column of H, namely the column that corresponds to the error. If the column is the binary column representation of j, then the error was in position j of the received word, and can be corrected immediately.

The code we have constructed is called a $(2^n - 1, 2^n - 1 - n)$-*Hamming code*. These codes were first discovered by Richard Hamming. The Hamming codes are perfect single-error-correcting linear codes. So there is a perfect single-error-correcting code of length $2^n - 1$ for positive integer $n \geq 2$. (When $n = 1$, the message has length 0, so it has no content.)

Example 12.9: Describe the $(3, 1)$-Hamming code.

This is the case $n = 2$. The encoding matrix is

$$H = \begin{bmatrix} 0 & 1 & 1 \\ 1 & 0 & 1 \end{bmatrix}$$

and the two possible messages are 0 and 1. 0 encodes as $0\,0\,0$ and 1 encodes as $1\,1\,1$. □

The Venn diagram code is essentially the $(7, 4)$-Hamming code, with the positions in the codeword permuted to make encoding and decoding by diagram easier.

12.6   Codes and the Hat Problem

Suppose a code has minimum distance 3. If you are presented with a string U that is not a codeword, there is at most one codeword distance 1 from U, because if V and W are codewords such that $d(U, V) = 1$ and $d(U, W) = 1$ then $d(V, W) \leq 2$; the only possibility is that $V = W$. As a consequence, if you are given a word that is not a codeword, there is at most one place where changing the symbol will result in a codeword. If the code is perfect, there will be exactly one such place.

We now go back to the Hat Game that we discussed in Section 1.1. If we look at the three hat colors in some order (for example, order the wearers' names alphabetically), the contestants could interpret what they are wearing as a binary string, like a codeword, with red and black replacing the binary symbols 0 and 1.

Maybe the three contestants are obsessed with Hamming codes. Maybe they learned coding theory from Dr. Walter Johns, and they hated his teaching style so much that they never want to see a code again. They say, "Why don't we assume that, whatever happens, the string we are wearing is *not* a word in the $(3, 1)$-Hamming code!"

The code consists of the two words 000 and 111, and every possible string is a codeword or else differs from a codeword in exactly one place. Suppose the string that occurs is not a codeword. There is exactly one place where a change will produce a codeword. If you were the contestant wearing the hat that corresponded to that position, you would *know* the color of your hat: your reasoning would be something like, "If my hat were red, our string would be a codeword, and we decided that isn't going to happen." In that case you would know that your hat was black. Of course, your assumption (that the string is not a codeword) might be wrong, and you would lose. But only two of the eight possible strings are codewords, so you would win the game in two out of every eight cases, or 75% of the time.

This same sort of reasoning applies whenever there is a perfect code available. So, if n is any positive integer greater than 1, the $(2^n - 1, 2^n - 1 - n)$-Hamming code can be used. There are $2^{[2^n-1-n]}$ codewords, and since they are of length $2^n - 1$ there are $2^{[2^n-1]}$ possible binary strings. If the hat colors are chosen at random (as is true of the Hat Game), the chance that the string representing them will be a codeword is

$$\frac{2^{[2^n-1-n]}}{2^{[2^n-1]}} = \frac{1}{2^n},$$

so the probability that the strategy we described will be succesful approaches 1 as the number of players goes to infinity.

12.7 Variable-Length Codes and Data Compression

So far we have looked at *block codes* or *fixed-length codes*, where every word is the same length. Many codes are like this. But some codes use different length words. They are called *variable-length codes*.

Perhaps the best-known example is the *Morse code*. This was designed for transmission where there can be a break between codewords. Beginning in 1836, Samuel F. B. Morse and Alfred Vail developed the electric telegraph, and the code was developed by Morse to transmit numbers and expanded by Vail to an alphabetic code in 1844. The code uses short and long transmissions ("dots" and "dashes") to represent letters. A dot is formed by transmitting continuously for one time period and a dash by three periods. There is a codeword for each letter in the English alphabet. The break between two letters is denoted by seven periods with no transmission. For example, to transmit "dot dash" followed by a new letter, one might depress a key for one-tenth of a second, let it go then immediately hold it down for three-tenths of a second, let the key go for seven-tenths of a second, then start on the next letter.

The Morse code is shown in Figure 12.4. For example the "dot dash" we mentioned above would translate as the letter A. Notice that letters commonly used in the English language were assigned short sequences; the most common

A	● ■	J	● ■ ■ ■	R	● ■ ●
B	■ ● ● ●	K	■ ● ■	S	● ● ●
C	■ ● ■ ●	L	● ■ ● ●	T	■
D	■ ● ●	M	■ ■	U	● ● ■
E	●	N	■ ●	V	● ● ● ■
F	● ● ■ ●	O	■ ■ ■	W	● ■ ■
G	■ ■ ●	P	● ■ ■ ●	X	■ ● ● ■
H	● ● ● ●	Q	■ ■ ● ■	Y	■ ● ■ ■
I	● ●			Z	■ ■ ● ●

FIGURE 12.4:   The Morse code

vowel, E, is represented by a single dot and the most common consonant, T, is a single dash. Encoding so that the most common letters (or words or phrases) get the fewest symbols is called *data compression*.

Morse code was designed for manual transmission, and the break between codewords is provided by the person doing the transmitting. Electronically, variable-length codes can either use the "end-of-line" symbol for a break or else the codewords can include information that signals the end of a codeword. For example, (00 at end).

The genetic code provides an interesting example. Genes are strings of four nucleotides, adenine, thymine, guanine and cytosine, with repetitions possible. The names are usually abbreviated A, T, G and C, so a gene can be specified by a sequence of these four symbols.

Suppose we want to use a binary code (that is, 0s and 1s) to represent a genetic string. One possibility is to use codewords

00 for A
01 for C
10 for T
11 for G.

For example, the string AACACAGTATACA would be represented by

000001000100110010000100,

which uses 26 symbols.

It happens that A is most common in genetic strings, then C, then T and G. If we use the following (data compression) encoding:

0   for A
10   for C
110 for T
111 for G

then AACACAGTATACA is represented by 0010010011111001100100, which only takes 22 symbols. When very large collections of very large strings are to be transmitted, the shorter version (the *compressed genetic code*) is useful.

In order to decode the symbol for a string of nucleotides, notice that:

- Every time a 0 occurs, it is the end of a codeword;
- Every time 111 occurs, it's the codeword for G;
- The only way a codeword can end is with 0 or 111.

Example 12.10: Decode 0011001011101101100001000.

First put commas at the ends of words: 0011001011101101100000010 becomes 0, 0, 110, 0, 10, 111, 0, 110, 110, 0, 0, 10, 0, 0. Now decode symbol by symbol, to obtain AATACGATTAAAACAA.

Exercises 12A

1. What is the distance between the following strings?

(i) 101110111 and 101100110

(ii) 10101010 and 01010101

(iii) 000010000 and 001010110

(iv) 0101101011 and 0011100110

2. Prove (Theorem 12.1) the triangle inequality for distance between binary strings.

3. Consider the five-word code given in Example 12.3. How would you interpret the following received messages, using nearest neighbor decoding?

(i) 1011101 (ii) 0001010

(iii) 1101010 (iv) 0000110

4. Consider a code consisting of the seven words 0000000, 0010111, 0101110, 1011101, 1110010, 1100101, 1001011. What are the weight and distance of this code? What are its error-correcting and -detecting properties? If the word 0111001 were added, what changes are made to these parameters?

5. The *parity-check code* of length 6 contains words of six binary symbols. The first five characters represent the message, and the sixth is chosen such that the sum of the symbols is even.

(i) How many messages are possible?

(ii) How many errors will this code detect? How many will it correct?

6. Show that there is a set of four binary strings of length 5 with minimum distance 3. Is there such a set with five members?

7. Encode the following messages using Venn diagram coding.

   (i) 1010   (ii) 0011
   (iii) 1001   (iv) 0010

8. Decode the following messages using Venn diagram coding.

   (i) 1001011   (ii) 0011111
   (iii) 1000101   (iv) 1101000
   (v) 0011010   (vi) 1110011

9. To transmit the eight possible messages of three binary symbols, a code appends three further symbols: if the message is ABC, the code transmits $ABC(A+B)(A+C)(B+C)$. Write down the encoding matrix for this code. How many errors can it detect and correct?

10. Use the compressed genetic code to encode the following:
    (i) AAACAATAACTAGACGAA
    (ii) CACAAAACAATACCAAAA

11. Use the compressed genetic code to decode the following:
    (i) 010001010000011100011000
    (ii) 000100011000111010010000010

Exercises 12B

1. What is the distance between the following strings?
   (i) 101110101 and 101100110
   (ii) 10101010 and 01100010
   (iii) 000011100 and 001010100
   (iv) 010110100101 and 101011100110

2. Consider a code consisting of eight words 00000000, 00101110, 01011100, 10111010, 11100101, 11001011, 01110011, 10010111. What are the weight and distance of this code?

3. Encode the following messages using Venn diagram coding.

   (i) 1111   (ii) 0110

  (iii) 1100   (iv) 1101

4. Decode the following messages using Venn diagram coding.

   (i) 1001101   (ii) 0010111

  (iii) 1011001   (iv) 1100011

   (v) 0001101   (vi) 0101010

5. Write down the encoding matrix for the $(15,11)$-Hamming code.

   (i) How does this code encode the message 11001011111?

  (ii) How does this code encode the message 00001110000?

  (iii) How does it decode 110010101100111?

  (iv) How does it decode 011110001010001?

(Assume the check digits are in positions 1, 2, 4 and 8.)

6. The *binary repetition code of length* n consists of the two codewords $00\ldots0$ and $11\ldots1$ of length n.

   (i) Show that the binary repetition code of length 5 is 2-error-correcting (that is, given any error in at most two bits the nearest neighbor decoding provides the correct codeword).

  (ii) Generalize to the binary repetition code of length $2n + 1$.

7. Use the compressed genetic code to encode the following:

   (i) AAGAACAACATAAACCACTAA

  (ii) ACAAAACAAGAAACTAACAA

8. Use the compressed genetic code to decode the following:

   (i) 000100011101000100001110001100010

  (ii) 00000011011010010110001010

Problems

Problem 1: The n-dimensional hypercube is a graph with 2^n vertices corresponding to the $(0, 1)$-vectors of length n. Vertices are adjacent if and only if their vectors differ in exactly one position. Explain how block codes correspond to subgraphs of hypercubes.

Problem 2: Write out a formal proof of Theorem 12.2.

Problem 3: Suppose you construct a code from the Venn diagram code by adding an eighth symbol to each codeword so as to make the sum of the symbols even. That is, if the code had an odd number of 1s, the last symbol is 1; otherwise it is 0. How many errors can this code detect or correct?

Problem 4: You need to send messages in the five symbols A, B, C, D, E. Among the possible messages, A occurs most often, B second most, then C, then D, and E is least frequent. Devise a code that will make the average length of messages as short as possible.

Problem 5: Prove that a code can correct n errors if and only if the minimum distance between codewords is at least $2n + 1$.

Chapter 13

Latin Squares

13.1 Introduction

"There are thirty-six military officers, six each from six different regiments. The officers from any one regiment include one of each of six ranks (a colonel, a lieutenant-colonel, a major, a captain, a lieutenant, and a sub-lieutenant). They wish to parade in a square formation, so that every row and every column contains one member of each regiment and one officer of each rank. Is this possible?"

This problem fascinated Euler in the eighteenth century [33], but he could not solve it. In fact, it was not until 1900 that Tarry [114] proved, by an exhaustive search, that no solution was possible. (See Appendix C for biographies of Euler and Tarry.)

Suppose we are given an arrangement of the officers in which every row and every column contains one representative of each regiment. It can be represented as a 6×6 array; if the officer in row i, column j belongs to regiment k, then the array has (i, j) entry k. For example, one possible arrangement of thirty-six officers from regiments 1, 2, 3, 4, 5, 6 could be represented by

1	3	6	2	4	5
5	1	3	6	2	4
4	5	1	3	6	2
3	6	2	4	5	1
6	2	4	5	1	3
2	4	5	1	3	6

Euler discussed these arrays at length in [33], and gave (essentially) the following definition:

Definition: A *Latin square* of *side* (or *order*) n is an $n \times n$ array based on some set S of n symbols (treatments), with the property that every row and every column contains every symbol exactly once. In other words, every row and every column is a permutation of S.

The arithmetical properties of the symbols in a Latin square are not relevant to the definition, so their nature is often immaterial; unless otherwise specified, we assume a Latin square of order n to be based on $\{1, 2, \ldots, n\}$. (Other symbol sets will be used when appropriate.)

Here are some small Latin squares:

$$\boxed{1} \qquad \begin{array}{|c|c|} \hline 1 & 2 \\ \hline 2 & 1 \\ \hline \end{array} \qquad \begin{array}{|c|c|c|} \hline 1 & 2 & 3 \\ \hline 2 & 3 & 1 \\ \hline 3 & 1 & 2 \\ \hline \end{array}.$$

There are Latin squares of side 3 that look quite different from the one just given, for example

$$\begin{array}{|c|c|c|} \hline 1 & 2 & 3 \\ \hline 3 & 1 & 2 \\ \hline 2 & 3 & 1 \\ \hline \end{array}.$$

but this can be converted into the given 3×3 square by exchanging rows 2 and 3.

As we said, the arithmetical properties of the symbols are not considered in the definition of a Latin square. However, many examples *arise* from simple arithmetic and algebra.

Example 13.1: Use modular arithmetic to construct a Latin square of order 5.

Consider addition modulo 5. If we write $\mathbb{Z}_5 = \{0, 1, 2, 3, 4\}$ for the set of integers modulo 5, the addition table looks like

	0	1	2	3	4
0	0	1	2	3	4
1	1	2	3	4	0
2	2	3	4	0	1
3	3	4	0	1	2
4	4	0	1	2	3

The core of this,

0	1	2	3	4
1	2	3	4	0
2	3	4	0	1
3	4	0	1	2
4	0	1	2	3

is a Latin square of order 5. □

This same method works if 5 is replaced by any positive integer n, so:

THEOREM 13.1 (Existence of Latin Squares)
There exists a Latin square of every positive integer order.

Multiplication in modular arithmetic can sometimes give rise to a Latin square. In this case the square is of order one less than the modulus, because the zero element must be deleted, or else there would be a row and column of all zeroes.

Example 13.2: Use multiplication modulo 5 to construct a Latin square of order 4.

The multiplication table is

	0	1	2	3	4
0	0	0	0	0	0
1	0	1	2	3	4
2	0	2	4	1	3
3	0	3	1	4	2
4	0	4	3	2	1

and if we delete the labels and the row and column corresponding to 0, we obtain

1	2	3	4
2	4	1	3
3	1	4	2
4	3	2	1

which is called the *core* of the table. It is a Latin square, and symbol 0 never occurs. □

This construction is not available for all positive integers; for example, replacing 5 by 4, the 0 occurs in an inappropriate place:

	0	1	2	3
0	0	0	0	0
1	0	1	2	3
2	0	2	0	2
3	0	3	2	1

$\Rightarrow$

1	2	3
2	0	2
3	2	1

.

If we replace 5 by n, the "multiplication table" construction yields a Latin square if and only if n is prime.

These considerations lead us to treat Latin squares as the tables of binary operations.

Definition: A *quasigroup* $(Q, *)$ consists of a set Q together with a binary operation $*$ such that, for each pair x and y of elements of Q there exist unique elements u and v of Q for which

$$x * u = y = v * x.$$

THEOREM 13.2 (Latin Squares and Quasigroups)

Suppose $Q = \{x_1, x_2, \ldots, x_n\}$ is an n-set and $*$ is a binary operation on Q such that $(Q, *)$ is a quasigroup. Define an $n \times n$ array A by setting $a_{ij} = k$ when $x_i * x_j = x_k$. Then A is a Latin square.

Proof. Suppose some row, say row i of A, contains a repetition. Then $a_{ij} = a_{ih} = k$ for some h, j and k. Then $x_i * x_j = x_k = x_i * x_h$. Therefore x_j is the unique element such that $x_i * x_j = x_k$; and *so is* x_h. So $x_j = x_h$. This implies that $j = h$, so k only appears once in row i. Since i and k were arbitrary, this means each element appears exactly once in each row. Similarly, no column can contain a repeated element. As every cell of A is occupied by an element of $\{1, 2, \ldots, n\}$, each element occurs precisely once, and A is a Latin square, as required. $\qquad\square$

It is clear that a Latin square on symbol-set S will serve as the operation table for a quasigroup based on S, so the two concepts are equivalent. Theorem 13.1 is a restatement of the fact that the set $\mathbb{Z}_n$ of integers modulo n form a quasigroup under addition for any positive integer n. Other quasigroups include the integers $\mathbb{Z}$, under addition or under subtraction, the reals $\mathbb{R}$ under addition, subtraction or multiplication, and the non-zero reals (often denoted $\mathbb{R}^*$) under multiplication or division. The set $\mathbb{Z}_n^*$, the integers modulo n with zero omitted, form a quasigroup under multiplication (or under division) if and only if n is prime.

A quasigroup containing an identity element for $*$, an element e such that $x * e = x = e * x$ for every element x, is called a *loop*. If the underlying set is ordered so that e is the first element, then the corresponding Latin square has its first row and first column identical.

Two special features of quasigroups can be interpreted as properties of Latin squares. If the operation $*$ on the quasigroup is *commutative* in the ordinary algebraic sense then the quasigroup is called *Abelian* and the Latin square is *symmetric* (when considered as a matrix). If the quasigroup is *idempotent* ($x * x = x$ for all $x \in Q$) then the entries on the main diagonal of the Latin square are all different. A set of elements of a Latin square is called a *transversal* if it contains exactly one member of each row, one member of each column, and exactly one copy of each symbol. So the Latin square from a quasigroup contains a rather special transversal, one that lies on the main diagonal of the square. We shall discuss transversals further in the next section.

It is also significant if the operation of a quasigroup is *associative* ($x * (y * z) = (x * y) * z$ for all x, y, z). An associative quasigroup is a *semigroup* and an associative loop is a *group*. Groups are widely studied objects, and most readers will be familiar with group theory. However, this significance is more algebraic than combinatorial.

13.2 Orthogonality

We now return to Euler's problem of the 36 officers. As we said, the problem cannot be solved. However, if we replace 36 by 9, and have three regiments a, b, c with three ranks α, β, γ represented in each, there is a solution. An example is

$a\alpha$	$b\beta$	$c\gamma$
$b\gamma$	$c\alpha$	$a\beta$
$c\beta$	$a\gamma$	$b\alpha$

This array was constructed by superposing two 3×3 Latin squares, namely

a	b	c
b	c	a
c	a	b

and

α	β	γ
γ	α	β
β	γ	α

The two squares have the following property: The positions occupied by a fixed symbol in the first square form a transversal in the second square. (For example, the positions containing a in rows 1, 2 and 3 contain α, β, and γ, respectively in the second.) We say the second Latin square is *orthogonal* to the first.

Definition: Suppose A and B are Latin squares of side n based on symbol sets S_A and S_B, respectively. B is *orthogonal* to A (written $B \perp A$) if, for every $x \in S_A$, the set of n positions in A occupied by x contain every member of S_B.

It is clear that each member of S_B will occur precisely once in the cells occupied by a fixed element in A.

Equivalently, suppose we construct from A and B an $n \times n$ array (A, B) of ordered pairs, where (x, y) occurs in position (i, j) if and only if x occurs in position (i, j) of A and y occurs in position (i, j) of B, then B is orthogonal to A if every possible ordered pair with first element in S_A and second element in S_B occurs in the new array.

THEOREM 13.3 (Symmetry of Orthogonality)
Orthogonality is symmetric; that is, B is orthogonal to A implies A is orthogonal to B.

Proof. Suppose A and B are Latin squares of side n based on symbol sets S_A and S_B, respectively, and B is orthogonal to A. Write $S_A = \{x_1, x_2, \ldots, x_n,$ and select $y \in S_B$. There is exactly one position in which A contains x_1 and B contains y (put $x = x_1$ in the definition). There is also one position where A contains x_2 and B contains y (put $x = x_2$). In this way we see that every member of S_A occurs once with y. This proof can be applied to every member y of S_B. So A is orthogonal to B. $\square$

The array formed by superposing two orthogonal Latin squares is sometimes called a *Graeco-Latin square*. (In fact, the name "Latin square" comes from Euler's habit of using the Roman (that is, Latin) alphabet for the symbols of a square; and when representing a pair of orthogonal squares, he used the Greek alphabet for the second square, just as we did.) But it is more common to speak of "a pair of orthogonal Latin squares," and to use the symbols $\{1, 2, \ldots, n\}$ for each. We have exhibited a pair of orthogonal Latin squares of order 3, and stated that there is no pair of orthogonal Latin squares of order 6. One can ask: for what orders do there exist a pair of orthogonal Latin squares? Can we find three or more Latin squares that are *mutually orthogonal*—each one is orthogonal to all the others?

Before we discuss this problem, we define a *standardized* Latin square to be a Latin square whose symbols are the first n positive integers, and whose first row is

$$\boxed{1\ 2\ 3\ \ldots\ n}. \tag{13.1}$$

If A is any Latin square, we can convert it to standardized form by renaming the symbols. If A and B are orthogonal, we can assume that both are standardized; we rename the symbols in each separately.

THEOREM 13.4 (Maximum Number of MOLS)
If there are k mutually orthogonal Latin squares (MOLS) of side n, $n > 1$, then $k < n$.

Proof. Suppose $A_1, A_2, \ldots, A_k$ are Latin squares of side $n, n > 1$, each of which is orthogonal to each other one. Without loss of generality, assume that each has been standardized, so that each has first row (13.1). Write a_i for the $(2, 1)$ entry in A_i. No a_i can equal 1 (since the first columns can contain no repetition), and the a_i must be different (if $a_i = a_j$, then the n cells that contain a_i in A_i must contain a repetition in A_j; both the $(1, a_i)$ and $(2, 1)$ cells of A_i and A_j contain a_i). So $\{a_1, a_2, \ldots, a_k\}$ contains k distinct elements of $\{2, 3, \ldots, n\}$, and $k < n$. $\square$

Let us write $N(n)$ for the number of squares in the largest possible set of mutually orthogonal Latin squares of side n. In this notation, we have just shown

$$N(n) \leq n - 1 \text{ if } n > 1.$$

(If $n = 1$, we can take $A_1 = A_2 = \ldots$ and it makes sense to write "$N(1) = \infty$" in some situations. Whenever a theorem requires the existence of at least k mutually orthogonal Latin squares of side n, the conditions are satisfied by $n = 1$ for any k. In other applications it is more appropriate to say $N(1) = 1$. To avoid this problem, many authors only define orthogonality for side greater than 1.)

For example, $N(4) \leq 3$. In fact, $N(4) = 3$; one set of three mutually orthogonal Latin squares of side 4 is

1	2	3	4
2	1	4	3
3	4	1	2
4	3	2	1

1	2	3	4
3	4	1	2
4	3	2	1
2	1	4	3

1	2	3	4
4	3	2	1
2	1	4	3
3	4	1	2

On the other hand, the theorem tells us only that $N(6) \leq 5$, but we know from Tarry's result that $N(6) = 1$.

THEOREM 13.5 (Number of MOLS, Prime Power Order)
$N(n) = n - 1$ *whenever* n *is a prime power.*

Proof. Suppose n is prime. We define a set of $n - 1$ mutually orthogonal Latin squares $A_1, A_2, \ldots, A_{n-1}$ of order n. Let us write $A_h = (a_{ij}^h)$. We define

$$a_{ij}^h = i + hj,$$

where i, h and j are integers modulo n (remainder 0 is written as n, not 0). Then A_h is easily seen to be a Latin square, and the $n - 1$ Latin squares are orthogonal.

The case where n is a prime power is similar, but uses the existence of a finite field with n elements. $\square$

The *Cartesian product* of two Latin squares A and B is defined as follows. If a is any symbol in A, write (a, B) for the array derived from B by replacing each entry x by the ordered pair (a, x). Then replace every occurrence of a in A by (a, B).

For example, suppose A and B are Latin squares of order 2:

$$A = \begin{array}{|cc|} \hline 1 & 2 \\ 2 & 1 \\ \hline \end{array} \qquad B = \begin{array}{|cc|} \hline x & y \\ y & x \\ \hline \end{array}$$

Then the Cartesian product $A \times B$ is

$$A \times B = \begin{array}{|cccc|} \hline (1,x) & (1,y) & (2,x) & (2,y) \\ (1,y) & (1,x) & (2,y) & (2,x) \\ (2,x) & (2,y) & (1,x) & (1,y) \\ (2,y) & (2,x) & (1,y) & (1,x) \\ \hline \end{array}$$

In general, if A is a Latin square of order n and B is a Latin square of order r then $A \times$ is a Latin square of order nr; if A is based on symbol-set S and B is based on symbol-set T, then the symbol-set for $A \times B$ is the Cartesian product $S \times T$, as defined in Section 1.2 (which is why the name *Cartesian product* is used for these Latin squares). A similar idea—the Cartesian product of matrices—is defined in Appendix B.

It is easy to see that the Cartesian product of Latin squares is a Latin square, and moreover the taking of Cartesian products preserves orthogonality: if A_1 is orthogonal to A_2 and B_1 is orthogonal to B_2, then $A_1 \times B_1$ is orthogonal to $A_2 \times B_2$. Therefore

$$N(nr) \geq \min\{N(n), N(r)\}. \tag{13.2}$$

So we have the following result:

THEOREM 13.6 (MacNeish's Theorem) [70]
Suppose $n = p_1^{a_1} p_2^{a_2} \ldots p_r^{a_r}$, where the p_i are distinct primes and each $a_i \geq 1$. Then

$$n - 1 \geq N(n) \geq \min(p_i^{a_i}).$$

We know that there cannot be a pair of orthogonal Latin squares of order 6, and order 2 is also impossible. Are there any further values n for which $N(n) = 1$? Euler thought so; he conjectured that $N(n) = 1$ whenever $n \equiv 2$ modulo 4. However, the opposite is true:

THEOREM 13.7 (Falsity of Euler's Conjecture) [11]
$N(n) \geq 2$ *for every integer $n > 6$.*

The proof can be found in most books on design theory (such as [123]). The history of this problem and its solution can be found in [39].

Further results are known. For example, many authors have contributed to the proof that:

THEOREM 13.8 (Three MOLS)
There exist three mutually orthogonal Latin squares of every side except 2, 3, 6, and possibly 10.

A proof appears in [123]. In general, for any given constant k there exists a lower bound n_k such that $N(n) \geq k$ whenever $n > n_k$, but the only known value is $n_2 = 6$; we do not know $N(10)$, so n_3 may be either 10 or 6. For the best known lower bounds on n_k, as well as a table of the largest known numbers of Latin squares of orders less than 10,000, see [22].

A Latin square with *no* orthogonal mate is called a *bachelor square*. We already know of bachelor squares of sides 2 and 6. (In fact, any Latin square of these orders must be a bachelor.) At side 3, there is no bachelor.

Norton [78] (see also [95]) listed the isomorphism classes of Latin squares of side 7; there are 147 in all. There is one set of six mutually orthogonal squares (all six squares are isomorphic); no other set of three or more mutually orthogonal square exists, and only five of the remaining classes contain squares with mates. So there are 141 isomorphism classes of bachelor squares. This led to speculation that bachelor squares might be very common, but empirical evidence suggests that for orders larger than 7, almost all Latin squares have mates. However, there are bachelor squares of all orders other than 3 and 7. The proof for odd orders may be found in [72] and [124].

To discuss even orders we define a function

$$\Delta_A(i,j) \equiv i + j - a_{ij} \pmod{n}.$$

LEMMA 13.9

If T is a transversal in A, the sum of $\Delta_A(i,j)$ over the entries a_{ij} of T is $\frac{1}{2}n$ if n is even and 0 if n is odd.

Proof. Suppose cells $(1, j_1), (2, j_2), \ldots, (n, j_n)$ form a transversal. As they form a transversal, their entries are $\{1, 2, \ldots, n\}$ in some order. So

$$
\begin{aligned}
\sum_{i=1}^{n} \Delta_A(i, j_i) &= (1 + 2 + \ldots + n) + (1 + 2 + \ldots + n) - (1 + 2 + \ldots + n) \\
&= (1 + 2 + \ldots + n) \\
&= \tfrac{1}{2}n(n+1).
\end{aligned}
$$
$\square$

THEOREM 13.10 (Latin Squares Without Transversals)

Let A be the Latin square of side $2t$ defined by

$$a_{ij} \equiv i + j \pmod{2t}, 1 \le a_{ij} \le 2t.$$

Then A contains no transversal.

Proof. Suppose cells $(1, j_1), (2, j_2), \ldots, (2t, j_{2t})$ form a transversal. From Lemma 13.9

$$\sum \Delta_A(i, j_i) = t.$$

However, $a_{ij_i} \equiv i + j_i$, so

$$\sum \Delta_A(i, j_i) \equiv 2(1 + 2 + \ldots + 2t) \equiv 0 \pmod{2t},$$

a contradiction. $\square$

Obviously these squares have no orthogonal mates.

We conclude this section by exhibiting some small examples of orthogonal Latin squares whose orders are not prime powers.

The transpose of a Latin square is a Latin square of the same order, so it is possible to ask whether the two squares are orthogonal. A Latin square that is orthogonal to its transpose is called *self-orthogonal*. It is in fact not difficult to construct self-orthogonal Latin squares.

We have not yet seen a pair of orthogonal Latin squares of order 10, so we now exhibit a self-orthogonal square of that order:

1	2	3	4	5	6	7	8	9	X
2	X	4	8	6	3	5	9	7	1
3	6	7	1	8	9	2	5	X	4
4	7	X	5	2	1	9	6	8	3
5	3	8	X	9	4	6	1	2	7
6	9	1	3	7	2	8	X	4	5
7	8	6	9	3	X	1	4	5	2
8	5	9	2	X	7	4	3	1	6
9	4	2	6	1	5	X	7	3	8
X	1	5	7	4	8	3	2	6	9

Another interesting example of orthogonal Latin squares is the following:

THEOREM 13.11 (Latin Squares of Side 12) [54]
There is a set of five mutually orthogonal Latin squares of side 12.

Proof. One of the Latin squares is

1	2	3	4	5	6	7	8	9	X	E	D
2	3	4	5	6	1	8	9	X	E	D	7
3	4	5	6	1	2	9	X	E	D	7	8
4	5	6	1	2	3	X	E	D	7	8	9
5	6	1	2	3	4	E	D	7	8	9	X
6	1	2	3	4	5	D	7	8	9	X	E
7	8	9	X	E	D	1	2	3	4	5	6
8	9	X	E	D	7	2	3	4	5	6	1
9	X	E	D	7	8	3	4	5	6	1	2
X	E	D	7	8	9	4	5	6	1	2	3
E	D	7	8	9	X	5	6	1	2	3	4
D	7	8	9	X	E	6	1	2	3	4	5

Each of the other squares is obtained from this by column permutation; the four permutations give the following four first rows:

$$[1 \quad 7 \quad 9 \quad 3 \quad 8 \quad 2 \quad X \quad D \quad 5 \quad E \quad 6 \quad 4],$$
$$[1 \quad 4 \quad 7 \quad 2 \quad X \quad D \quad 3 \quad 9 \quad 6 \quad 5 \quad 8 \quad E],$$
$$[1 \quad 9 \quad 2 \quad D \quad 6 \quad X \quad 4 \quad E \quad 3 \quad 8 \quad 7 \quad 5],$$
$$[1 \quad 5 \quad D \quad E \quad 3 \quad 8 \quad 9 \quad 7 \quad X \quad 2 \quad 4 \quad 6]. \qquad \square$$

THEOREM 13.12 (Latin Squares of Side 15) [97]

$$N(15) \geq 4.$$

Proof. The first rows of the squares are

$$
\begin{array}{*{16}{c}}
[\,1 & 15 & 2 & 14 & 3 & 13 & 4 & 12 & 5 & 11 & 6 & 10 & 7 & 9 & 8\,], \\
[\,1 & 14 & 3 & 11 & 6 & 9 & 8 & 7 & 10 & 4 & 13 & 12 & 5 & 15 & 2\,], \\
[\,1 & 10 & 7 & 13 & 4 & 2 & 15 & 6 & 11 & 9 & 8 & 3 & 14 & 12 & 5\,], \\
[\,1 & 6 & 11 & 10 & 7 & 15 & 2 & 5 & 12 & 14 & 3 & 9 & 8 & 4 & 13\,].
\end{array}
$$

In each square the later rows are formed by developing the first row modulo 15. □

13.3   Idempotent Latin Squares

An $n \times n$ Latin square is called a *transversal square* if its diagonal entries form a transversal—that is, they are all different. In particular a transversal square whose main diagonal is

$$(1, 2, \ldots, n) \qquad (13.3)$$

is called an *idempotent Latin square*.

LEMMA 13.13
There exists a set of $N(n) - 1$ mutually orthogonal idempotent Latin squares of side n.

Proof. Suppose $A_1, A_2, \ldots, A_{N(n)}$ are mutually orthogonal Latin squares of side n based on $1, 2, \ldots, n$. By permuting the columns, transform $A_{N(n)}$ so that the element 1 appears in all the diagonal cells. Carry out the same column permutation on the other squares. The diagonals of the first $N(n) - 1$ new squares will be transversals. Now, in each square, permute the names of the elements so as to produce the diagonal (13.3). We have the required orthogonal idempotent Latin squares. □

The above proof embodies a simple but useful fact. If you need an idempotent Latin square, it is sufficient to construct a transversal square. The names of entries can then be permuted to make it idempotent.

We shall write $I(n)$ for the maximal cardinality of a set of mutually orthogonal idempotent Latin squares of side n. Using Lemma 13.13, we see that for every n,

$$N(n) - 1 \leq I(n).$$

On the other hand, $I(n)$ cannot possibly be larger than $N(n)$. So

$$N(n) - 1 \leq I(n) \leq N(n). \tag{13.4}$$

As easy generalizations of earlier results, we have:

THEOREM 13.14 (Orthogonal Idempotent Latin Squares)
If $n > 1$ then $I(n) \leq n - 2$, with equality when n is a prime power.

The proof that $I(n) \leq n - 2$ is similar to that of Theorem 13.4, and is left as an exercise. The other part follows because all but one of the squares constructed in Theorem 13.5 are in fact idempotent.

The only known cases where $I(n) = N(n)$ are the trivial case $n = 1$ (some authors would omit this case, saying that neither $N(1)$ nor $I(1)$ is defined) and

$$I(6) = N(6) = 1.$$

(We know that $N(6) = 1$, so $I(6) = 1$ or 0. Proving the existence of an idempotent Latin square of side 6 is again an exercise.)

THEOREM 13.15 (Orthogonal Idempotent Latin Squares, Composite Order)
For every m and n, $I(mn) \geq \min(I(m), I(n))$. If n has prime power decomposition

$$n = p_1^{a_1} p_2^{a_2} \ldots p_k^{a_k},$$

then $I(n) \geq \min(p_i^{a_i} - 1) - 1$.

Proof. If A and B are idempotent Latin squares, then the Cartesian product of A and B is a transversal square. The proof now follows that of Theorem 13.6. $\square$

As a consequence of Theorem 13.8, we have:

THEOREM 13.16 (Orthogonal Idempotent Latin Squares, Lower Bound)
There is a pair of orthogonal idempotent Latin squares of every order except 2, 3 and 6.

Proof. The only case to be investigated is order 10. But the square of order 10 given in the preceding section is a transversal square and is orthogonal to its transpose, so these two squares are a pair orthogonal idempotent Latin squares of that order. $\square$

13.4 Partial Latin Squares and Subsquares

A *partial Latin square* of side n is an $n \times n$ array in which some cells are filled with members of some n-set while others are empty, such that no row or column contains a repeated element. In 1960, Evans [34] conjectured that any partial Latin square of side n that has $n-1$ or fewer cells occupied can be *completed*, that is, the empty cells can be filled in so that the result is a Latin square. This turned out to be a difficult problem, but in 1981 Smetaniuk [105] showed that the Evans conjecture was true. This is the best possible result. It is easy to find an n-element example that cannot be completed.

Probably the partial Latin square most familiar to our readers is the Sudoku puzzle, which we mentioned in Chapter 1. This consists of a 9×9 array, partitioned into 3×3 subarrays, in which some cells contain entries chosen from $\{1, 2, 3, 4, 5, 6, 7, 8, 9\}$. The array is in fact a partial Latin square, and the object is to complete the Latin square. However, there is a further condition: each subarray must contain the nine numbers precisely once each.

We defined a *Sudoku square* to be a Latin square that could be the solution to a Sudoku puzzle: a 9×9 Latin square partitioned into 3×3 subarrays, where each subarray contains each of $\{1, 2, \ldots, 9\}$ precisely once. In a well-posed Sudoku puzzle, there is one and only one Sudoku square that contains the original partial Latin square.

There is no reason to restrict Sudoku to the 9×9 case. We define a (p, q)-square to be a Latin square of side pq partitioned into $p \times q$ subarrays, such that each subarray contains all integers from 1 to pq once each. For example, here is a very simple $(2, 3)$-square:

1	2	3	4	5	6
4	5	6	1	2	3
2	3	1	5	6	4
5	6	4	2	3	1
3	1	2	5	6	4
5	6	4	3	1	2

Among partial Latin squares, the *Latin rectangle* is of special interest. A Latin rectangle of size $k \times n$ is a $k \times n$ array with entries from $\{1, 2, \ldots, n\}$ such that every row is a permutation and the columns contain no repetitions. For example, the first k rows of an $n \times n$ Latin square form a Latin rectangle. Clearly, k can be no larger than n.

The following important theorem is a consequence of the results on systems of distinct representatives (SDRs), which we discussed in Section 1.1.5.

THEOREM 13.17 (Completing Idempotent Latin Rectangles)
If A is a $k \times n$ Latin rectangle, then one can append $(n - k)$ further rows to

A so that the resulting array is a Latin square.

Proof. If $k = n$, the result is trivial. So assume that $k < n$. Write S for $\{1, 2, \ldots, n\}$ and define S_j to be the set of members of S not occurring in column j of A. Each S_j has k elements. Now consider any element i of S. The k occurrences of i in A are in distinct columns, so i belongs to exactly $(n - k)$ of the sets S_j.

Suppose there were r of the sets S_j whose union contained fewer than r members, for some r. If we wrote a list of all members of each of these r sets, we would write down $r(n - k)$ symbols, since each S_j is an $(n - k)$-set. On the other hand, we would write at most $r - 1$ different symbols, and each symbol occurs in exactly $n - k$ sets, so there are at most $(r-1)(n-k)$ symbols written. Since $k < n, (r - 1)(n - k) < r(n - k)$, a contradiction. So there is no set of r of the S_j that contain between them fewer than r symbols. By Theorem 14.7, there is a system of distinct representatives for the S_i. Say the representative of S_j is i_j. Then $(i_1, i_2, \ldots, i_n)$ is a permutation of S. If we append this to A as row $k + 1$, we have a $(k + 1) \times n$ Latin rectangle. The process may be repeated. After $(n - k)$ iterations we have a Latin square with A as its first k rows. $\square$

A *subsquare* of side s in a Latin square of side n is an $s \times s$ subarray that is itself a Latin square of side s. A necessary and sufficient condition for an $s \times s$ subarray to be a subsquare is that it should contain only s different symbols.

COROLLARY 13.17.1

If L is a Latin square of side s and $n \geq 2s$, then there is a Latin square of side n with L as a subsquare.

Proof. Say $n = s + t$. Define M to be the $s \times t$ array

$$M = \begin{bmatrix} s+1 & s+2 & \cdots & s+t \\ s+2 & s+3 & \cdots & s+1 \\ \vdots & \vdots & & \vdots \\ s+s & s+s+1 & \cdots & s+s-1 \end{bmatrix}$$

with entries reduced modulo t, where necessary, to ensure that they lie in the range $(s+1, s+2, \ldots, s+t)$. Write A for the array constructed by laying row i of M on the right of row i of L, for $1 \leq i \leq s$:

$$A = \boxed{L \ \ M}.$$

Then A is an $s \times n$ Latin rectangle; if we embed it in an $n \times n$ Latin square, that square has L as a subsquare. $\square$

This is the best possible result (see the exercises), provided one ignores the trivial case $s = n$.

One can also ask whether or not a given array can be embedded in a Latin square.

Example 13.3: In each case, can the given array be embedded in a 7×7 Latin square? If so, give an example.

$$A = \begin{array}{|ccccc|} \hline 1 & 2 & 5 & 6 & 7 \\ 5 & 1 & 2 & 7 & 3 \\ 3 & 7 & 1 & 4 & 2 \\ 2 & 5 & 6 & 1 & 4 \\ 6 & 3 & 4 & 2 & 1 \\ \hline \end{array} \qquad B = \begin{array}{|ccccc|} \hline 1 & 2 & 5 & 6 & 7 \\ 5 & 1 & 2 & 7 & 3 \\ 3 & 7 & 1 & 4 & 2 \\ 2 & 5 & 6 & 1 & 4 \\ 7 & 3 & 4 & 2 & 1 \\ \hline \end{array}$$

We can assume the given array will appear in the top left corner of the Latin square, if one exists. Symbols 1 and 2 appear five times each in A. That means there can be only two more occurrences of each; the only way will be for them to appear in the bottom right corner of the square, forming a 2×2 subsquare. One example is

$$A = \begin{array}{|ccccc|cc|} \hline 1 & 2 & 5 & 6 & 7 & 3 & 4 \\ 5 & 1 & 2 & 7 & 3 & 4 & 6 \\ 3 & 7 & 1 & 4 & 2 & 6 & 5 \\ 2 & 5 & 6 & 1 & 4 & 7 & 3 \\ 6 & 3 & 4 & 2 & 1 & 5 & 7 \\ \hline 4 & 6 & 7 & 3 & 5 & 1 & 2 \\ 7 & 4 & 3 & 5 & 6 & 2 & 1 \\ \hline \end{array}.$$

For B there is no solution: 6 occurs only twice in B, so it would have to appear five more times. But the maximum is four (2 rows, 2 columns). □

13.5 Applications

Designing Experiments Using Latin Squares

Latin squares provide a model for experiments where there are several different factors. As an example, consider a motor oil manufacturer who claims that using its product improves gasoline mileage in cars. Suppose you wish to test this claim by comparing the effect of four different oils on gas mileage. However, because the car model and the drivers' habits greatly influence the mileage obtained, the effect of an oil may vary from car to car and from driver to driver. To obtain results of general interest, you must compare the oils in several different cars and with several different drivers. If you choose four car models and four drivers, there are 16 car-driver combinations. The type of car and driving habits are so influential that you must consider each of these

16 combinations as a separate experimental unit. If each of the four oils is used in each unit, 4×16, or 64, test drives are needed.

In this case there are three factors; the oil to be used in a test run is one, and so are the driver and the type of car.

An experiment is designed using a 4×4 Latin square. Suppose the cars are denoted X_1, X_2, X_3, X_4, and the drivers are Y_1, Y_2, Y_3, Y_4. Call the oils Z_1, Z_2, Z_3, Z_4. We call these the four *levels* of the factors. We assign one oil to each of the 16 car-driver combinations in the following arrangement:

	Y_1	Y_2	Y_3	Y_4
X_1	Z_1	Z_2	Z_3	Z_4
X_2	Z_2	Z_1	Z_4	Z_3
X_3	Z_3	Z_4	Z_1	Z_2
X_4	Z_4	Z_3	Z_2	Z_1

For example, the cell in the row marked X_2 and the column marked Y_3 contains entry Z_4. This means, in the test drive when car X_2 is driven by Y_3, oil Z_4 is to be used. The diagram is a plan for 16 test drives, one for each combination of car and driver. It tells us which oil to use in each case.

The array underlying the experiment is a Latin square, so each oil appears four times, exactly once in each row (for each car) and also exactly once in each column (for each driver). This setup can show how each oil performs with each car and with each driver, while still requiring only 16 test drives, rather than 64. If 64 test drives are available, one technique is to use a Latin square to schedule 16 tests, and then running each test four times. This enables the experimenters to make a more reliable estimate of the underlying variability due to uncontrollable factors (sometimes called "error"), which improves the statistical reliability of the experiment.

This technique can be used with any size Latin square. Suppose there are three types of treatment, X, Y and Z, and each is to be applied at n levels (X at levels $X_1, X_2, \ldots, X_n$, and so on). An $n \times n$ Latin square provides n^3 comparisons with n^2 tests; if the (i, j) element is k, then treatment levels X_i, Y_j and Z_k are applied in one test.

Further savings can be achieved by using orthogonal Latin squares. In the motor oil example, one might wonder whether automobile performance depends on the type of roads. So four different test routes could be chosen: one primarily urban, one about two-thirds urban and one-third interstate, one two-thirds interstate and one all interstate. If these routes are denoted T_1, T_2, T_3 and T_4, then the following design allows us to test all pairs of levels of the four factors in 16, rather than 256, tests:

	Y_1	Y_2	Y_3	Y_4
X_1	Z_1T_1	Z_2T_2	Z_3T_3	Z_4T_4
X_2	Z_2T_3	Z_1T_4	Z_4T_1	Z_3T_2
X_3	Z_3T_4	Z_4T_3	Z_1T_2	Z_2T_1
X_4	Z_4T_2	Z_3T_1	Z_2T_4	Z_1T_3

The design was constructed from a pair of orthogonal Latin squares.

In experiments where interaction between the different factors is unlikely, orthogonal Latin squares allow a large amount of information to be obtained with relatively little effort. This idea has been applied in a number of areas other than experimentation. One interesting application is the use of orthogonal Latin squares in compiler testing; for details, see [71].

Golf Designs

We conclude with an application to tournament scheduling. The teams representing $2n + 1$ golf associations wish to play a round robin tournament. There are $2n+1$ rounds; each team serves as "host" for the one round in which they do not play, and the n matches in that round are played (simultaneously) on the course of the host team.

The teams wish to play $2n - 1$ complete tournaments in such a way that every pair of teams play exactly once on each of the $2n - 1$ available courses. (We shall call such an arrangement a *golf design*.) When is this possible?

This question arose in planning an actual golf tournament in New Zealand. It was reported by Robinson [89], who pointed out that the problem of constructing one round robin is equivalent to finding an idempotent symmetric Latin square; the square has (i, j) entry k, when $i \neq j$, if and only if i plays j at k. So the whole problem is equivalent to finding $2n - 1$ symmetric idempotent Latin squares such that the $2n - 1$ (i, j) entries cover $\{1, 2, \ldots, 2n + 1\}$ with i and j deleted, for all $i \neq j$. He proved that case $n = 5$ has no solution and exhibited a solution for case $n = 7$ (see Figure 13.1).

We can use a Steiner triple system of order v to construct a $v \times v$ Latin square $L = (l_{ij})$. We set $l_{ii} = i$ for all i. If $i \neq j$, find the unique triple that contains both i and j. If it is $\{i, j, k\}$, then set $l_{ij} = k$. The resultant square L is idempotent and symmetric.

Tierlinck [115, 116] proved that if a golf design exists for order 11, then it exists for all odd orders other than 5 and possibly 41. Subsequently, Colbourn and Nonay [23] found an example of order 11, and recently Chang [20] has constructed one of order 41. So the problem is completely solved.

To solve particular cases, one could try to construct symmetric idempotent Latin squares cyclically. If the first row of such a square is $0, a_1, a_2, \ldots, a_{2n}$

```
0 2 4 5 1 6 3        0 3 6 1 5 4 2        0 4 5 6 2 3 1
2 1 6 4 5 3 0        3 1 5 0 6 2 4        4 1 3 2 0 6 5
4 6 2 0 3 1 5        6 5 2 4 1 0 3        5 3 2 1 6 4 0
5 4 0 3 6 2 1        1 0 4 3 2 6 5        6 2 1 3 5 0 4
1 5 3 6 4 0 2        5 6 1 2 4 3 0        2 0 6 5 4 1 3
6 3 1 2 0 5 4        4 2 0 6 3 5 1        3 6 4 0 1 5 2
3 0 5 1 2 4 6        2 4 3 5 0 1 6        1 5 0 4 3 2 6
```

```
    0 5 3 2 6 1 4        0 6 1 4 3 2 5
    5 1 4 6 3 0 2        6 1 0 5 2 4 3
    3 4 2 5 0 6 1        1 0 2 6 5 3 4
    2 6 5 3 1 4 0        4 5 6 3 0 1 2
    6 3 0 1 4 2 5        3 2 5 0 4 6 1
    1 0 6 4 2 5 3        2 4 3 1 6 5 0
    4 2 1 0 5 3 6        5 3 4 2 1 0 6
```

FIGURE 13.1:   A golf design for seven clubs.

then the square would be

$$
\begin{array}{cccccc}
0 & a_1 & a_2 & a_3 & \cdots & a_{2n} \\
a_{2n}+1 & 1 & a_1+1 & a_2+1 & \cdots & a_{2n-1}+1 \\
a_{2n-1}+2 & a_{2n}+2 & 2 & a_1+2 & \cdots & a_{2n-2}+2 \\
& \cdots & & & & \cdots
\end{array}\; .
$$

For symmetry we would require $a_{2n+1-j} = a_j - j \pmod{2n+1}$. In order that the rows be Latin we need $0, a_1, a_2, \ldots, a_n, a_n - n, a_{n-1} - (n-1), \ldots, a_1 - 1$ to be distinct modulo $2n+1$; a similar condition on columns exists. For a golf design we need $2n - 1$ such sequences with different entries in all nondiagonal positions. This property can be derived from the starter rows. We have the following:

THEOREM 13.18 (Circulant Golf Arrays) [121]
Suppose there exists a $(2n-1) \times (2n+1)$ array A with the following properties:
  (a) *when $1 \le k \le n$, $a_{i,n+k} = a_{i,n+1-k} + n + k$;*
  (b) *$a_{i0} = 0$;*
  (c) *$a_{ij} + k \ne a_{ik} + j$ when $j \ne k$;*
  (d) *each row of A is a permutation of $\{0, 1, \ldots, 2n\}$;*
  (e) *column j of A is a permutation of $\{1, 2, \ldots, 2n\} \setminus \{j\}$.*
(All arithmetic is carried out modulo $2n + 1$.) Then a golf design for $2n + 1$ teams exists.

2	5	9	14	16	13	15	12
3	1	10	12	11	15	4	13
4	7	12	2	13	16	1	14
5	8	2	13	15	1	14	11
6	9	15	3	1	14	11	10
7	16	13	8	3	11	2	9
8	14	16	9	6	5	10	2
9	4	14	16	10	7	3	6
10	13	4	7	2	12	5	16
11	15	7	5	14	8	6	3
12	10	5	1	4	9	13	15
13	6	8	15	7	3	16	1
14	11	1	6	8	10	12	7
15	3	11	10	12	5	9	4
16	12	6	4	9	2	8	5

FIGURE 13.2: Array to generate a golf design for 17 teams.

The $2n - 1$ Latin squares are derived by developing the rows of A modulo $2n + 1$.

The point of this theorem is that it yields a relatively easy computer construction. To construct row i of A it is only necessary to select the n entries $a_{i1}, a_{i2}, \ldots, a_{in}$. Then the row is

$$0, a_{i1}, a_{i2}, \ldots, a_{in}, (a_{in} + n + 1), \ldots, (a_{i2} + 2n - 1), (a_{i1} + 2n)$$

(reduced mod $2n + 1$). Moreover, you may as well fix a_{i1} to be $i + 2$. Then properties (c) through (e) restrict the search considerably.

No solutions exist for $2n + 1 = 5, 7$ or 11. It is found that there is a unique solution for $2n + 1 = 9$, and exactly four solutions exist for $2n + 1 = 13$.

A solution for $2n + 1 = 17$ is illustrated in Figure 13.2. Only columns 1 through 8 of A are shown. Column 0 consists of all zeros; columns 9 through 16 are derived from columns 8 through 1 by adding $(9, 10, 11, 12, 13, 14, 15, 16)$ modulo 17.

Exercises 13A

1. Write down the Latin square derived from the addition of integers modulo 6.

2. Verify that the multiplication table of integers modulo 6 does not produce a 5×5 Latin square.

3. A *reduced Latin square* of side n is one in which the first row and first column are $(1, 2, \ldots, n)$ in that order.

   (i) Prove that every Latin square is equivalent to a reduced Latin square.

   (ii) Say there are r_n different reduced Latin squares of side n. Prove that the number of Latin squares of side n is

   $$n!(n-1)!r_n.$$

   (iii) Find r_n for $n = 3$ and 4.

   (iv) Exhibit all reduced Latin squares of order 4.

4. Prove that the Cartesian product of Latin squares is a Latin square, and that the Cartesian product of transversal squares is a transversal square.

5. Recall that a Latin square is called *self-orthogonal* if it is orthogonal to its own transpose.

   (i) Prove that the diagonal elements of a self-orthogonal Latin square of side n must be $1, 2, \ldots, n$ in some order.

   (ii) Prove that there is no self-orthogonal Latin square of side 3.

   (iii) Find self-orthogonal Latin squares of sides 4 and 5.

6. Walter Johns and John Walters wish to invite 25 colleagues to an inter-disciplinary conference. These colleagues will include five people from each of five universities A, B, C, D, and E; each school will send one person from each of five disciplines, d_1 through d_5. For the introductory talks, they wish to seat these people in a five-by-five grid so that no row or column contains two people from the same discipline or from the same college.

   (i) Find a seating plan that accomplishes this.

   (ii) The football teams from College A and College B are blood rivals. Can you come up with a seating plan in which no member of College A sits next to a member of College B (to avoid fights)?

   (iii) Can you find a seating plan in which each member of discipline d_1 is sitting next to a member of discipline d_2?

   (iv) Can you find a seating plan as required in part (iii) if there are only four universities and four disciplines involved? Can you find one that also satisfies the condition of part (ii)?

   (In this question "next to" refers only to rows, not columns.)

7. Prove that if a Latin square of side n has a subsquare of side s, where $s < n$, then $n \geq 2s$.

8. Without using the result of Theorem 13.17, prove directly that any $(n-1) \times n$ Latin rectangle can be completed to give an $n \times n$ Latin square.

9. Use SDRs to complete the following Latin square.

1	2	3	4	5
5	1	2	3	4
4	5	1	2	3

10. In each case, can the given array be embedded in a 6×6 Latin square? If so, give an example.

(i)
1	5	3	4
3	2	1	5
2	3	6	1
6	1	4	2

(ii)
1	2	3	4
2	1	6	5
3	6	4	1
4	3	2	6

11. In each case, is there a value for x such that the given array can be embedded in a 6×6 Latin square? If so, what is that value? Give an example of an embedding.

(i)
1	2	3	4
5	6	1	2
3	4	5	1
4	1	2	x

(ii)
1	2	3	4
4	6	1	2
3	1	6	5
6	4	2	x

12. Find a completion of the partial Latin square

	2	3		4	
3		2	4		
	4			2	3
2	3		1		4
	1	4	2	3	
4			3		2

.

13. Consider golf designs of order 5:

   (i) Show that there is a unique symmetric Latin square of side 5, up to isomorphism.

   (ii) Prove that there is no golf design of order 5.

14. Suppose A is a Latin square of side n, and is *symmetric* (that is, $a_{ij} = a_{ji}$ for all i and j)

   (i) Say n is odd. Show that the diagonal entries are $\{1, 2, \ldots, n\}$ in some order.

   (ii) Show that the above statement is false when n is even.

Exercises 13B

1. Write down the Latin square derived from the multiplication of integers modulo 7.

2. Prove that $\mathbb{Z}_n^*$ is not a quasigroup under multiplication when n is composite.

3. (For those with sufficient algebraic background.) Prove Theorem 13.5 in the case where n is a prime power but not a prime.

4. Prove that $I(n) \le n - 2$ for every integer $n > 1$.

5. Show that the array

1			
		2	
		3	

can be completed to a Latin square.

6. Construct a $(2,2)$-square.

7. Use SDRs to complete the following Latin square.

1	2	3	4	5	6
6	1	4	3	2	5
4	5	1	6	3	2

8. Consider the problem in Exercise 11A.6. Suppose Drs. Johns and Walters wish to invite n^2 colleagues from n universities, each of which sends one person from each of n disciplines. They wish to seat these people in a square grid so that no row or column contains two people from the same discipline or from the same university.

   (i) For what n is there a seating plan that accomplishes this?

   (ii) For what n is there a plan in which no member of College A sits next to a member of College B?

   (iii) For what n is there a plan in which each member of discipline d_1 is sitting next to a member of discipline d_2?

9. Verify that the array in Figure 13.2 yields a golf design for 17 teams.

Problems

Problem 1: Prove that there is no idempotent Latin square of side 2.

Problem 2: Construct an idempotent Latin square of side 6.

Problem 3: Prove that there exist at least

$$n!(n-1)!\ldots(n-k+1)!$$

$k \times n$ Latin rectangles and therefore at least

$$n!(n-1)!\ldots 2!1!$$

Latin squares of side n.

Problem 4: Find a 4×4 array containing four entries, one each of 1, 2, 3, 4, that cannot be completed to a Latin square. Generalize your result to show that the Evans conjecture is the best possible result.

Problem 5: Verify that the construction in Theorem 13.5 produces Latin squares, and that the squares are orthogonal. (If you have not studied finite fields, restrict yourself to the case of prime order.)

Problem 6: Suppose we have t orthogonal Latin squares of side n; the kth square has (i,j) entry a_{ij}^k. Form a new array of order $(t+2) \times n^2$ as follows: column $(i-1)n+j$ is the transpose of

$$\left(i, j\, a_{ij}^1, a_{ij}^2, \vdots, a_{ij}^t \right).$$

This array is called a $(t+2) \times n^2$ *orthogonal array*.

(i) Construct a 4×9 orthogonal array.

(ii) Is there a 4×36 orthogonal array?

(iii) Construct a set of 16 codewords of length 5 based on the alphabet $\{1,2,3,4\}$ such that, given any two words, there is exactly one position in which they have the same symbol.

Problem 7: (i) Show that the partial Latin square

1		3		5
	1		3	
4	5		2	
3	4		1	
		4	5	

has a unique completion.

(ii)   Find all completions of the partial Latin square

$$
\begin{array}{|ccccc|}
\hline
1 & & 3 & & 5 \\
& 1 & & 3 & \\
& & 1 & & \\
& 4 & & & \\
& & & 5 & \\
\hline
\end{array}
$$

.

Chapter 14

Balanced Incomplete Block Designs

Suppose you want to compare six varieties of corn, varieties x_1, x_2, x_3, x_4, x_5 and x_6, in order to decide which variety yields the most grain in your area. You could find a large field, divide it into six smaller areas (*plots*) and plant one variety in each plot. Eventually you could weigh the different yields and make your comparisons.

However, suppose one part of the field is more fertile than another. The variety planted there would seem better than the other varieties, but you are only observing the quality of the soil, not the corn. The way round this is to partition the field into four midsized areas—*blocks*—so that the fertility of each plot is close to uniform. You could then plant six plots, one for each variety, in each block, for a total of 24.

But now there is another problem. Not all seeds germinate equally, and not all plants grow equally well, even in identical environments. Living things are not uniform, and do not develop uniformly. If the plots are too small, this random variation could affect the yields enough to outweigh the differences you are trying to study. Maybe 24 plots are too many, the individual plots are too small.

The solution is to make the plots larger; not every variety will appear in each block. For example, you might restrict each block to three varieties. If the first block contains x_1, x_2 and x_3, it makes sense to compare the yields for x_1 and x_2, in that block, and in every block where both appear.

For conciseness, let us call the blocks $B_1, B_2, \ldots, B_4$. We identify the block with the set of varieties, so $B_1 = \{x_1, x_2, x_3\}$. One possible selection of four blocks is

$$
\begin{aligned}
B_1 &= \{x_1, x_2, x_3\} \\
B_2 &= \{x_4, x_5, x_6\} \\
B_3 &= \{x_1, x_2, x_5\} \\
B_4 &= \{x_3, x_4, x_6\}.
\end{aligned}
$$

We could compare the yield for varieties x_1 and x_2 twice (using blocks B_1 and B_3).

Such an arrangement of varieties into blocks is called an *experimental design*, because they were first studied as tools for conducting experiments, such as the agricultural experiment we have outlined, or *designs* for short. In the particular case when all the blocks are the same size (like our example) the term *block design* is used.

In the 24-plot example that we mentioned first, each block is a complete set of all the varieties. This is called a *complete design*. Designs like our second example, whose blocks are proper subsets of the varieties, are called *incomplete*.

Formally, we define a *block design with parameters* v, b, k to be a way of choosing b subsets of size k (called *blocks*) from a set V of v objects called *varieties* (or *treatments*). We say the design is *based on V*. A block design is *regular* if every variety occurs in the same number of blocks. This is often a desirable property when the data is analyzed, because we have the same amount of information about each variety. If a block design is regular, it is also called a *1-design*; the number of blocks in which a treatment occurs is called the *replication number* or *frequency* of the design, and is usually denoted r.

The word *incidence* is used to describe the relationship between blocks and varieties in a design. We say block B is incident with treatment t, or treatment t is incident with block B, to mean that t is a member of B. When we study designs, we shall primarily be involved with incidence and other structural properties of a collection of subsets of the varieties. It makes no difference whether the varieties are labels representing types of experimental material (such as corn), or positive integers, or whatever. We are purely interested in the combinatorics of the situation. For this reason, the objects we are discussing are often called *combinatorial designs*.

14.1   Design Parameters

The four design parameters we have defined are not independent:

THEOREM 14.1 (First Parameter Relation)
In a regular block design,

$$bk = vr. \tag{14.1}$$

Proof. We count, in two different ways, all the ordered pairs (x, y) such that variety x belongs to block y. Since every variety belongs to r blocks, there are r ordered pairs for each variety, so the number is vr. Similarly, each block contributes k ordered pairs, so the summation yields bk. Therefore $bk = vr$. $\square$

If x and y are any two different varieties in a block design, we refer to the number of blocks that contain both x and y as the *covalency* of x and y, and write it as λ_{xy}. Many important properties of block designs are concerned with this covalency function. The one that has most frequently been studied is the property of *balance:* a *balanced incomplete block design,* or BIBD, is a regular incomplete design in which λ_{xy} is a constant, independent of the choice of x and y. This constant covalency is called the *index* of the design.

Balanced incomplete block designs were defined (essentially as a puzzle) by Woolhouse [140], in the annual *Lady's and Gentleman's Diary,* a "Collection of mathematical puzzles and aenigmas" that he edited. The Reverend T. P. Kirkman [57, 58, 59, 60] studied the case of block size 3; in particular he introduced his famous "schoolgirl problem" [59]:

> A schoolmistress has 15 girl pupils and she wishes to take them on a daily walk. The girls are to walk in five rows of three girls each. It is required that no two girls should walk in the same row more than once per week. Can this be done?

The solution is a balanced incomplete block design with $v = 15, k = 3, \lambda = 1$ with additional properties. Yates [142] first studied balanced incomplete block designs from a statistical viewpoint.

A balanced design with $\lambda = 0$, or a null design, is often called *trivial,* and so is a complete design. We shall demand that a balanced incomplete block design be not trivial; since completeness is already outlawed, the added restriction is that $\lambda > 0$.

It is usual to write λ for the index of a balanced incomplete block design. We often refer to a balanced incomplete block design by using the five parameters (v, b, r, k, λ), and call it a (v, b, r, k, λ)-*design* or (v, b, r, k, λ)-*BIBD*.

As an example, here is a $(7, 7, 4, 4, 2)$-BIBD:

$$
\begin{aligned}
B_1 &= \{x_1, x_2, x_3, x_4\} \\
B_2 &= \{x_1, x_4, x_5, x_6\} \\
B_3 &= \{x_1, x_2, x_5, x_7\} \\
B_4 &= \{x_1, x_3, x_6, x_7\} \\
B_5 &= \{x_3, x_4, x_5, x_7\} \\
B_6 &= \{x_2, x_3, x_5, x_6\} \\
B_7 &= \{x_2, x_4, x_6, x_7\}.
\end{aligned}
$$

There is a further relation between the parameters of a balanced incomplete block design:

THEOREM 14.2 (Second Parameter Relation)
In a (v, b, r, k, λ)-*BIBD,*

$$r(k - 1) = \lambda(v - 1). \tag{14.2}$$

Proof. Consider the blocks of the design that contain a given variety, x say. There are r such blocks. Because of the balance property, every variety other than x must occur in λ of them. So if we list all entries of the blocks, variety x is listed r times and we list every other treatment λ times.

The list contains rk entries. So:

$$rk = r + \lambda(v-1),$$
$$r(k-1) = \lambda(v-1). \qquad \square$$

Relations (14.1) and (14.2) allow us to find all parameters of a BIBD when only three are given. For example, if we know v, k and λ, we can calculate r from (14.2) and then deduce b using (14.1). As an example, if $v = 13$, $k = 3$ and $\lambda = 1$, we have $r = \lambda(v-1)/(k-1) = 12/2 = 6$ and $b = vr/k = 13 \times 6/3 = 26$, and a $(13, 26, 6, 3, 1)$-design exists. But not all sets $\{v, k, \lambda\}$ give a design. First, all the parameters must be integers; for example, if $v = 11$, $k = 3$ and $\lambda = 1$, we would get $r = 5$ and $b = 55/3$; you cannot have a fractional number of blocks, so this is impossible. And even when the parameters are all integers, there might be no corresponding design; for example, there is no $(22, 22, 7, 7, 2)$-design (see Section 14.3), even though both equations are satisfied. In fact, whether there exists a design corresponding to a given set of parameters is not known in general.

There is no requirement that all the blocks in a design should be different. If two blocks have the same set of elements, we say there is a "repeated block." A design that has no repeated block is called *simple*.

If a design has b blocks $B_1, B_2, \ldots, B_b$ and v varieties $t_1, t_2, \ldots, t_v$, we define a $v \times b$ matrix A with (i, j) entry a_{ij} as follows:

$$a_{ij} = \begin{cases} 1 & \text{if } t_i \in B_j; \\ 0 & \text{otherwise.} \end{cases}$$

This matrix A is called the *incidence matrix of the design*. The definition means that each block corresponds to a column of the incidence matrix, and each variety corresponds to a row.

Example 14.1: Exhibit the incidence matrix of the regular design with varieties t_1, t_2, t_3, t_4 and blocks $B_1 = \{t_1, t_2\}$, $B_2 = \{t_3, t_4\}$, $B_3 = \{t_1, t_4\}$, $B_4 = \{t_2, t_3\}$.

This design has incidence matrix $A = \begin{bmatrix} 1 & 0 & 1 & 0 \\ 1 & 0 & 0 & 1 \\ 0 & 1 & 0 & 1 \\ 0 & 1 & 1 & 0 \end{bmatrix}$. $\qquad \square$

When we define designs, the order of the varieties does not matter. Similarly, the order of the set of blocks is not important. So we do not usually distinguish between two designs if the incidence matrix of one can be obtained by reordering the rows and/or columns of the incidence matrix of the other. Two such designs are called *isomorphic*.

Suppose we interchange blocks B_2 and B_3 in the preceding example, and also reorder the treatments t_1, t_4, t_2, t_3. The new design has incidence matrix

$$B = \begin{bmatrix} 1 & 1 & 0 & 0 \\ 0 & 1 & 1 & 0 \\ 1 & 0 & 0 & 1 \\ 0 & 0 & 1 & 1 \end{bmatrix}.$$

These matrices are related by:

$$\begin{bmatrix} 1 & 1 & 0 & 0 \\ 0 & 1 & 1 & 0 \\ 1 & 0 & 0 & 1 \\ 0 & 0 & 1 & 1 \end{bmatrix} = \begin{bmatrix} 1 & 0 & 0 & 0 \\ 0 & 0 & 0 & 1 \\ 0 & 1 & 0 & 0 \\ 0 & 0 & 1 & 0 \end{bmatrix} \begin{bmatrix} 1 & 0 & 1 & 0 \\ 1 & 0 & 0 & 1 \\ 0 & 1 & 0 & 1 \\ 0 & 1 & 1 & 0 \end{bmatrix} \begin{bmatrix} 1 & 0 & 0 & 0 \\ 0 & 0 & 1 & 0 \\ 0 & 1 & 0 & 0 \\ 0 & 0 & 0 & 1 \end{bmatrix}.$$

The above equation could be written

$$B = PAQ,$$

where

$$P = \begin{bmatrix} 1 & 0 & 0 & 0 \\ 0 & 0 & 0 & 1 \\ 0 & 1 & 0 & 0 \\ 0 & 0 & 1 & 0 \end{bmatrix}, \quad Q = \begin{bmatrix} 1 & 0 & 0 & 0 \\ 0 & 0 & 1 & 0 \\ 0 & 1 & 0 & 0 \\ 0 & 0 & 0 & 1 \end{bmatrix}.$$

The permutation matrix P represents the reordering of the varieties, and could be obtained by performing the same reordering on the rows of the 4×4 identity matrix. Matrix Q represents the block reordering.

As usual, we write J_{mn} for an $m \times n$ matrix with every entry 1. (If $m = n$, we simply write J_n; and if the order is obvious from the context, we omit subscripts altogether.) Also, I_n (or simply I) is an $n \times n$ identity matrix.

THEOREM 14.3 (Incidence Matrix Equations)
If A is the incidence matrix of a balanced incomplete block design with parameters (v, b, r, k, λ), then

$$J_v A = k J_{vb} \tag{14.3}$$

and

$$AA^T = (r - \lambda)I_v + \lambda J_v. \tag{14.4}$$

Conversely, if a matrix A, with every entry 0 or 1, satisfies (14.3) and (14.4) with $k < v$ and $\lambda > 0$, then A is the incidence matrix of a (v, b, r, k, λ)-BIBD and

$$v = \frac{r(k-1)}{\lambda} + 1, \quad b = \frac{vr}{k}. \tag{14.5}$$

Proof. First, assume A is the incidence matrix of a (v, b, r, k, λ)-BIBD.

For any i, the (i, j) entry of $J_v A$ counts the number of 1s in column j of A, which equals the number of varieties in block j. So it equals k. Since $J_v A$ is the product of a $v \times v$ and a $v \times b$ matrix, it is of size $v \times b$, so (14.3) holds.

The (i, i) entry of AA^T is

$$\sum_{p=1}^{b} a_{ip} a_{ip} = \sum_{p=1}^{b} a_{ip}^2;$$

a_{ip}^2 is 1 when variety i belongs to block p, which happens for r values of p (variety i is in r blocks), and 0 otherwise. So the sum is r. When $i \neq j$, the (i, j) entry of AA^T is

$$\sum_{p=1}^{b} a_{ip} a_{jp};$$

$a_{ip} a_{jp}$ is 1 when varieties i and j both belong to block p, which happens λ times for given i and j, and 0 otherwise. So the sum is λ. Therefore AA^T has diagonal entries r and off-diagonal entries λ; that is,

$$AA^T = rI_v + \lambda(I_v - J_v) = (r - \lambda)I_v + \lambda J_v.$$

Conversely, suppose A is a $v \times b$ $(0, 1)$-matrix that satisfies the equations. We define a block design with varieties $x_1, x_2, \ldots, x_v$ and blocks $Y_1, Y_2, \ldots, Y_b$ by the rule

$$x_i \in Y_j \text{ if and only if } a_{ij} = 1.$$

Equation (14.3) implies that every block has size k. From (14.4) the design has constant replication number r (because of the diagonal entries of AA^T) and constant covalency λ (from the off-diagonal elements). So the design is a (v, b, r, k, λ)-BIBD with incidence matrix A. The equalities (14.5) now follow from (14.1) and (14.2). $\qquad\square$

Another useful equation for balanced incomplete block designs,

$$AJ_v = rJ_{vb},\tag{14.6}$$

follows from the fact that every variety belongs to r blocks, so each row of A contains r 1's.

14.2   Fisher's Inequality

In this section we need to know the determinant of the matrix AA^T.

LEMMA 14.4
The determinant of

$$M = \begin{pmatrix} r & \lambda & \lambda & \cdots & \lambda \\ \lambda & r & \lambda & \cdots & \lambda \\ \lambda & \lambda & r & \cdots & \lambda \\ & \cdots & & & \cdots \\ \lambda & \lambda & \lambda & \cdots & r \end{pmatrix}$$

is $(r-1)^{v-1}[r + \lambda(v-1)]$.

Proof. If we subtract the first column of M from every other column we obtain

$$\begin{pmatrix} r & \lambda - r & \lambda - r & \cdots & \lambda - r \\ \lambda & r - \lambda & 0 & \cdots & 0 \\ \lambda & 0 & r - \lambda & \cdots & 0 \\ & \cdots & & & \cdots \\ \lambda & 0 & 0 & \cdots & r - \lambda \end{pmatrix}.$$

Now add row 2 to row 1, then add row 3 to row 1, and so on. The resulting matrix,

$$\begin{pmatrix} r + (v-1)\lambda & 0 & 0 & \cdots & 0 \\ \lambda & r - \lambda & 0 & \cdots & 0 \\ \lambda & 0 & r - \lambda & \cdots & 0 \\ & \cdots & & & \cdots \\ \lambda & 0 & 0 & \cdots & r - \lambda \end{pmatrix}.$$

This matrix has the same determinant as M; and, as all entries above its diagonal are zero, the determinant is the product of the diagonal elements, $(r-\lambda)^{v-1}[r + \lambda(v-1)]$. □

THEOREM 14.5 (Fisher's Inequality) [35]
In any balanced incomplete block design, $b \geq v$.

Proof. Suppose A is the incidence matrix of a (v, b, r, k, λ)-BIBD. Then $\det(AA^T) = (r-1)^{v-1}[r + \lambda(v-1)]$. From (14.2) we can replace $\lambda(v-1)$ by $r(k-1)$, so

$$\det(AA^T) = (r-1)^{v-1}[r + r(k-1)] = (r-1)^{v-1}rk.$$

On the other hand, (14.2) can be rewritten as

$$r = \frac{v-1}{k-1}\lambda;$$

by incompleteness, $k < v$, so $r > \lambda$, and $(r - \lambda)$ is non-zero. Therefore $\det(AA^T)$ is non-zero, so the $v \times v$ matrix AA^T has rank v. That is

$$v = \text{rank}(AA^T) \leq \text{rank}(A) \leq \min\{v, b\},$$

because A is $v \times b$. Therefore $v \leq b$. $\square$

14.3   Symmetric Balanced Incomplete Block Designs

Designs that only just satisfy Fisher's inequality, those with $b = v$, are called *symmetric*. From (14.1), such a design also satisfies $r = k$, so only three parameters need to be specified; the common phrase is "(v, k, λ)-design." Another abbreviation is SBIBD. For these designs, Equation (14.2) takes on the simpler form

$$\lambda(v - 1) = k(k - 1). \tag{14.7}$$

In the following theorem, the matrices are all $v \times v$, so we simply write I and J instead of I_v and J_v.

THEOREM 14.6 (Incidence Matrix Equations, Symmetric Designs)

If A is the incidence matrix of a (v, k, λ)-design, then

$$\lambda J A^T = kJ \tag{14.8}$$

and

$$\lambda A^T A = (k - \lambda)I + \lambda J. \tag{14.9}$$

So A^T is also the incidence matrix of a (v, k, λ)-design.

Proof. For a symmetric design, (14.3) and (14.4) become

$$JA = kJ, \tag{14.10}$$

$$AJ = kJ, \tag{14.11}$$

and

$$AA^T = (k - \lambda)I + \lambda J. \tag{14.12}$$

$J^T = J$, so (14.8) is just the transpose of (14.11).

Since A is square, $\det(A)$ and $\det(A^T)$ are defined and are equal; $\det(A) = \sqrt{\det(AA^T)}$ is non-zero. So A has an inverse. Multiplying (14.11) on the left by A^{-1}, $A^{-1}AJ = kA^{-1}J$, so $A^{-1}J = k^{-1}J$. Now multiply Equation (14.12) on the left by A^{-1} and on the right by A:

$$
\begin{aligned}
A^{-1}AA^TA &= (k-\lambda)A^{-1}A + \lambda A^{-1}JA \\
&= (k-\lambda)I + \lambda A^{-1}kJ \\
&= (k-\lambda)I + \lambda k A^{-1}J \\
&= (k-\lambda)I + \lambda k k^{-1}J \\
&= (k-\lambda)I + \lambda J
\end{aligned}
$$

as required. It follows from Theorem 14.3 that A^T is the incidence matrix of a (v, k, λ)-design. $\qquad\square$

COROLLARY 14.6.1
In a symmetric balanced incomplete block design, any two blocks intersect in λ blocks.

A block design whose blocks have constant-size intersection is called *linked*. So the corollary tells us that a symmetric balanced incomplete block design is also a linked design.

In general, if a design $\mathcal{D}$ has incidence matrix A, the design with incidence matrix A^T is called the *dual* of $\mathcal{D}$, and is often denoted $\mathcal{D}^*$. The dual of any balanced incomplete block design is linked, but the only incomplete block designs that are both symmetric and linked are SBIBDs.

The following important theorem is called the *Bruck–Chowla–Ryser Theorem*.

THEOREM 14.7 (Bruck–Chowla–Ryser Theorem) [15, 21, 99]
If there exists a symmetric balanced incomplete block design with parameters (v, k, λ), then:

(i) *if v is even, $k - \lambda$ must be a perfect square;*

(ii) *if v is odd, there must exist integers x, y and z, not all zero, such that*

$$
x^2 = (k-\lambda)y^2 + (-1)^{(v-1)/2}\lambda z^2. \tag{14.13}
$$

Proof. Part (i) only: Suppose v is even, and suppose there exists a (v, k, λ)-design with incidence matrix A. Since A is square it has a determinant, denoted $\det(A)$, where $\det(A^T) = \det(A)$, so

$$
\det(A) = \sqrt{\det(AA^T)}.
$$

But from Lemma 14.4,

$$
\det(AA^T) = k^2(k-\lambda)^{v-1},
$$

so

$$\det(A) = \pm k(k - \lambda)^{\frac{1}{2}(v-1)}.$$

Now A is an integer matrix, so $\det(A)$ is an integer. Since k and λ are integers, and v is even, it follows that $\sqrt{k - \lambda}$ is an integer also.

The proof of part (ii) requires more mathematical background than we are assuming. See, for example, Chapter 7 of [123]. □

For example, consider the case $k = 7, \lambda = 2$. From (14.7), v must equal 22; but this is even, so no $(22, 7, 2)$-design exists.

Some applications require some harder results from number theory, but we shall give one example where all you need to know is that 2 is not a square in the arithmetic modulo 3.

Example 14.2: Show that no $(141, 21, 3)$-design exists.

Suppose a $(141, 21, 3)$-design exists. Then there are integers x, y and z, not all zero, such that

$$x^2 = 18y^2 + 3z^2.$$

Without loss of generality we can assume that x, y, z have no mutual common factor. Clearly, 3 must divide x^2, so 3 divides x: say $x = 3e$. Then

$$9e^2 = 18y^2 + 3z^2$$

and z must also be divisible by 3; say $z = 3f$. We have

$$
\begin{aligned}
9e^2 &= 18y^2 + 27f^2, \\
e^2 &= 2y^2 + 3f^2.
\end{aligned}
$$

Reducing modulo 3, we find

$$e^2 \equiv 2y^2 \; (\mathrm{mod} \; 3).$$

If y is not divisible by 3, $2y^2$ is congruent to 2, which is not a perfect square in the arithmetic modulo 3, a contradiction. So 3 divides y, and x, y and z have common factor 3, another contradiction. Therefore no $(141, 21, 3)$-design can exist. □

14.4 New Designs from Old

If B_0 is any block of a balanced incomplete block design with index λ, then any two varieties that do not belong to B_0 must occur together in λ of the remaining blocks, while any two members of B_0 must be together in $\lambda - 1$ of the remaining blocks. It follows that the blocks $B \backslash B_0$ form a balanced design of index λ when B ranges through the remaining blocks, and the blocks $B \cap B_0$ form a balanced design of index $\lambda - 1$. We shall refer to these as the *residual* and *derived* designs of the original design with respect to the block B_0.

In general the replication number of a residual or derived design is not constant, and the block sizes follow no particular pattern. If each variety in the original design occurs r times, its residual and derived designs also have constant replication numbers, r and $r - 1$ respectively.

The block pattern can be predicted if the original design is linked, that is the size of $B \cap B_0$ is constant; the blocks of the derived design will be of that constant size, and the blocks of the residual design will be $|B \cap B_0|$ smaller than the blocks of the original. In particular, if we start with a symmetric balanced incomplete block design, we obtain constant replication numbers and block sizes.

THEOREM 14.8 (Derived and Residual Design Parameters)
The residual design of a (v, k, λ)-SBIBD is a balanced incomplete block design with parameters

$$(v - k, v - 1, k, k - \lambda, \lambda), \qquad (14.14)$$

provided $\lambda \neq k - 1$. The derived design of a (v, k, λ)-SBIBD is a balanced incomplete block design with parameters

$$(k, v - 1, k - 1, \lambda, \lambda - 1), \qquad (14.15)$$

provided $\lambda \neq 1$.

The two exceptions are made so as to exclude "trivial" designs.

Example 14.3: Exhibit derived and residual designs of a $(11, 6, 3)$-design.

Figure 14.1 shows the designs. The first column shows the blocks of a $(11, 6, 3)$-design $\mathcal{B}$. The other columns show the blocks of the residual design $\mathcal{R}$ and the derived design $\mathcal{D}$ with respect to the first block (block $\{1, 2, 3, 4, 5, 6\}$). Designs $\mathcal{R}$ and $\mathcal{D}$ have parameters $(5, 10, 6, 3, 3)$ and $(6, 10, 5, 3, 2)$, respectively. $\qquad\square$

Suppose a balanced incomplete block design $\mathcal{D}$ has parameters $(v - k, v - 1, k, k - \lambda, \lambda)$ for some integers v, k and λ. One could ask whether there is a (v, k, λ)-design for which $\mathcal{D}$ is the residual. The answer is "not necessarily," but in most cases there must be such a symmetric design. The exact situation is given by the following theorem, which is proven in various texts (for example, [50]).

THEOREM 14.9 (Existence of Residual Designs) [51, 103, 104]
Suppose $\mathcal{D}$ is a balanced incomplete block design with parameters (14.14).

$\mathcal{B}$						$\mathcal{R}$			$\mathcal{D}$		
1	2	3	4	5	6						
2	5	6	7	10	11	7	10	11	2	5	6
1	4	6	7	8	10	7	8	10	1	4	6
2	4	5	7	8	9	7	8	9	2	4	5
3	5	6	8	9	10	8	9	10	3	5	6
3	4	6	7	9	11	7	9	11	3	4	6
1	3	5	7	8	11	7	8	11	1	3	5
1	2	6	8	9	11	8	9	11	1	2	6
1	2	3	7	9	10	7	9	10	1	2	3
2	3	4	8	10	11	8	10	11	2	3	4
1	4	5	9	10	11	9	10	11	1	4	5

FIGURE 14.1: Residual and derived designs.

(i) *If* $\lambda = 1$ *or* $\lambda = 2$, *there is a* (v, k, λ)-*design of which* $\mathcal{D}$ *is the residual.*

(ii) *There is a number* $f(\lambda)$, *depending only on* λ, *such that if* $k \geq f(\lambda)$, *there is a* (v, k, λ)-*design of which* $\mathcal{D}$ *is the residual.*

A design with parameters (14.14) is called *quasi-residual*; thus Theorem 14.9(ii) says that "all sufficiently large quasi-residual designs are residual." For example, $f(3)$ is at most 90.

The first example of a quasi-residual design that is not residual was given by Bhattacharya [9] in 1947. It had parameters $(16, 24, 9, 6, 3)$.

14.5   Difference Methods

Suppose you decide a $(7, 3, 1)$-design is what you need for an experiment, so you try to construct one. If you took a very simple-minded approach, you would observe that there are $\binom{7}{3} = 35$ possible blocks, so there are $\binom{7}{3}^7$, more than 64 billion, ways to select seven blocks. With a little thought, you will realize that (up to isomorphism) you can choose 123 to be a block; moreover, the design will be simple, so the remaining six blocks can be chosen in at most $\binom{34}{6}$, or 12.7 million, ways. This number can be reduced significantly by using more and more facts about block designs, and finally you will find that there is only one solution, up to isomorphism. But the point is that there are a large number of possibilities to be eliminated. It comes as no surprise that the computer searches involved in looking for block designs can be very long.

In view of this, it is very useful to have techniques for constructing block designs quickly, even if success is not guaranteed. In this section we look at a method that has produced a large number of BIBDs.

Consider the following blocks based on varieties $\{0, 1, 2, 3, 4, 5, 6\}$, the integers modulo 7. The blocks are $124, 235, 346, 450, 561, 602, 013$. The second block is formed by adding 1 to every member of the first, the third by adding 1 to every member of the second, and so on. The defining properties of a block design can be deduced from this additive structure: clearly every block has the same size, and every variety appears equally often—once in the first position in the blocks, once in the second, and so on.

There are seven pairs of varieties whose labels differ by ± 1, namely 01, 12, 23, 34, 45, 56, 60. The first block contains one of these pairs (12), and the additive property means that each of the others appears exactly once. Similarly, there is one pair with each of the other two possible differences, ± 2 and ± 3 (pairs 24 and 14, respectively). This is enough to prove balance. The properties of a balanced incomplete block design can all be deduced from the choice of the first block and the additive structure.

The important property we used was that the first block contained exactly one pair of entries with difference ± 1, one with difference ± 2 and one with difference ± 3. If every difference had occurred twice, we could have generated a design with $\lambda = 2$, and so on. This leads us to the following definition.

Definition: A (v, k, λ)-*difference set* B is a k-element set of integers modulo v with the property that given any integer $d, 0 < d < v$, there are precisely λ ordered pairs of elements of B whose difference is d.

The definition is written in terms of ordered pairs. When we looked at the introductory example, we talked about unordered pairs and differences like ± 1. Using ordered pairs gives the same result (after all, if we say elements x and y have difference $\pm d$, it is the same as saying the two ordered pairs (x, y) and (y, x) have differences d and $-d$), but avoids any confusion in the case where v is odd, and the specific difference $\frac{1}{2}v$ is being discussed.

Given a set B, we write $B + i$ for the set obtained by adding i to every element of B. Then we have the following theorem.

THEOREM 14.10 (Difference Sets and Designs)

If B is a (v, k, λ)-difference set then the sets $B + i$, for $0 \leq i < v$, are the blocks of a (v, k, λ)-design.

For example, in the case $v = 7$, we saw that $\{1, 2, 4\}$ is a $(7, 3, 1)$-difference set, and the corresponding design is

$$124 \quad 235 \quad 346 \quad 450 \quad 561 \quad 602 \quad 013.$$

Notice that instead of $\{1, 2, 4\}$ we could have used $\{0, 1, 3\}$ or $\{2, 3, 5\}$ or any of the seven blocks. (These other blocks are called *shifts* of the original block.)

v	k	λ	Difference set
15	7	3	$1, 2, 3, 5, 6, 9, 11$
21	5	1	$1, 2, 7, 9, 19$
73	9	1	$1, 2, 4, 8, 16, 32, 37, 55, 64$
37	9	2	$1, 7, 9, 10, 12, 16, 26, 33, 34$

FIGURE 14.2: Some difference sets.

COROLLARY 14.10.1

If there is a (v, k, λ)-difference set, then $\lambda(v - 1) = k(k - 1)$.

Example 14.4: Construct a $(7, 4, 2)$-difference set D.

There will be exactly two pairs of elements whose difference is 1; without loss of generality we might as well take one of them as 1 and 2 (if we took values t and $t+1$, we would obtain a shift of a block containing 0 and 1). So the set is $\{1, 2, x, y\}$ for some x and y. Now we need another pair with difference 1; the possibilities are that D contains $\{0, 1\}$ ($D = \{0, 1, 2, x\}$ for some x), D contains $\{2, 3\}$ ($D = \{1, 2, 3, x\}$), or $y = x + 1$ ($D = \{1, 2, x, x + 1\}$). In the first case $x \neq 3$ or 6, because these both give a third pair with difference 1. So the possibilities are $\{0, 1, 2, 4\}$ and $\{0, 1, 2, 5\}$. To test the first one, we see that the differences of the elements are

$$\{\pm(1 - 0), \pm(2 - 0), \pm(4 - 0), \pm(2 - 1), \pm(4 - 1), \pm(4 - 2)\}$$
$$= \{\pm 1, \pm 2, \pm 4, \pm 1, \pm 3, \pm 2\},$$

so it is a difference set (remember that $\pm 4 = \pm 3$ modulo 7). Similarly, the second set works. The second two cases give solutions $\{1, 2, 3, 5\}$ and $\{1, 2, 3, 6\}$, which are shifts of the first two. In the third case, the possibilities are $\{1, 2, 4, 5\}$ and $\{1, 2, 5, 6\}$. Set $\{1, 2, 4, 5\}$ yields differences

$$\{\pm(2 - 1), \pm(4 - 1), \pm(5 - 1), \pm(4 - 2), \pm(5 - 2), \pm(5 - 4)\}$$
$$= \{\pm 1, \pm 3, \pm 4, \pm 2, \pm 3, \pm 1\},$$

and there is only one difference ± 2, so this is not a difference set. The other example does not work either. So there are 14 candidates for D—the two examples that work and their shifts. □

Figure 14.2 shows some further examples of difference sets.

The process of starting with one block, adding 1 to each member to form a second block, then continuing, is called *developing* the block. We are only dealing with cases where the elements are integers modulo v (called *developing mod v*), but the same process can be carried out in other arithmetic structures, such as groups. It is also possible to develop more than one block. For example, consider the blocks $\{1, 2, 3, 5\}$ and $\{1, 2, 5, 7\}$ modulo 9. The first block contains differences ± 1 twice, ± 2 twice, ± 3 once and ± 4 once. So developing that block yields nine blocks in which every pair with difference ± 1—pairs 01, 12, 23, ..., 78, 80—occur twice. Block $\{1, 2, 5, 7\}$ contains differences ± 1 once, ± 2 once, ± 3 twice and ± 4 twice; pairs 01, and so on, will appear once in the nine blocks developed from it. Similarly, every other pair will appear exactly three times in the 18 blocks obtained by developing both initial blocks. A quick check shows that the blocks together form a $(9, 18, 8, 4, 3)$-BIBD. Blocks like this are called *supplementary difference sets*. Further examples are given in Figure 14.3.

v	b	r	k	λ	Initial blocks
41	82	10	5	1	$\{1, 10, 16, 18, 37\}; \{5, 8, 9, 21, 39\}$
13	26	12	6	5	$\{1, 2, 4, 7, 8, 12\}; \{1, 2, 3, 4, 8, 12\}$
16	80	15	3	2	$\{1, 2, 4\}; \{1, 2, 8\}; \{1, 3, 13\}; \{1, 4, 9\}; \{1, 5, 10\}$
22	44	14	7	4	$\{1, 7, 12, 16, 19, 21, 22\}; \{1, 6, 8, 9, 10, 14, 20\}$

FIGURE 14.3: Some supplementary difference sets.

Exercises 14A

1. You need to construct a block design with five blocks of size 2, based on variety-set $\{1, 2, 3\}$, in which every possible pair of varieties occurs at least once. Prove that there are exactly two non-isomorphic ways to do this. Give examples.

2. Suppose v is any positive integer greater than 2. Prove that a $(v, v, v - 1, v - 1, v - 2)$-design exists.

3. Prove that the following blocks form a $(6, 10, 5, 3, 2)$-design:

$$123, \quad 124, \quad 135, \quad 146, \quad 156,$$
$$236, \quad 245, \quad 256, \quad 345, \quad 346.$$

4. Write down the blocks of a $(7, 14, 6, 3, 2)$-design.

5. Prove that, in any balanced incomplete block design, $\lambda < r$.

6. Suppose S is a set with v elements, $v > 2$. Write $2(S)$ for the set of all unordered pairs of elements of S. Show that if we take the members of S as varieties and the members of $2(S)$ as blocks, the result is a balanced incomplete block design. What are its parameters?

7. In each row of the following table, fill in the blanks so that the parameters are possible parameters for a balanced incomplete block design, or else show that this is impossible.

v	b	r	k	λ
	35		3	1
21	28		3	
11		5	5	
14	7	4		
12		6	3	
49			7	1

8. Show that there do not exist symmetric balanced incomplete block designs with the following parameters: $(46, 10, 2)$, $(52, 18, 6)$.

9. Suppose $\mathcal{D}$ is a balanced incomplete block design with parameters (v, b, r, k, λ). The design $t\mathcal{D}$ is defined to have tb blocks: each block B of $\mathcal{D}$ gives rise to precisely t blocks, each with the same set of varieties as B. Prove that $t\mathcal{D}$ is a balanced incomplete block design. What are its parameters? (The design $t\mathcal{D}$ is called the *t-multiple* of $\mathcal{D}$.)

10. The balanced incomplete block design $\mathcal{D}$ has parameters $(6, 10, 5, 3, 2)$ and blocks

$$
\begin{array}{rclrcl}
B_0 &=& \{0,1,2\} & B_5 &=& \{1,2,5\} \\
B_1 &=& \{0,1,3\} & B_6 &=& \{1,3,4\} \\
B_2 &=& \{0,2,4\} & B_7 &=& \{1,4,5\} \\
B_3 &=& \{0,3,5\} & B_8 &=& \{2,3,4\} \\
B_4 &=& \{0,4,5\} & B_9 &=& \{2,3,5\}.
\end{array}
$$

Construct the blocks of the dual of $\mathcal{D}$, and verify that it is not balanced.

11. For the following parameter sets, what are the residual and derived parameters?
  (i) $(23, 11, 5)$
  (ii) $(15, 7, 3)$
  (iii) $(45, 12, 3)$
  (iv) $(13, 9, 6)$
  (v) $(69, 17, 4)$

12. Find an $(11, 5, 2)$-difference set.

13. Find two initial blocks whose development modulo 13 yields a $(13, 26, 6, 3, 1)$-design.

Exercises 14B

1. A combinatorial design has six varieties, $\{1, 2, 3, 4, 5, 6\}$, and nine blocks of size 2. Every variety occurs in three blocks, and the design is simple (no two blocks are identical). Prove that there are exactly two non-isomorphic ways to do this.

2. Prove that any two $(4, 4, 3, 3, 2)$-designs must be isomorphic.

3. In each row of the following table, fill in the blanks so that the parameters are possible parameters for a balanced incomplete block design, or else show that this is impossible.

v	b	r	k	λ
7	14		3	
		6	4	2
		13	6	1
17		8	5	
	30		7	
33	44		6	

4. Show that, in a symmetric balanced incomplete block design, λ cannot be odd when v is even.

5. Suppose $\mathcal{D}$ is a (v, b, r, k, λ)-BIBD with variety-set V. Write T for the set of all subsets of V of the form $V \backslash B$, where B is a block of $\mathcal{D}$. Prove that the members of T, as blocks, form a balanced incomplete block design based on V, unless $b - 2r + \lambda$ is zero. (This design is called the *complement* of $\mathcal{D}$.) What are its parameters?

6. The complement of a design was defined in the previous exercise. What are the parameters of the complement of a design with the parameters shown?

   (i) $(7, 7, 3, 3, 1)$ (iii) $(16, 24, 9, 6, 3)$

   (ii) $(13, 26, 6, 3, 1)$ (iv) $(12, 44, 33, 9, 24)$

7. Show that a balanced incomplete block design for which $b - 2r + \lambda$ is zero must have parameters $(v, v, v - 1, v - 1, v - 2)$, provided $r \leq v$.

8. Prove that the dual of any balanced incomplete block design is a linked design.

9. For the following parameter sets, what are the residual and derived parameters?

   (i) $(16, 6, 2)$

   (ii) $(22, 7, 2)$

   (iii) $(15, 8, 4)$

   (iv) $(31, 10, 3)$

   (v) $(41, 16, 6)$

10. Find a $(13, 4, 1)$-difference set.

11. Verify that the examples given in Figure 14.2 are, in fact, difference sets with the stated parameters.

12. Suppose D is a (v, k, λ)-difference set. Write E for the set of all integers modulo v that *do not* belong to D. Prove that E is a $(v, v-k, v-2k+\lambda)$-difference set.

13. Verify that $\{0, 1, 2, 5\}$ is a $(7, 4, 2)$-difference set, but $\{1, 2, 5, 6\}$ is not.

Problems

Problem 1: Prove that any two $(7, 7, 4, 4, 2)$-designs must be isomorphic.

Problem 2: Can you find a simple $(7, 14, 6, 3, 2)$-design, that is, one with no repeated blocks? (See Exercise 12A.4.)

Problem 3: Suppose there were a symmetric balanced incomplete block design with $\lambda = 3$ and $k = 21 + 27t$ for some positive integer t.

   (i) According to (14.1) and (14.2), what is the corresponding value of v?

   (ii) Prove that no such design exists.

Problem 4: Suppose $\mathcal{B}$ is a balanced incomplete block design with $v < b$ (that is, $\mathcal{B}$ is *not* symmetric). Show that $\mathcal{B}$ is not a linked design.

Problem 5: (Generalization of Exercise 12A.6.) Suppose v and k are positive integers, with $v > k \geq 2$, and S is a set with v elements. Write $k(S)$ for the set of all unordered k-sets of elements of S. Show that if we take the members of S as varieties and the members of $k(S)$ as blocks, the result is a balanced incomplete block design. What are its parameters?

Problem 6: See the definition of the *complement* of a design in Exercise 14B.5. Suppose $\mathcal{D}$ is a symmetric balanced incomplete block design with support set S, and suppose $\mathcal{E}$ is its complement. Select any block B of $\mathcal{D}$.

Prove that the residual design of $\mathcal{E}$ with respect to $S \backslash B$ is the complement of the derived design of $\mathcal{D}$ with respect to B. Select any block B of $\mathcal{D}$. Prove that the residual design of $\mathcal{E}$ with respect to $S \backslash B$ is the complement of the derived design of $\mathcal{D}$ with respect to B.

Problem 7: $\mathcal{D}$ is the $(15, 15, 7, 7, 3)$-design with blocks

$$
\begin{array}{llll}
B_0 & = & \{0,1,2,3,4,5,6\} & B_1 & = & \{0,3,4,9,10,13,14\} \\
B_2 & = & \{0,1,2,7,8,9,10\} & B_3 & = & \{0,5,6,7,8,13,14\} \\
B_4 & = & \{0,1,2,11,12,13,14\} & B_5 & = & \{0,5,6,9,10,11,12\} \\
B_6 & = & \{0,3,4,7,8,11,12\} & B_7 & = & \{1,3,5,7,9,11,13\} \\
B_8 & = & \{1,3,6,7,10,12,14\} & B_9 & = & \{2,3,6,8,9,11,14\} \\
B_{10} & = & \{1,4,5,8,10,11,14\} & B_{11} & = & \{2,4,5,7,9,12,14\} \\
B_{12} & = & \{1,4,6,8,9,12,13\} & B_{13} & = & \{2,4,6,7,10,11,13\} \\
B_{14} & = & \{2,3,5,8,10,12,13\}
\end{array}
$$

(i) Write down the blocks of the dual design $\mathcal{D}^*$ of $\mathcal{D}$.

(ii) A *triplet* is a set of three blocks whose mutual intersection has size 3. (For example, $\{B_0, B_1, B_6\}$ is a triplet in D.) By considering triplets, prove that $\mathcal{D}$ and $\mathcal{D}^*$ are not isomorphic.

Problem 8: Consider an SBIBD with parameters $(n^2 + n + 1, n + 1, 1)$. (This is called a *finite projective plane*.)

(i) What are the parameters of the complement of this design (as defined in Exercise 12A.5)?

(ii) What are the parameters of the residual of this complementary design?

(iii) Prove that a design with these residual parameters always exists.

Problem 9: A *finite binary plane* consists of a finite, non-empty set of *points* (the varieties) and a set of *lines*. Each line is a subset of the points, and two lines are different if and only if they have different point-sets. Moreover, the following axioms are satisfied:

  B1 any two lines have precisely one common point;

  B2 any point lies in precisely two lines.

(i) Show that every line must contain the same number of points;

(ii) Construct all finite binary planes.

Problem 10: Prove that no (v, k, λ)-difference set satisfies $v = 2k$.

Problem 11: Suppose D is a (v, k, λ)-difference set and m is a positive integer such that m and v are coprime (that is, they have no common factor greater than 1). By mD we mean the $\{mx : x \in D\}$, where all numbers are treated as integers modulo v. Prove that mD is a (v, k, λ)-difference set. Is this true if m and v are not coprime?

Problem 12: Fourteen riders take part in a motorcycle sprint championship. The qualifying consists of fourteen heats, of four riders each. Each competitor races in four heats, and no two meet in more than one heat.

  (i) Find a schedule for the competition.

 (ii) Is there a schedule where the qualifying is divided into seven rounds of two heats each, and no rider appears in both heats of a round?

Chapter 15

Linear Algebra Methods in Combinatorics

Every mathematical discipline is enriched by and enriches other mathematical disciplines. The reader may be familiar with such areas as analytic number theory, for instance, in which the techniques of complex analysis are applied to questions from number theory. Combinatorics, as we have seen in earlier chapters, uses calculus and group theory (among many other disciplines) for generating functions and Pólya counting. In this chapter, we take a deeper look at some topics within combinatorics that rely especially upon results and ideas from linear algebra.

15.1 Recurrences Revisited

We saw in Chapter 6 that there are two ways (generating functions and the characteristic equation) to solve a linear homogeneous recurrence relation. A third way involving linear algebra provides a different perspective. We illustrate with an example.

Example 15.1: Let A denote the matrix $\begin{bmatrix} 0 & 1 \\ 1 & 1 \end{bmatrix}$. Find $A^n \begin{bmatrix} 1 \\ 1 \end{bmatrix}$.

Let the row matrix $[x_n, y_n]$ be the result of the matrix multiplication for a given n. Then because matrix multiplication is associative, we have

$$[x_{n+1}, y_{n+1}] = A \times A^n \times [x, y]^T = A \times [x_n, y_n]^T = [y_n, x_n + y_n].$$

Given $[x_0, y_0] = [1, 1] = [F_1, F_2]$ (the first two Fibonacci numbers, as outlined in Chapter 6), it follows that $[x_1, y_1] = [1, 2] = [F_2, F_3]$, so we see $[x_n, y_n] = [F_{n+1}, F_{n+2}]$. □

This gives us a faster method of computing a given Fibonacci number than using the recurrence. To find F_{100}, we would take the matrix A in the example and square it, then square the square to get A^4, and continue to A^{64}. Then calculate $A^{64} \times A^{32} \times A^2 \times [1, 1]^T = [x_{97}, x_{98}] = [F_{99}, F_{100}]$. This would be nine matrix multiplications, as opposed to 98 iterations of the recurrence.

Considering this process, we see that the matrix A^n consists only of Fibonacci numbers. In fact, $A^n = \begin{bmatrix} F_{n-1} & F_n \\ F_n & F_{n+1} \end{bmatrix}$. We use this observation to obtain a Fibonacci identity. Since $A^k \times A^m = A^{k+m}$, we look at the $(2, 2)$-entry in the product to see $F_k F_m + F_{k+1} F_{m+1} = F_{k+m+1}$. In this way, the matrix identity allows us to see relationships among the Fibonacci numbers that would be more tedious to prove without matrix methods. Other Fibonacci identities are given easy matrix proofs in [106], and in the problems.

Other two-term recurrences may be treated similarly. (In fact, the difference between a "method" and a "trick" in mathematics is often said to be that a trick only works in one case, where a method works in many; the matrix technique with this convention is a "method," because it works for many recurrences.) Suppose we look at the recurrence $a_{n+1} = ca_n + da_{n-1}$, with specified initial values a_1, a_2. We seek a matrix A for which $A^n[a_1, a_2]^T = [a_{n+1}, a_{n+2}]$. It is not hard to see that the first row will be $[0, 1]$ and the second row (where the recurrence "lives," so to speak) will be $[d, c]$ (note the reversal of the coefficients!).

Although we have seen two ways in Chapter 6 to find an explicit formula for a_n in a two-term linear homogeneous recurrence, we will explore a third way that uses diagonalization. Recall that some square matrices have the property that there exists a matrix P where $P^{-1}AP = D$ is a diagonal matrix (all entries a_{ij} are 0 unless $i = j$). If A has this property, we say A is *diagonalizable*, and D's non-zero entries are the eigenvalues of A and the columns of P are the corresponding eigenvectors.

Example 15.2: Solve the recurrence $a_{n+1} = a_n + 6a_{n-1}$ with initial conditions $a_0 = 2$, $a_1 = 1$.

We want eigenvalues and eigenvectors of the matrix $A = \begin{bmatrix} 0 & 1 \\ 6 & 1 \end{bmatrix}$. The characteristic equation $\det(\lambda I - A)$ is $\lambda(\lambda - 1) - 6 = \lambda^2 - \lambda - 6$. The roots are $\lambda_1 = 3$ and $\lambda_2 = -2$. (The characteristic equation of the matrix is the same as the characteristic equation of the recurrence that we discussed in Chapter 6.) The eigenvector corresponding to λ_1 is the solution to $A\vec{v} = 3\vec{v}$. We write $\vec{v} = [v_1, v_2]^T$. There are infinitely many eigenvectors, so we arbitrarily let $v_1 = 1$; then $A\vec{v} = 3\vec{v}$ gives us $v_2 = 3v_1$, or $\vec{v} = [1, 3]^T$.

Similarly, the eigenvector $[1, -2]$ has eigenvalue -2. We conclude that

$$P = \begin{bmatrix} 1 & 1 \\ -2 & 3 \end{bmatrix}; P^{-1} = \begin{bmatrix} .6 & -.2 \\ .4 & .2 \end{bmatrix}; D = P^{-1}AP = \begin{bmatrix} -2 & 0 \\ 0 & 3 \end{bmatrix}.$$

To solve the recurrence, we use the fact that a diagonal matrix may be easily raised to any power.

$$[a_n, a_{n+1}] = A^n \begin{bmatrix} a_0 \\ a_1 \end{bmatrix} = (PDP^{-1})^n \begin{bmatrix} 2 \\ 1 \end{bmatrix} = PD^n P^{-1} \begin{bmatrix} 2 \\ 1 \end{bmatrix}$$

The resulting matrix multiplication yields

$$\begin{bmatrix} 1 & 1 \\ -2 & 3 \end{bmatrix} \begin{bmatrix} (-2)^n & 0 \\ 0 & 3^n \end{bmatrix} \begin{bmatrix} .6 & -.2 \\ .4 & .2 \end{bmatrix} \begin{bmatrix} 2 \\ 1 \end{bmatrix} = \begin{bmatrix} 3^n + (-2)^n \\ 3^{n+1} + (-2)^{n+1} \end{bmatrix}.$$

Thus, we arrive at the solution $a_n = 3^n + (-2)^n$. □

15.2   State Graphs and the Transfer Matrix Method

Occasionally we look at systems that can be in any of a finite number of states, and can under certain circumstances change from one state to another. For a simple example, consider an automobile with a standard transmission. It can be in several states (gears); first, second, third, fourth, stopped or reverse. (We omit neutral for the sake of simplicity.)

We cannot shift gears entirely at will; for instance, it would be a bad idea to shift directly from fourth gear to reverse. For simplicity, we agree that we may shift a car from a particular gear to another that is only one higher or lower.

We will use a graph to designate the possible changes of state. Our graph will have a vertex for each legal state and an edge between two vertices indicates that we may safely shift between the corresponding states. This graph is shown in Figure 15.1.

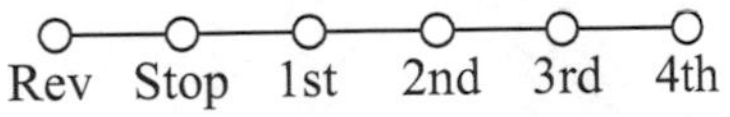

$$A = \begin{bmatrix} 0 & 1 & 0 & 0 & 0 & 0 \\ 1 & 0 & 1 & 0 & 0 & 0 \\ 0 & 1 & 0 & 1 & 0 & 0 \\ 0 & 0 & 1 & 0 & 1 & 0 \\ 0 & 0 & 0 & 1 & 0 & 1 \\ 0 & 0 & 0 & 0 & 1 & 0 \end{bmatrix}$$

FIGURE 15.1:   Gear shift graph.

Now consider the question: We have shifted gears five times since we began with the car in the initial Stop state. How many ways can we be in 3rd gear now? Of course, this problem is easily solved by simply going through all possible sequences of changes that might end in 3rd gear. But it becomes significantly more difficult if we specify 50 or 200 (or n) changes in state. We could also observe that to get to 3rd gear in n steps we must get to 2nd gear or 4th gear in $n-1$ steps, and develop a recurrence, although this method would fail if our graph were more complex, with more edges. Instead we consider a 6×6 matrix, the *adjacency matrix* A of the graph, which was introduced in

Chapter 10. This matrix has a 1 in the (i, j) position if there is an edge from vertex i to vertex j, and a 0 otherwise. There is a theorem concerning the adjacency matrix that is useful to us.

THEOREM 15.1 (Counting Walks)
If A is the adjacency matrix of a graph G, then the (i, j) entry of A^n is the number of ways to walk from i to j along edges in exactly n steps.

The proof is left to Problem 15.2.

The application of the theorem is clear; we compute the matrix A for the given graph (also shown in Figure 15.1). The calculation of A^5 may be left to a sophisticated calculator or computer algebra system. The $(2, 5)$-entry (corresponding to changes from state 2, or *stop*, to state 5, or *3rd gear*), is 5.

Some changes in state are irreversible. Consider the change of state of an egg from whole (raw, unbroken) to broken (but uncooked) to scrambled. In such a case, we use directed edges and the adjacency matrix will not be symmetric. However, Theorem 15.1 still applies.

Example 15.3: How many ways are there to move from the vertex "Start" to the vertex "Stop" of the graph in Figure 15.2 below?

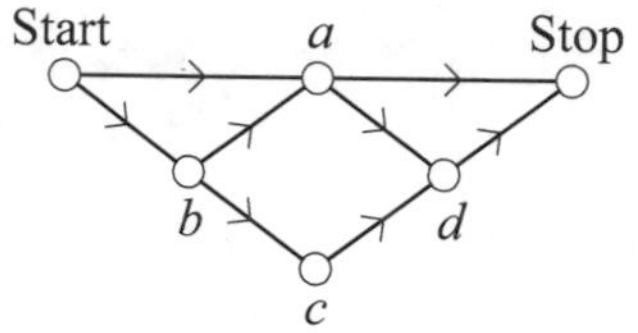

$$A = \begin{bmatrix} 0 & 1 & 1 & 0 & 0 & 0 \\ 0 & 0 & 1 & 1 & 0 & 0 \\ 0 & 0 & 0 & 0 & 1 & 1 \\ 0 & 0 & 0 & 0 & 1 & 0 \\ 0 & 0 & 0 & 0 & 0 & 1 \\ 0 & 0 & 0 & 0 & 0 & 0 \end{bmatrix}$$

FIGURE 15.2: State graph and adjacency matrix.

In this case, we see a 0 in row 1, column 6, indicating that there is no edge directly from Start (the vertex of row 1) to Stop (the vertex of column 6). The two 1s in row 2 (representing vertex a) indicate the edges from a to b (column 3) and to c (column 4), and the two 1s in column 5 represent the two edges from b and c to d.

To solve the given problem, we need the various powers of A. Since A^2 has a 1 in the $(1, 6)$ position, we know that there is one path of length 2 from Start to Stop. The 2 in the $(1, 6)$ position of A^3 indicates two paths of length 3, namely Start-a-b-Stop and Start-b-d-Stop. Because the graph has only six vertices, we don't need powers beyond A^6, and in fact we may stop at A^4. We find that there are only 5 ways to walk from Start to Stop.   □

To calculate the matrix $A + A^2 + A^3 + A^4$ is somewhat time-consuming by hand, but that is what computers are for. Still, there is an easier method.

THEOREM 15.2 (Sum of Powers)
$I + A + A^2 + \cdots + A^n = (A^{n+1} - I)(A - I)^{-1}$ *if the inverse exists. Here I denotes the identity matrix with the same dimensions as A.*

Proof. Multiply the left-hand side by $A - I$; see Problem 15.3 for details. $\square$

This theorem is the matrix analogue of the well-known formula for the sum of a finite geometric progression. There is a similar analog for the sum of an infinite geometric series, and that will apply in the case of the foregoing example.

COROLLARY 15.2.1
$\sum_{i=0}^{\infty} A^i = (I - A)^{-1}$, *where* $A^0 = I$, *and* $\lim_{n \to \infty} A^n = 0$.
Proof. The proof of this corollary is a simple generalization of the proof from calculus of the sum of an ordinary geometric series, applied to Theorem 15.2.
$\square$

It is worth noting that the hypothesis that $A^n \to 0$ is equivalent to the assertion that all the eigenvalues of A are less than 1 in absolute value (or modulus, if the eigenvalues are complex). The proof is beyond the scope of this text.

Now applying the corollary to the matrix of our example, we find $(I - A)^{-1}$ has a $(1,6)$ entry of 5, so there are a total of five paths from Start to Stop.

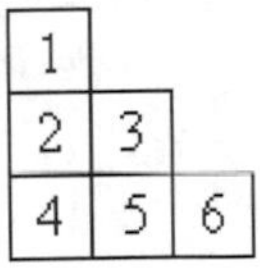

	1	2	3	4	5	6
1	0	1	0	0	0	0
2	1	0	1	1	0	0
3	0	1	0	0	1	0
4	0	1	0	0	1	0
5	0	0	1	1	0	1
6	0	0	0	0	1	0

	1	2	3	4	5	6
1	0	1	1	0	0	0
2	1	0	1	1	1	0
3	1	1	0	1	1	1
4	0	1	1	0	1	0
5	0	1	1	1	0	1
6	0	0	1	0	1	0

FIGURE 15.3: A diagram and two adjacency matrices.

Example 15.4: A marker is placed in the square labeled 1 in Figure 15.3. In how many ways can we move the marker to square 6 in n steps if a step is a move from a square to another square with which it shares an edge? In how many ways if a step is a move to any square sharing an edge or a corner?

The first of the two matrices shown is the matrix that defines two squares to be adjacent if they have a side in common. The answer is clearly the $(1,6)$ entry of the nth power of this matrix, which we denote A, and that may be easily computed for a specific n. If we choose $n = 6$, we find that the $(1,6)$ entry is 12, so that there are exactly 12 paths of six steps in length from square

1 to square 6. If we wish to compute the number of ways in *at most n* steps, we find that there is no shortcut; Theorem 15.2 will not help. This is because the matrix $I - A$ is singular, so we must sum the matrices individually. We do this using a programmable calculator or computer algebra system. It is also possible, but tedious, to use Microsoft Excel or another spreadsheet; Excel has the built-in function MMULT to multiply two matrices, but does not have a function to raise a matrix to a power. We find, for instance, $I + A + \cdots + A^6$ has a $(1,6)$ entry of 14, indicating that there are 14 ways to walk from square 1 to square 6 in six or fewer steps.

The second matrix is the "side or corner" matrix; so that, for example, the $(1,3)$ entry is 1 because square 1 shares a corner (but not a side) with square 3. Call this matrix B; we find the $(1,6)$ entry of B^6 to be 162, which reflects the greater number of permissible steps. It happens that in this case the matrix $I - B$ is invertible; we may compute the sum $I + B + \cdots + B^6$ or the product $(I - B^7)(I - B)^{-1}$. The two expressions give us exactly the same result; the $(1,6)$ entry is 222.

This works even though the series $\sum_{i=0}^{\infty} B^i$ does not converge. How could we tell that the series does not converge? Given any n, there is a walk from square 1 to square 6 of length n whenever there is such a walk of length $n-2$. To find such a walk, we simply retrace a step. For instance, if our walk begins by going from square 1 to square 2, we could simply insert two more steps in our walk by stepping from square 2 back to square 1, and then from square 1 to square 2, and follow the rest of the walk. As a result, the number of walks of length n from square 1 to square 6 never decreases as n goes to infinity so the sum can have no finite $(1,6)$ entry. □

Example 15.5: Suppose we move a marker along the squares of Figure 15.3 subject to the constraint that the marker may only move from a lower-number square to a higher-number square. In this case, we replace each non-zero entry below the main diagonal of either matrix with a zero to get two new matrices.

$$
\hat{A} = \begin{array}{c|cccccc}
 & 1 & 2 & 3 & 4 & 5 & 6 \\
\hline
1 & 0 & 1 & 0 & 0 & 0 & 0 \\
2 & 0 & 0 & 1 & 1 & 0 & 0 \\
3 & 0 & 0 & 0 & 0 & 1 & 0 \\
4 & 0 & 0 & 0 & 0 & 1 & 0 \\
5 & 0 & 0 & 0 & 0 & 0 & 1 \\
6 & 0 & 0 & 0 & 0 & 0 & 0 \\
\end{array}
\qquad
\hat{B} = \begin{array}{c|cccccc}
 & 1 & 2 & 3 & 4 & 5 & 6 \\
\hline
1 & 0 & 1 & 1 & 0 & 0 & 0 \\
2 & 0 & 0 & 1 & 1 & 1 & 0 \\
3 & 0 & 0 & 0 & 1 & 1 & 1 \\
4 & 0 & 0 & 0 & 0 & 1 & 0 \\
5 & 0 & 0 & 0 & 0 & 0 & 1 \\
6 & 0 & 0 & 0 & 0 & 0 & 0 \\
\end{array}
$$

Now, however, if we raise these matrices to the sixth power, we get a matrix of all zeros, because clearly we can move at most five steps before we come to a halt. We can raise either $\hat{A}$ or $\hat{B}$ to any lower power than 6; so if we find, for example, $\hat{B}^3$, its $(1,6)$ entry is 3, indicating three ways to go from square 1 to square 6 in three steps by crossing sides or corners and never moving to a lower-numbered square. If we compute $(I - \hat{B})^{-1}$, we find its $(1,6)$ entry is

8, indicating that there are a total of 8 ways to walk from square 1 to square 6. How do we know that $\sum_{i=0}^{\infty} \hat{B}^i$ converges? Simply because any power of $\hat{B}$ past the fifth power is a matrix of all zeros! Another way to obtain the same result would be to place a 1 in the $(6,6)$ entry of $\hat{B}$, and raise the resulting matrix to any power greater than 4. This would work because, for instance, a walk from square 1 to square 6 in three steps would give us a unique walk from square 1 to square 6 in four steps in which the last step was from square 6 to itself. The result is this matrix.

$$\tilde{B} = \begin{array}{c|cccccc} & 1 & 2 & 3 & 4 & 5 & 6 \\ \hline 1 & 0 & 1 & 1 & 0 & 0 & 0 \\ 2 & 0 & 0 & 1 & 1 & 1 & 0 \\ 3 & 0 & 0 & 0 & 1 & 1 & 1 \\ 4 & 0 & 0 & 0 & 0 & 1 & 0 \\ 5 & 0 & 0 & 0 & 0 & 0 & 1 \\ 6 & 0 & 0 & 0 & 0 & 0 & 1 \end{array} \qquad \tilde{B}^5 = \begin{array}{c|cccccc} & 1 & 2 & 3 & 4 & 5 & 6 \\ \hline 1 & 0 & 0 & 0 & 0 & 0 & 8 \\ 2 & 0 & 0 & 0 & 0 & 0 & 5 \\ 3 & 0 & 0 & 0 & 0 & 0 & 3 \\ 4 & 0 & 0 & 0 & 0 & 0 & 1 \\ 5 & 0 & 0 & 0 & 0 & 0 & 1 \\ 6 & 0 & 0 & 0 & 0 & 0 & 1 \end{array}$$

Notice that any power of $\tilde{B}$ greater than 4 gives us the same matrix. We get a matrix whose only non-zero entries are in column 6, and the $(1,6)$ entry is, as expected, 8. Similar remarks apply to the matrix $\hat{A}$. □

Example 15.6: In how many ways can we form a string of $n = 3$ letters from the set $\{a, b, c, d, e\}$ subject to the restrictions that e is the only letter that may appear twice in succession, the only letters that may follow b or c or d are a and e (that is, each consonant must be followed by a vowel), and c may not come directly after e?

This problem is similar to problems that we have done by means of inclusion-exclusion (Chapter 5) and by means of generating functions (Chapter 6), but more complicated. We form a directed multigraph whose vertices correspond to the letters, and a directed edge will go from one vertex to the next only when the corresponding letters may appear in succession in a legal string. The graph and its adjacency matrix appear in Figure 15.4.

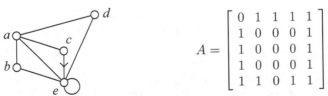

$$A = \begin{bmatrix} 0 & 1 & 1 & 1 & 1 \\ 1 & 0 & 0 & 0 & 1 \\ 1 & 0 & 0 & 0 & 1 \\ 1 & 0 & 0 & 0 & 1 \\ 1 & 1 & 0 & 1 & 1 \end{bmatrix}$$

FIGURE 15.4: Digraph and adjacency matrix for the string problem.

Notice that the $(5,3)$ entry is 0, but the $(3,5)$ entry is 1. This represents the existence of an edge from c to e but not from e to c, because we forbid strings containing *ec*. Now any legal string of length n corresponds to a walk along $n-1$ edges of the graph and because arbitrarily long walks are possible, there is no power n such that $A^n = 0$. An attempt to apply Corollary 15.2.1 will not work because the matrix $(I - A)^{-1}$ does not have integer entries.

To enumerate walks of length 2, we find A^2, which has entries from 1 to 4, indicating the number of walks with the corresponding first and last vertices. Since the $(5,4)$ entry is 2, there are two walks of length from d to e. We can see these walks from the diagram; they are d-e-e and d-a-e. The sum of all the matrix entries is 46, so there are 46 strings of three characters that satisfy the given constraints. $\qquad\square$

Example 15.7: Find the number of binary strings of length n that:
(a) Do not contain the string 11;
(b) Do not contain the string 111.

A solution to part (a) was done in Section 6.7. To use the transfer matrix, we take a graph with two vertices, $V = \{0, 1\}$ with an edge between the two vertices and a loop at 0 but not at 1. In this case the various non-zero powers of A form an infinite sequence, and we borrow a concept from Chapter 6 on generating functions. Look at the expression $\sum_{i \geq 0} A^i x^i$. Similarly to the result of Corollary 15.2.1, we may write the sum formally as $(I - xA)^{-1}$. Because this is a formal sum, the question of convergence is unimportant. Now, a string of length $n+1$ corresponds to a walk of length n on the graph. Each entry in the matrix $(I - xA)^{-1}$ is the generating function of the number of length-$(n+1)$ binary strings that begins at the vertex that corresponds to the row and ends at the vertex that corresponds to the column.

$$(I - xA)^{-1} = \left(\begin{bmatrix} 1 & 0 \\ 0 & 1 \end{bmatrix} - x \begin{bmatrix} 1 & 1 \\ 1 & 0 \end{bmatrix} \right)^{-1} = \frac{1}{1 - x - x^2} \begin{bmatrix} 1 & x \\ x & 1-x \end{bmatrix}$$

Thus, the number of walks beginning at 0 and ending at 0 is generated by $1/(1 - x - x^2)$, and so on. To find the total, we add all entries, and discover the generating function $(2 + x)/(1 - x - x^2)$. We can then prove rigorously if we like that the number of strings is the Fibonacci number f_{n+1}.

Part (b) will be a little trickier, and requires us to use a bigger graph. Here the vertices will be $\{00, 01, 10, 11\}$ and we will have a directed edge from ab to bc only if abc is a legitimate substring. The digraph that results is shown in Figure 15.5.

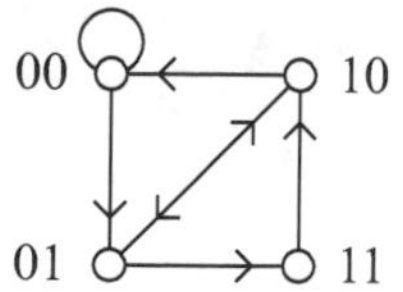

$$A = \begin{bmatrix} 1 & 1 & 0 & 0 \\ 0 & 0 & 1 & 1 \\ 1 & 1 & 0 & 0 \\ 0 & 0 & 1 & 0 \end{bmatrix}$$

FIGURE 15.5: Digraph and adjacency matrix.

In this case, a string of length n corresponds to $n-2$ steps along the digraph, because one step (say from 01 to 11) corresponds to the string 011. Thus the coefficient of x^n in our generating function will be the number of strings of length $n+2$. Now we find the matrix $(I-xA)^{-1}$ and add the entries. This is more easily done with a symbolic algebra program than by hand, and gives us the result $f(x) = (-2x^2 - 3x - 4)/(x^3 + x^2 + x - 1)$. Long division gives us $4 + 7x + 13x^2 + 24x^3 + 44x^4 + 81x^5 + \ldots$. For $n=2$ there are clearly four such strings, so we check for $n=3$ and find seven strings. For $n=4$ there are 13 strings, as we may check by enumeration, and we should be content that we have not made any mistakes since the coefficients agree. Checking the cases of $n=5$ and $n=6$ is a pleasant exercise in inclusion-exclusion. We may also deduce, using the methods of Chapter 6, that the sequence a_n we are looking at satisfies the recurrence $a_n = a_{n-1} + a_{n-2} + a_{n-3}$. $\square$

We conclude with a look at another application of the transfer-matrix method.

Example 15.8: In how many different ways may we cover all the squares of a $4 \times n$ chessboard with nonoverlapping dominos?

To solve this problem we represent it as a problem enumerating walks on an appropriate graph. Let each row of a $4 \times n$ chessboard be represented by a string of four bits, where a 1 represents a domino in the vertical position extending into the next row, and a 0 represents a domino in a horizontal position or extending from the row below. It follows that the topmost row must be represented by 0000. Of the 16 possible bit strings, we find only six may represent rows of a properly covered $4 \times n$ chessboard. Thus, for example, the string 0000 may represent a row containing two horizontal dominos, or one horizontal domino and two vertical dominos that rise from the lower row, or four vertical dominos from below. A little thought reveals that no binary string with an odd number of 1s can work since it can never "fit" with the topmost 0000 row. Similarly, a row represented by 1010 must be followed by a row of 0101, and the two alternate, never leading to any way to create a 0000 row. The other strings are 0011, 0110, 1001, 1100, and 1111. Now we make a graph of these six strings and make a directed edge from the first to the next if a row of the chessboard corresponding to the first can fit below one corresponding to the second. The graph and the matrix that corresponds to it are shown in Figure 15.6.

For any specific n, the $(2,2)$-entry of A^n will give the number of proper coverings of a $4 \times n$ chessboard by dominos. Thus, a walk of $n=3$ steps, say 0000-1100-0011-0000, corresponds to the domino tiling of the 4×3 board in which two vertical dominos occupy the lower-left four squares and two other vertical dominos occupy the upper-right four squares; and horizontal dominos occupy the upper-left and lower-right corners.

This matrix is large enough that to find $(I-xA)^{-1}$ we use a computer algebra system. In *Mathematica*, for instance, we define A and compute $(I-xA)^{-1}$ as follows.

	1111	0000	1100	0110	0011	1001
1111	0	1	0	0	0	0
0000	1	1	1	0	1	1
$A =$ 1100	0	1	0	0	1	0
0110	0	1	0	0	0	1
0011	0	1	1	0	0	0
1001	0	1	0	1	0	0

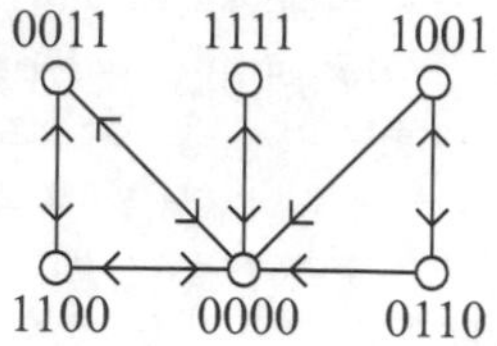

FIGURE 15.6: Digraph and adjacency matrix for domino tiling.

```
A = {{0, 1, 0, 0, 0, 0}, {1, 1, 1, 0, 1, 1}, {0, 1, 0, 0, 1, 0},
     {0, 1, 0, 0, 0, 1}, {0, 1, 1, 0, 0, 0}, {0, 1, 0, 1, 0, 0}};

B = Inverse[IdentityMatrix[6] - x*A];

B[[2,2]]
```

The semicolons at the end of each line suppress display of the results; the last line produces the $(2, 2)$-entry of B, namely,

$$\frac{1 - 2x^2 + x^4}{1 - x - 6x^2 - x^3 + 6x^4 + 2x^5 - x^6}.$$

We can expand this in a power series:

```
Series[%,{x,0,10}]
```

and we get $1 + x + 5x^2 + 12x^3 + 38x^4 + 107x^5 + 316x^6 + 915x^7 + 2671x^8 + 7771x^9 + 22640x^{10} + O[x]^{11}$.

The constant term of 1 indicates that there is one way of covering a 4×0 board with dominos (which is to do nothing). The x term says that there is one way to cover a 4×1 board with dominos, which we easily see is to place two dominos horizontally on the four squares of the board. That the coefficient of x^2 is 5 follows from the fact that a 4×2 board is also a 2×4 board, and the result from Chapter 6 about the number of domino tilings of a $2 \times n$ board. □

15.3   Kasteleyn's Permanent Method

The *permanent* of a matrix may be described, loosely, as "the determinant without the minus signs." More precisely, the matrix $A = \begin{bmatrix} a & b \\ c & d \end{bmatrix}$ has permanent $\text{Per}(A) = ad + bc$. For a 3×3 or larger matrix, the permanent may be computed by Laplace expansion, without the power of -1 that causes terms in the summation to alternate in sign. To put it another way, if the reader recalls that the determinant of a matrix, $\text{Det}(A)$, may be written as

$$\text{Det}(A) = \sum_{\sigma} (-1)^{\text{sgn}(\sigma)} a_{1,\sigma(1)} a_{2,\sigma(2)} \cdots a_{n,\sigma(n)}$$

then the permanent may be written as

$$\text{Per}(A) = \sum_{\sigma} a_{1,\sigma(1)} a_{2,\sigma(2)} \cdots a_{n,\sigma(n)}. \tag{15.1}$$

In each case the summation is over all permutations on n letters.

Why, the reader will wonder, this temporary digression on the permanent? The answer is that it has applications in enumeration. To see why, let's consider a balanced bipartite graph such as C_4. We may recall from Chapter 10 that there may exist a perfect matching, or one-factor. We may also see from the adjacency matrix that all the information about the graph's structure is preserved in only one quarter of the matrix, because the matrix of a balanced bipartite graph on $2n$ vertices is comprised of two $n \times n$ matrices of all zeroes together with two copies of an $n \times n$ matrix that describes the edges of the bipartite graph. This $n \times n$ matrix is sometimes called the *bipartite adjacency matrix*, and we shall denote it $\text{B}(G)$. More formally, $\text{B}(G)$ is the $n \times n$ matrix whose rows correspond to vertices in one part of the bipartite graph, and columns correspond to vertices in the other part of the bipartite graph, and the (i, j) position is 1 if i is adjacent to j in the bipartite graph, 0 otherwise.

THEOREM 15.3 (Matchings in Bipartites)
The balanced bipartite graph B has exactly Per(B(G)) perfect matchings.

Proof. We establish a bijection between matchings of G and permutations of n letters that contribute a 1 to the sum of Equation 15.1. Choose any perfect matching of the balanced bipartite graph G with parts X and Y, and form $A = \text{B}(G)$ by letting the rows of A correspond to vertices of the part X, and the columns of A correspond to the vertices of Y. Now let the permutation σ satisfy $\sigma(i) = j$ exactly when vertex x_i of X has an edge in the matching to vertex y_j of Y. Since each vertex of X has by definition exactly one neighbor in the matching, this defines a permutation on n letters, and corresponds to one summand of the sum of Equation 15.1. This summand must have the value 1, because every matrix entry of the form $a_{i,\sigma(i)}$ is a 1 that corresponds to an edge of the matching. In the same way, every non-zero summand of 15.1 corresponds to a permutation that arises from some perfect matching. It is easy to see that distinct perfect matchings give rise to distinct permutations and vice-versa, so the theorem is proven. $\qquad\square$

The foregoing discussion of the uses of the permanent may be generalized. Suppose we have a family of k subsets $\mathcal{D} = \{B_i : 1 \leq i \leq k\}$ of an n-set $\mathcal{U}$. We will recall from Chapter 1 that a *system of distinct representatives* or *SDR* is a subset $R = \{x_i : 1 \leq i \leq k\}$ of the n-set with the property that $x_i \in B_i$ for each i, and of course the elements are distinct (because R is a set, not a multiset). Theorem 1.2 gave a necessary and sufficient condition for an SDR to exist. However, this result, while it establishes the *existence* of an SDR, gives us no help in determining *how many* SDRs may exist.

We can modify the definition of the permanent so as to count the number of SDRs, or equivalently to count the number of matchings of an unbalanced bipartite graph that use all the vertices of the smaller part. The graph will have k vertices corresponding to the sets B_i in one part, and n vertices corresponding to the elements of the n-set $\mathcal{U}$ in the other part. We assume $k \leq n$ and look at the $k \times n$ bipartite adjacency matrix A of the graph; let V be the set of all k-permutations of the n-set $\mathcal{U}$. We define the permanent of A to be

$$\text{Per}(A) = \sum_{\sigma \in V} a_{1,\sigma(1)} a_{2,\sigma(2)} \cdots a_{n,\sigma(n)}. \tag{15.2}$$

Then the permanent of A again gives the number of SDRs of the associated family of subsets; the proof is exactly analogous to the proof of Theorem 15.3.

Equation 15.1 is unsatisfactory as a way to compute the permanent; we mentioned above that the equivalent of the familiar Laplace expansion for the determinant will work. Thus, the permanent of any 2×2 matrix was given at the beginning of this section as $ad + bc$; we can use this to compute the permanent of a larger matrix. So if A_i is the result of deleting row 1 and column i from the matrix A, we may write $\text{Per}(A) = \sum_{i=1}^{n} a_{1,i}\text{Per}(A_i)$. However, caution is required when using this technique. In particular, the traditional shortcut to computing the determinant with Laplace expansion is to perform some elementary row operations to create a row or column with many zero entries, and expand about that row or column. This will work because the determinant is invariant under elementary row or column operations; however, the permanent is not. Thus, this shortcut will yield an incorrect value for the permanent. It is permissible to expand the permanent about a different row of the matrix, as this merely corresponds to relabeling the vertices of the graph, or reordering the sets B_i. Also, because $\text{Per}(A^T) = \text{Per}(A)$, we may expand about any column when A is square.

Example 15.9: Three committees consist of five people altogether. Committee A is made up of 1, 2, and 3. Committee B contains 1, 3, and 5, and Committee C consists of 2, 3, 4, and 5. In how many ways can I choose an SDR consisting of one member of each committee?

The stated problem gives us a 3×5 matrix A; we will find its permanent.

$$\mathrm{Per}A = \mathrm{Per}\begin{bmatrix} & 1\ 2\ 3\ 4\ 5 \\ A & 1\ 1\ 1\ 0\ 0 \\ B & 1\ 0\ 1\ 0\ 1 \\ C & 0\ 1\ 1\ 1\ 1 \end{bmatrix}$$

$$= \mathrm{Per}\begin{bmatrix} 0\ 1\ 0\ 1 \\ 1\ 1\ 1\ 1 \end{bmatrix} + \mathrm{Per}\begin{bmatrix} 1\ 1\ 0\ 1 \\ 0\ 1\ 1\ 1 \end{bmatrix} + \mathrm{Per}\begin{bmatrix} 1\ 0\ 0\ 1 \\ 0\ 1\ 1\ 1 \end{bmatrix}$$

This equation came from expanding about the first row, corresponding to the set A. We continue the process; for instance,

$$\mathrm{Per}\begin{bmatrix} 0 & 1 & 0 & 1 \\ 1 & 1 & 1 & 1 \end{bmatrix} = \mathrm{Per}[1\ 1\ 1] + \mathrm{Per}[1\ 1\ 1].$$

Now, to find the permanent of a 1×3 matrix, we simply add the entries. In this case, each of these permanents is 3. We get 18 SDRs in all from these sets. ☐

Example 15.10: Use the permanent method to calculate the number of matchings in $K_{3,3}$.

$$A = \begin{bmatrix} 1 & 1 & 1 \\ 1 & 1 & 1 \\ 1 & 1 & 1 \end{bmatrix}, \text{ so } \mathrm{Per}(A) = 3 \cdot \mathrm{Per}\begin{bmatrix} 1 & 1 \\ 1 & 1 \end{bmatrix} = 3 \cdot 2 = 6$$

Or, as we knew before, there are 6 permutations on three letters. ☐

Example 15.11: The 4-rung ladder L_4 and B(L_4) are shown below. How many matchings does this graph have?

	2	4	6	8
1	1	0	0	1
3	1	1	0	1
5	0	1	1	0
7	0	1	1	1

We expand about the first row, as it has only two 1s; we obtain

$$\mathrm{Per}(A) = \mathrm{Per}\begin{bmatrix} 1 & 0 & 1 \\ 1 & 1 & 0 \\ 1 & 1 & 1 \end{bmatrix} + \mathrm{Per}\begin{bmatrix} 1 & 1 & 0 \\ 0 & 1 & 1 \\ 0 & 1 & 1 \end{bmatrix}.$$

The first permanent on the right-hand side may be expanded about the first row; the second, however, will most efficiently be expanded about the first column (which has only a single non-zero element). We get

$$\mathrm{Per}(A) = \mathrm{Per}\begin{bmatrix} 1 & 0 \\ 1 & 1 \end{bmatrix} + \mathrm{Per}\begin{bmatrix} 1 & 1 \\ 1 & 1 \end{bmatrix} + \mathrm{Per}\begin{bmatrix} 1 & 1 \\ 1 & 1 \end{bmatrix} = 5.$$

The graph has five perfect matchings, verifiable by listing them. ☐

By itself, the permanent is of little use in computing matchings for individual bipartite graphs; its real utility comes in working with families of graphs, using theorems about the permanent. We illustrate with two such theorems, provided here without proof.

For a specific $m \times n$ matrix $A = [a_{ij}]$, with $m \leq n$, let A_r denote one of the matrices formed by deleting r of the columns of A, where $n - m \leq r \leq n - 1$. We will add the elements of each row of A_r, and then multiply the resulting row sums together to get a quantity we denote ΠA_r. There are $\binom{n}{r}$ ways to form A_r by choosing which columns to delete; let $\Sigma\Pi A_r$ denote the sum over all $\binom{n}{r}$ possible A_r of the quantity ΠA_r.

THEOREM 15.4 (Ryser's Formula)

$$Per(A) = \Sigma\Pi A_{n-m} - \binom{n - m + 1}{1}\Sigma\Pi A_{n-m+1} +$$

$$\cdots + (-1)^{m-1}\binom{n - 1}{m - 1}\Sigma\Pi A_{n-1}$$

COROLLARY 15.4.1

For a square matrix A, we have $Per(A) = \sum_{i=0}^{n-1}(-1)^i\Sigma\Pi A_i$.

The proof of Theorem 15.4 may be found in [94]; Corollary 15.4.1 follows by setting $m = n$. For one use of this result, consider the matrix J_n, an $n \times n$ matrix with 1 in each entry. Clearly, $Per(J) = n!$ since J is $B(K_{n,n})$. Now, consider $J_n - I_n$; this is B of $K_{n,n}$ with a perfect matching removed, and its permanent is the number D_n of derangements on n letters. Now, applying the corollary, we note that the various matrices A_r all have row sums $n - r$ or $n - r - 1$. A little thought lets us see that a particular ΠA_r will be $(n - r)^r(n - r - 1)^{n-r}$, and there are $\binom{n}{r}$ such matrices A_r. We get

$$D_n = \sum (-1)^r \binom{n}{r}(n - r)^r(n - r - 1)^{n-r}.$$

This formula is different from the usual formula for derangements given in Chapter 5.

THEOREM 15.5 (van der Waerden's Inequality)

Let A be an $n \times n$ matrix with non-negative entries where the entries in each row sum to k. Then $Per(A) \geq n!(k/n)^n$.

Although Bartel Leendert van der Waerden conjectured that this inequality was true back in 1926, it was not until 1981 that it was first established by

G. P. Egorychev and, independently, by D. I. Falikman (see [69]). This result tells us more than just that a regular bipartite graph has a matching; it tells us that it must have many of them. Since the permanent must be an integer, we will round our answers up using the notation $\lceil x \rceil$ for the least integer greater than or equal to x.

Example 15.12: How many matchings are in a 3-regular bipartite graph on 12 vertices?

This graph will have a 6×6 bipartite adjacency matrix; because it is regular of degree 3, each row will have three 1s and three 0s, so $k = 3$. Theorem 15.5 tells us that the graph must have (using $n = 6$) at least $\lceil 6!/2^6 \rceil = 12$ matchings, no matter how the edges are arranged. □

For more on this subject, as well as for other applications of the permanent, the interested reader is referred to ([69], Chapter 8).

Exercises 15A

1. Explain how to find, for some square matrix A, the matrix A^{82} using only 8 matrix multiplications.

2. Solve each recurrence using the methods of this chapter:
   (i) $a_{n+1} = 2a_n + 3a_{n-1}$, $a_0 = 2, a_1 = 2$.
   (ii) $a_{n+1} = 2a_n + 3a_{n-1}$, $a_0 = 1, a_1 = 7$.
   (iii) $b_{n+1} = 2b_n + 8b_{n-1}$, $b_0 = -1, b_1 = 8$.
   (iv) $b_{n+1} = 2b_n + 8b_{n-1}$, $b_0 = 2, b_1 = 2$.

3. Suppose a machine has five states A, B, C, D, and *Halt*. From state A it may transfer to either state C or D; from B, it may go to D or *Halt*; from C it may transfer to B or D; from D it may go to A or *Halt*; and it cannot transfer out of state *Halt*.
   (i) Find the state digraph that represents this machine.
   (ii) Find the adjacency matrix M of the digraph of part (i).
   (iii) Compute M^4; in how many ways can the machine go from state A to *Halt* in exactly four steps?

4. Let M be the matrix from the preceding exercise, and suppose $\hat{M}$ is formed from M by changing the $(5, 5)$ entry from 0 to 1 (that is, allowing the machine to go from state *Halt* to *Halt*). Explain why this matrix would be useful in determining the number of ways that the machine can go from state A to state *Halt* in *at most* four steps. How will this compare to calculating the matrix $(I_5 - M^5)(I_5 - M)^{-1}$?

5. A marker starts in square 1 and may move from a square to any square sharing a side with that square. For each of the four diagrams shown, find the adjacency matrix A of the graph. Compute $(I - A)^{-1}$ if it exists, and find the number of walks of six or fewer steps from square 1 to square 6.

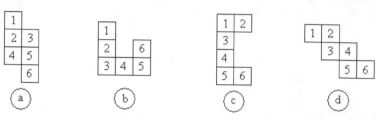

6. For the same diagrams as the previous exercise, a marker may move to a square with which its current square shares a side or a corner. As above, find the adjacency matrix A, the matrix $(I - A)^{-1}$ if it exists, and the number of walks from square 1 to square 6 in six or fewer steps.

7. For the same diagrams as Exercise 5, a marker may move only to a *higher-numbered* square with which its current square shares a side or a corner. Find the adjacency matrix A by replacing all non-zero entries below the main diagonal of the matrices from Exercise 6, find the matrix $(I - A)^{-1}$ if it exists, and compute the number of walks from square 1 to square 6 in six or fewer steps.

8. For the same diagrams, allow the marker to move as in Exercise 7, but also to move from square 6 to square 6. Find $(I - A)^{-1}$, $\sum_{i=1}^{6} A^i$, and the $(1,6)$ entry of A^6. What is the number of ways to move from square 1 to square 6 in six or fewer moves?

9. Suppose that the set A consists of the odd numbers between 1 and 6 inclusive; B consists of the even numbers in that range; and C consists of the numbers divisible by 3 in that range. Construct a matrix A and use this matrix to find the number of possible SDRs for this collection of sets.

10. Find the number of perfect matchings for the bipartite graph shown.

11. Use van der Waerden's inequality to determine the minimum number of perfect matchings of a 4-regular bipartite graph on 16 vertices (8 in each part).

Exercises 15B

1. Explain how to find, for some square matrix A, the matrix A^{61}, using at most nine matrix multiplications.

2. Solve each recurrence using the methods of this chapter:
   (i) $a_{n+1} = 2a_n + 3a_{n-1}$, $a_0 = 2, a_1 = 2$.
   (ii) $a_{n+1} = 2a_n + 3a_{n-1}$, $a_0 = 1, a_1 = 5$.
   (iii) $b_{n+1} = 2b_n + 15b_{n-1}$, $b_0 = 1, b_1 = 1$.
   (iv) $b_{n+1} = 2b_n + 15b_{n-1}$, $b_0 = 1, b_1 = 5$.

3. Suppose a machine has six states, A, B, C, D, E, and *Halt*. From state A we may go to B or C; from B, we may go to C, D, or *Halt*; from C we may only go to D; from D we may go to E or *Halt*; and from E we may only go to C.
   (i) Find the state digraph that represents this machine.
   (ii) Find the adjacency matrix M of the digraph of part (i).
   (iii) Compute M^6; in how many ways can the machine go from state A to *Halt* in exactly six steps?

4. Let M be the matrix from the preceding exercise, and suppose $\hat{M}$ is formed from M by changing the $(6,6)$ entry from 0 to 1 (that is, allowing the machine to go from state *Halt* to *Halt*). Explain why this matrix would be useful in determining the number of ways that the machine can go from state A to state *Halt* in *at most* six steps. How will this compare to calculating the matrix $(I_6 - M^6)(I_6 - M)^{-1}$?

5. A marker starts in square 1 and may move from a square to any square sharing a side with that square. For each of the four diagrams shown, find the adjacency matrix A of the graph. Compute $(I - A)^{-1}$ if it exists, and find the number of walks of six or fewer steps from square 1 to square 6.

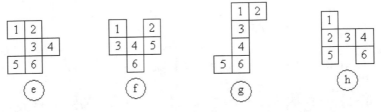

6. For the same diagrams as the previous exercise, a marker may move to a square with which its current square shares a side or a corner. As above, find the adjacency matrix A, the matrix $(I - A)^{-1}$ if it exists, and the number of walks from square 1 to square 6 in six or fewer steps.

7. For the same diagrams as the previous exercise, a marker may move only to a *higher-numbered* square with which its current square shares a side or a corner. Find the adjacency matrix A by replacing all non-zero entries below the main diagonal of the matrices from Exercise 6, find the matrix $(I-A)^{-1}$ if it exists, and compute the number of walks from square 1 to square 6 in six or fewer steps.

8. For the same diagrams, allow the marker to move as in Exercise 7, but also to move from square 6 to square 6. Find $(I-A)^{-1}$, $\sum_{i=1}^{6} A^i$, and the $(1,6)$ entry of A^6. What is the number of ways to move from square 1 to square 6 in six or fewer moves?

9. For the set $\{1,2,3,4,5,6,7,8\}$, there are three subsets $A = \{3,4,5,6\}$, $B = \{2,4,7\}$, and $C = \{1,4,7,8\}$. How many SDRs are possible?

10. Find the number of perfect matchings for the bipartite graph shown.

11. Use van der Waerden's inequality to determine the minimum number of perfect matchings of a 3-regular bipartite graph on 20 vertices (10 in each part).

Problems

Problem 1: We will modify the approach of Section 15.1 to deal with a three-term recurrence and solve the recurrence $a_{n+1} = 2a_n + a_{n-1} - 2a_{n-2}$ with $a_0 = 1$, $a_1 = 0$, and $a_2 = 4$.

(i) Find a matrix A such that $A \begin{bmatrix} a_{n-2} \\ a_{n-1} \\ a_n \end{bmatrix} = \begin{bmatrix} a_{n-1} \\ a_n \\ a_{n+1} \end{bmatrix}$.

(ii) Find the eigenvalues of the matrix A and corresponding eigenvectors (you should have three distinct eigenvalues).

(iii) Form the matrix P whose columns are the eigenvectors and find P^{-1}.

(iv) Find the diagonal matrix D and use it to solve the recurrence.

Problem 2: Use induction to prove Theorem 15.1.

Problem 3: Use the distributive law to find the matrix product $(I + A + A^2 + \cdots + A^n)(I - A)$.

Problem 4: Dr. John Walters and Dr. Walter Johns assist the biology department in a mathematical analysis of the life cycle of an obscure creature, the *nerdus academicus*. The creature goes through five different phases of its life cycle, colloquially referred to as *grad*, *lecturer*, *assistant*, *associate*, and *full*. However, the progression through the different phases (which may occur every three to seven years) is not regular. Thus, the *grad* stage may only develop into the *lecturer* stage, from which the creature may remain in the *lecturer* state or advance to *assistant*. From *assistant* three possible outcomes exist; the creature may revert to *lecturer*, remain as *assistant*, or progress to *associate*. It is virtually unheard-of for the creature to revert to the *assistant* phase from *associate*, so the possibilities from there are to remain *associate* or to progress to the final *full* stage. From that stage, no development occurs, so the creature remains in that position for the remainder of its life, after which its body (or *oeuvre*, to use the specialized term) will be used to feed the next generation. Draw a state graph representing the life cycle of this bizarre creature, and find its adjacency matrix. In how many ways can the creature progress from *grad* to *full* in six steps?

Problem 5: Prove or find a counterexample: If a matrix has a permanent of 0 it must also have a determinant of 0.

Problem 6: Use *Stirling's approximation* that says $n! \simeq n^n e^{-n} \sqrt{2\pi n}$ and van der Waerden's inequality to get an approximation to a lower bound for the number of perfect matchings of a bipartite graph that is regular of degree $k = 2$. Then use L'Hopital's rule to find the limit of this expression as n goes to infinity.

Problem 7: Construct a family of 2-regular bipartite graphs on $2n$ vertices for which the number of matchings increases exponentially as n increases.

Appendix A: Sets and Proof Techniques

AA.1 Sets and Basic Set Operations

A *set* is a collection of objects with a well-defined rule, called the *membership law*, for determining whether a given object belongs to the set. The individual objects in the set are called its *elements* or *members* and are said to belong to the set. If S is a set and s is one of its elements, we denote this fact by writing

$$s \in S$$

which is read as "s belongs to S" or "s is an element of S."

A set may be defined by listing all the elements, usually between braces; if S is the set consisting of the numbers 3 and 5, then $S = \{3, 5\}$. We write $\{1, 2, \ldots, 10\}$ to mean the set of all integers from 1 to 10. Another method is the use of the membership law of the set: for example, since 3 and 5 are precisely the solutions of the equation $x^2 + 8x + 15 = 0$, we could write

$$S = \{x : x^2 + 8x + 15 = 0\}$$

("the set of all x such that $x^2 + 8x + 15 = 0$"). The set of integers from 1 to 10 is

$$\{x : x \text{ integral}, 1 \le x \le 10\}.$$

Sometimes a vertical line is used instead of a colon, so we might instead write

$$\{x | x \text{ integral}, 1 \le x \le 10\}.$$

The set of all integers is denoted $\mathbb{Z}$, so we could also say

$$\{1, 2, \ldots, 1\} = \{x : x \in \mathbb{Z}, 1 \le x \le 10\}.$$

The *positive integers* or *natural numbers* are denoted $\mathbb{Z}^+$ or $\mathbb{N}$. $\mathbb{Z}^0$ denotes the *non-negative* integers. The *rational numbers*, *real numbers* and *complex numbers* are denoted $\mathbb{Q}$, $\mathbb{R}$ and $\mathbb{C}$ respectively. These sets are *infinite* (as opposed to *finite* sets like $\{0, 1, 3\}$).

Example AA.1: Find three ways to describe the set consisting of the three numbers 0, 1 and 2.

One way is to list the elements: $\{0, 1, 2\}$. The listing could also be expressed as $\{x : x \text{ integral}, 0 \leq x \leq 2\}$ (or $\{x : x \in \mathbb{Z}, 0 \leq x \leq 2\}$). Or we could note that 1, 2 and 3 are the solutions of $x^3 + 3x^2 + 2x = 0$, and write $\{x : x^3 + 3x^2 + 2x = 0\}$. $\qquad\square$

The number of elements in a set is called its *order* or *size*. We shall denote the order of S by $|S|$ (another common notation is $\#(S)$). For finite sets, the order will be a non-negative integer; for infinite sets we usually use the infinity symbol ∞.

Suppose n is a positive integer. The *residue classes modulo n* are the n sets $[0], [1], \ldots, [n-1]$, where $[i]$ is the set of all integers that leave remainder i on division by n:

$$[n] = \{\ldots, i - kn, \ldots, i - n, i, i + n, \ldots, i + kn, \ldots\}.$$

The set of residue classes modulo n is denoted $\mathbb{Z}_n$. Members x and y of the same residue classes are called *congruent modulo n* (see Section 15) and we write $y \equiv x (\bmod\, n)$. The operations of addition and multiplication are defined on $\mathbb{Z}_n$ in the obvious way:

$$[x] + [y] = [x + y]\,;\; [x] \times [y] = [x \times y].$$

When no confusion arises we simply write x instead of $[x]$.

It is sometimes useful to discuss number systems with the number 0 omitted from them, especially when division is involved. We denote this with an asterisk: for example, $\mathbb{Z}^*$ is the set of non-zero integers.

The notation $S \subseteq T$ means that every member of S is a member of T:

$$x \in S \Rightarrow x \in T.$$

Then S is called a *subset* of T. We also say that T contains S or T is a *superset* of S, and write $T \supseteq S$. Sets S and T are equal, $S = T$, if and only if $S \subseteq T$ and $T \subseteq S$ are both true. We can represent the situation where S is a subset of T but S is not equal to T—there is at least one member of T that is not a member of S—by writing $S \subset T$.

The number systems satisfy $\mathbb{Z}^+ \subseteq \mathbb{Z} \subseteq \mathbb{Q} \subseteq \mathbb{R}$. Rational numbers that are not integers are called *proper fractions*, and real numbers that are not rational are called *irrational numbers*.

The definition of a set does not allow for ordering of its elements, or for repetition of its elements. Thus $\{1, 2, 3\}, \{1, 3, 2\}$ and $\{1, 2, 3, 1\}$ all represent the same set (which could be written $\{x \mid x \in \mathbb{Z}^* \text{ and } x \leq 3\}$, or $\{x \in \mathbb{Z}^* \mid x \leq 3\}$). To handle problems that involve ordering, we define a *sequence* to be an ordered set. Sequences can be denoted by parentheses; $(1, 3, 2)$ is the sequence with first element 1, second element 3 and third element 2, and is different from $(1, 2, 3)$. Sequences may contain repetitions, and $(1, 2, 1, 3)$ is quite different from $(1, 2, 3)$; the two occurrences of object 1 are distinguished by the fact that they lie in different positions in the ordering.

Another way to deal with multiple elements is to use *multisets*. This is particularly useful when the objects are distinct but we choose to ignore distinctions between them. Thus, we might say that a set of items from the store consists of twelve eggs, three tomatoes, and four onions, and not describe the set as consisting of 19 elements, but as a set of three elements in which each element is repeated. This multiset might be described as $\{\text{egg} \times 12, \text{tomato} \times 3, \text{onion} \times 4\}$. The number of times that an element may be repeated in a multiset is variously referred to as its *multiplicity* or its *repetition number*; we shall use either as seems appropriate. A multiset S is said to be a *multisubset* of a multiset T if the multiplicity of each element of S is less than or equal to the multiplicity of that element in T. Equivalently, $S \subseteq T$ provided that each element has at least as large a repetition number in T as in S.

Example AA.2: Which of these are *not* multisubsets of the multiset $S = \{a \times 3, b \times 2, x \times 1, y \times 2, z \times 3\}$?

$A = \{a \times 1, b \times 2\}$.
$B = \{a \times 2, b \times 2, x \times 2, y \times 2, z \times 2\}$.
$C = \{a \times 1, b \times 1, c \times 1\}$.

Clearly, A is a multisubset of S because the elements of A do not appear more times in A than in S. B and C are not multisubsets of S because each has an element that appears more times than in S. B has two copies of b where S has one; C has a copy of c where S has none. □

An important concept is the *empty* set, or *null* set, which has no elements. This set, denoted by $\emptyset$, is unique and is a subset of every other set.

In all the discussions of sets in this book, we shall assume (usually without bothering to mention the fact) that all the sets we are dealing with are subsets of some given universal set U. U may be chosen to be as large as necessary in any problem we deal with; in most of our discussion so far we could have chosen $U = \mathbb{Z}$ or $U = \mathbb{R}$. U can often be chosen to be a finite set.

The *power set* of any set S consists of all the subsets of S (including S itself and $\emptyset$), and is denoted by $\mathcal{P}(S)$:

$$\mathcal{P}(S) = \{T : T \subseteq S\}. \tag{1}$$

The power set is a set whose elements are *themselves* sets.

Given sets S and T, the *union* of S and T is the set

$$S \cup T = \{x : x \in S \text{ or } x \in T \text{ (or both) }\},$$

the *intersection* of S and T is the set

$$S \cap T = \{x : x \in S \text{ and } x \in T\},$$

and the *relative complement* of T with respect to S (or alternatively the *set-theoretic difference* or *relative difference* between S and T) is the set

$$S \backslash T = \{x : x \in S \text{ and } x \notin T\}.$$

So $S \subseteq T$ if and only if $S \backslash T = \emptyset$.

The relative complement of T with respect to the universal set is written $\overline{T}$ and called the *complement* of T. With this notation $S \backslash T = S \cap \overline{T}$, since each of these sets consists of the elements belonging to S but not to T.

If two sets, S and T, have no common element, then S and T are called *disjoint*. The sets $S \backslash T$ and T are necessarily disjoint, and in particular, S and $\overline{S}$ are always disjoint. A *partition* of a set S is a collection of non-empty subsets $S_1, S_2, \ldots, S_n$ of S such that each pair of them are disjoint; we say that these sets are *mutually* or *pairwise disjoint*.

Both union and intersection are commutative; in other words

$$S \cup T = T \cup S$$

and

$$S \cap T = T \cap S,$$

for any sets S and T. Similarly, the associative laws

$$R \cup (S \cup T) = (R \cup S) \cup T$$

and

$$R \cap (S \cap T) = (R \cap S) \cap T$$

hold. The associative law means that we can omit brackets in a string of unions (or a string of intersections); expressions like $(A \cup B) \cup (C \cup D)$, $((A \cup B) \cup C) \cup D$ and $(A \cup (B \cup C)) \cup D$, are all equal, and we usually omit all the parentheses and simply write $A \cup B \cup C \cup D$. Combining the commutative and associative laws, we see that any string of unions can be rewritten in any order: for example,

$$(D \cup B) \cup (C \cup A) = C \cup (B \cup (A \cup D)) = (A \cup B \cup C \cup D).$$

The following distributive laws hold:

$$
\begin{array}{rcll}
R \cup (S \cap T) & = & (R \cup S) \cap (R \cup T); & \quad (2) \\
R \cap (S \cup T) & = & (R \cap S) \cup (R \cap T); & \quad (3) \\
(R \cup S) \backslash T & = & (R \backslash T) \cup (S \backslash T). & \quad (4)
\end{array}
$$

We have also the equation

$$R \backslash (S \cup T) = (R \backslash S) \cap (R \backslash T), \qquad (5)$$

and the analogous

$$R \backslash (S \cap T) = (R \backslash S) \cup (R \backslash T). \qquad (6)$$

When we take the particular case where R is the universal set in (5) and (6), those two equations become *de Morgan's laws*:

$$\overline{S \cup T} = \overline{S} \cap \overline{T}, \qquad (7)$$

$$\overline{S \cap T} = \overline{S} \cup \overline{T}. \qquad (8)$$

These laws extend to three or more sets in the obvious way: for example,

$$\overline{R \cup S \cup T} = \overline{R} \cap \overline{S} \cap \overline{T}.$$

Another property is that both union and intersection are *idempotent*. That is,

$$S \cup S = S \cap S = S$$

for any set S.

We often illustrate sets and operations on sets by diagrams. A set R is represented by a circle, and the elements of R correspond to points inside the circle. If we need to show a universal set (for example, if complements are involved), it is shown as a rectangle enclosing all the other sets. These illustrations are called *Venn diagrams*, because they were popularized by George Venn [118, 119].

Here are Venn diagrams representing $R \cup S$, $R \cap S$, $\overline{R}$, $R \backslash S$ and $R \cap S \cap T$; the set represented is shown by the shaded area, and a universal set is also shown.

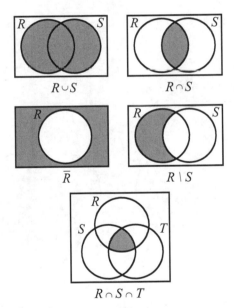

Two sets are equal if and only if they have the same Venn diagram. In order to illustrate this, we again consider the distributive law

$$R \cap (S \cup T) = (R \cap S) \cup (R \cap T). \tag{2}$$

The Venn diagram for $R \cap (S \cup T)$ is constructed in the upper half of Figure A.1, and that for $(R \cap S) \cup (R \cap T)$ is constructed in the lower half. The two are obviously identical.

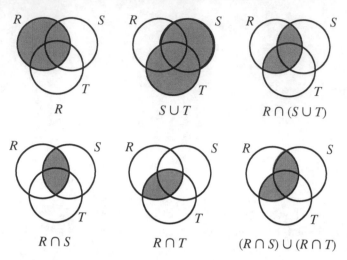

FIGURE A.1: $R \cap (S \cup T) = (R \cap S) \cup (R \cap T)$.

Example AA.3: Consider the set-theoretic identity

$$(R \cup S) \cap C \subseteq (R \cup T) \cap (S \cup T).$$

This can be established using Venn diagrams.

The diagram is:

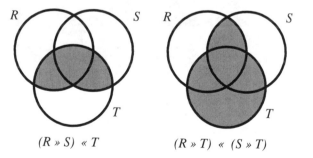

$(R » S) « T$ $\qquad\qquad$ $(R » T) « (S » T)$ $\qquad$ □

Sometimes we draw a Venn diagram in order to represent some properties of sets. For example, if A and B are disjoint sets, the diagram can be drawn with A and B shown as disjoint circles. If $A \subseteq B$, the circle for A is entirely inside the circle for B.

Example AA.4: Consider the argument:

All planets directly circle stars; our Moon does not directly circle the relevant star (the Sun); therefore the Moon is not a planet.

To examine this, we take the universal set to be the set of all heavenly bodies, A to be the set of all those bodies that directly circle stars, and B to be the set of all planets; m represents the Moon. We know $B \subseteq A$, so the sets look like the diagram. As $m \notin A$, it must be in the outer region, so it is certainly not in B. So the argument is valid. □

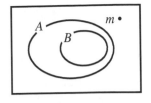

Some arguments that look logical at first sight turn out not to be valid. For example, consider:

All planets directly circle stars; Halley's Comet directly circles the relevant star (the Sun); therefore Halley's Comet is a planet.

We label the sets as in the example. In the diagram shown, Halley's Comet could be represented either by x or by y, so we can draw no conclusion. The argument is not valid. (In fact, x is correct.)

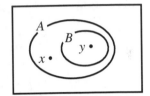

The *symmetric difference* of two sets consists of all elements in one or other but not both. That is, the symmetric difference of R and S is

$$R + S = \{x : x \in R \text{ or } x \in S \text{ but } x \notin R \cap S\}. \tag{9}$$

This definition could be stated as

$$\begin{aligned} R + S &= (R \cup S) \backslash (R \cap S) \\ &= (R \cup S) \cap \overline{(R \cap S)} \\ &= (R \cup S) \cap (\overline{R} \cup \overline{S}), \end{aligned} \tag{10}$$

using (7). We could also consider $A + B$ to be the union of the difference between A and B with the difference between B and A. This implies that

$$A + B = (A \backslash B) \cup (B \backslash A),$$

and hence that

$$A + B = (A \cap \overline{B}) \cup (A \cap \overline{B}). \tag{11}$$

It is easy to see that symmetric difference satisfies the commutative and associative laws.

We define the *Cartesian product* (or *cross product*) $S \times T$ of sets S and T to be the set of all ordered pairs (s, t) where $s \in S$ and $t \in T$:

$$S \times T = \{(s, t) : s \in S, t \in T\}.$$

There is no requirement that S and T be disjoint; in fact, it is often useful to consider $S \times S$.

The number of elements of $S \times T$ is $|S| \times |T|$. (This is one reason why the symbol $\times$ was chosen for Cartesian product.) As an example, consider $S = \{0, 1\}$ and $T = \{1, 2\}$. Then

$$S \times T = \{(0, 1), (0, 2), (1, 1), (1, 2)\},$$

the set of all four of the possible ordered pairs; $4 = 2 \times 2$.

The sets $(R \times S) \times T$ and $R \times (S \times T)$ are not equal; one consists of an ordered pair whose *first* element is itself an ordered pair, and the other of pairs in which the *second* is an ordered pair. So there is no associative law, and no natural meaning for $R \times S \times T$. On the other hand, it is sometimes natural to talk about ordered triples of elements, so we slightly abuse notation and write

$$R \times S \times T = \{(r, s, t) : r \in R, s \in S, t \in T\}.$$

This notation can be extended to ordered sets of any length.

There are several distributive laws involving the Cartesian product:

THEOREM AA.1 (Distributive Laws)
If R, S and T are any sets, then

(i) $R \times (S \cup T) = (R \times S) \cup (R \times T)$;
(ii) $R \times (S \cap T) = (R \times S) \cap (R \times T)$.

Proof. (i) We prove that every element of $R \times (S \cup T)$ is a member of $(R \times S) \cup (R \times T)$, and conversely.

First observe that

$$
\begin{aligned}
R \times (S \cup T) &= \{(r, s) \mid r \in R \text{ and } s \in S \cup T\} \\
&= \{(r, s) \mid r \in R \text{ and } (s \in S \text{ or } s \in T)\}.
\end{aligned}
$$

On the other hand,

$$
\begin{aligned}
(R \times S) \cup (R \times T) &= \{(r, s) \mid (r, s) \in R \times S \text{ or } (r, s) \in R \times T\} \\
&= \{(r, s) \mid (r \in R \text{ and } s \in S) \text{ or } (r \in R \text{ and } s \in T)\} \\
&= \{(r, s) \mid r \in R \text{ and } (s \in S \text{ or } s \in T)\}.
\end{aligned}
$$

It is now clear that $(r, s) \in R \times (S \cup T)$ and $(r, s) \in (R \times S) \cup (R \times T)$ are equivalent.

The proof of part (ii) is left as an exercise.   □

AA.2 The Principle of Mathematical Induction

In working with finite sets or with sets of integers, we shall repeatedly use a technique of proof known as the method of *mathematical induction*. The general idea is as follows: suppose we want to prove that every positive integer n has a property $P(n)$. We first prove $P(1)$ to be true. Then we prove that, for any n, the truth of $P(n)$ implies that of $P(n+1)$; in symbols:

$$P(n) \text{ true} \Rightarrow P(n+1) \text{ true.} \tag{12}$$

Intuitively, we would like to say:

$$
\begin{aligned}
&P(1) \text{ true} \\
&P(1) \text{ true} \quad \Rightarrow \quad P(2) \text{ true , by 12,} \\
\therefore \ &P(2) \text{ true;} \\
&P(2) \text{ true} \\
&P(2) \text{ true} \quad \Rightarrow \quad P(3) \text{ true , by 12,} \\
\therefore \ &P(3) \text{ true;}
\end{aligned}
$$

and so on. There is a difficulty, however: given any positive integer k, we can select an integer n such that the proof of $P(n)$ requires at least k steps, so the proof can be arbitrarily long. As "unbounded" proofs present logical difficulties in mathematics—who could ever finish writing one down?—we need an axiom or theorem that states that induction is a valid procedure. This is the *principle of mathematical induction*, and may be stated as follows.

Principle of mathematical induction: *Suppose the proposition $P(n)$ satisfies*

(i) *$P(1)$ is true; and*

(ii) *For every positive integer n, whenever $P(n)$ is true, then $P(n+1)$ is true.*

Then $P(n)$ is true for all positive integers n.

This principle is sometimes called *weak induction*. Another form is:

Strong induction: *Suppose the proposition $P(n)$ satisfies*

(i) *$P(1)$ is true; and*

(ii) *For every positive integer n, if $P(k)$ is true whenever $1 \le k < n$, then $P(n)$ is true.*

Then $P(n)$ is true for all positive integers n.

At first sight, the second statement looks as though we have assumed more than in the first statement. However these two forms are equivalent.

Induction can also be stated in strictly set-theoretic form. Suppose S is the set of integers n such that $P(n)$ is true. Then the principle of mathematical induction (weak induction form) is:

> *Let S be a subset of $\mathbb{Z}^+$ such that:*
> (i) $1 \in S$*; and*
> (ii) *whenever $n \in S$, then $n+1 \in S$.*
>   *Then $S = \mathbb{Z}^+$.*

One could equally well state the principle in terms of non-negative integers, instead of positive integers, by changing the case $P(0)$ to $P(1)$, and converting references from "positive integers" to "non-negative integers." It can in fact be stated in terms of any starting point: if S is a set of integers for which
   (i) $t \in S$,
where t is any integer, and
   (ii) for every integer $n \geq t$, if $n \in S$, then $n+1 \in S$,
then the principle can be used to prove that S contains all integers equal to or greater than t. This form is sometimes called "induction from t."

The Well-Ordering Principle

Another principle that is very useful in proving results about sets of positive integers is:

The well-ordering principle: *Suppose S is any non-empty set of positive integers. Then S has a smallest member.*

Any set with the property that every non-empty subset has a least member is called being *well-ordered*; so the principle says "the positive integers are well-ordered." Many number-sets, such as the real numbers, are not well-ordered.

At first the result seems obvious. If S is not empty, then it must contain some member, x say. In order to find the smallest member of S, one need only check to see whether or not $x-1, x-2, \ldots, 1$ are members of S, and this requires only a finite number of steps. However, this "proof" contains the same problem as the "proof" of the principle of mathematical induction: the starting value, x, can be arbitrarily large.

In fact, the induction principle can be proved from the well-ordering principle. To see this, let us assume the well-ordering principle is true and suppose P is any proposition about positive integers such that $P(1)$ is true, and for every positive integer n, whenever $P(n)$ is true, then $P(n+1)$ is true. (These are the requirements for induction.) Write S for the set of positive integers n such that $P(n)$ is not true. In order to prove that induction works, we need to show that S is empty.

If S is not empty, then by well-ordering S has a smallest member, x say. So $P(x)$ is false. We know that $P(1)$ is true, so $x \neq 1$. But x is a positive integer. So $x - 1$ is a positive integer, and $P(x)$ must be true—otherwise $x - 1$ would be a member of S, and smaller than x. But this means $P((x-1)+1)$ is true: that is, $P(x)$ is both true and false! This cannot happen, so our original assumption, that S is not empty, must have been wrong. So induction is proved.

We can also work the other way. If you assume induction is true, you can show that the positive integers are well-ordered.

However, there is no absolute proof here. It is necessary to assume that either induction or well-ordering is a property of the positive integers. So we assume these as axioms about numbers.

AA.3 Some Applications of Induction

THEOREM AA.2 (Power Set Order)
Let A be a set with $|A| = n$. Then $|\mathcal{P}(A)| = 2^n$.

Proof. To apply induction, we can rephrase the statement as: *Let $P(n)$ be the statement "$|\mathcal{P}(A)| = 2^n$ for any n-element set A"; then $P(n)$ is true for all positive integers n.* Since the elements of A do not matter, we may as well assume $A = \{a - 1, a_2, \dots, a_n\}$.

First we consider the case $n = 1$, so $A = \{a_1\}$. Then $\mathcal{P}(A) = \{\emptyset, A\}$, so $|\mathcal{P}(A)| = 2^1$ and in this case the theorem is true. Now suppose the result has been proved for $n = k - 1$ and assume $n = k$; we might as well say $A = A' \cup \{a_k\}$ where $A' = \{a - 1, a_2, \dots, a_{k-1}\}$. By the induction hypothesis $|\mathcal{P}(A')| = 2^{k-1}$. Any subset of A is either a subset of A' or a subset of A' with the element s_k adjoined to it, so to each subset of A' there correspond two subsets of A. Hence $|\mathcal{P}(A)| = 2^{k-1} \cdot 2$ and the theorem is proved. $\qquad\square$

THEOREM AA.3 (Order of Cartesian Product)
If $|S| = m$ and $|T| = n$, then $|S \times T| = mn$.

Proof. We proceed by induction on m. If $m = 1$, then $S = \{s_1\}$ and an ordered pair with s_1 as its first element may be constructed in n ways, giving $(s_1, t_1), (s_1, t_2), \dots, (s_1, t_n)$, so the theorem is true for $m = 1$. Now suppose the statement is true for $m = k - 1$, and consider the case $m = k$.

Let $S' = \{s_1, s_2, \dots, s_{k-1}\}$, so $|S' \times T| = (k - 1)n$. Also, $|\{s_k\} \times T| = n$. But $S \times T = (S' \times T) \cup (\{s_k\} \times T)$ and since the two products on the right-hand side of this equation are disjoint, we know that $|SxT| = (k - 1)n + n$, proving the theorem. $\qquad\square$

Mathematical induction is very often used in proving general algebraic formulas.

Example AA.5: Prove by induction that the sum of the first n positive integers is

$$1 + 2 + \cdots + n = \tfrac{1}{2}n(n+1).$$

The case $n = 1$ is $\tfrac{1}{2} \cdot 1(1+1) = 1$, which is obviously true, so the formula gives the correct answer when $n = 1$. Suppose it is true when $n = k - 1$; therefore

$$1 + 2 + \cdots + (k-1) = \tfrac{1}{2}(k-1)k.$$

Then

$$
\begin{aligned}
1 + 2 + \cdots + (k-1) + k &= \tfrac{1}{2}(k-1)k + k \\
&= \tfrac{1}{2}(k^2 - k + 2k) \\
&= \tfrac{1}{2}k(k+1),
\end{aligned}
$$

and the formula is proved correct when $n = k$. So, by induction, we have the required result. □

The following example looks geometrical, but it also yields to induction.

Example AA.6: Suppose n straight lines are drawn in two-dimensional space, in such a way that no three lines have a common point and no two lines are parallel. Prove that the lines divide the plane into $\tfrac{1}{2}(n^2 + n + 2)$ regions.

We proceed by induction. If $n = 1$, the formula yields $\tfrac{1}{2}(1 + 1 + 2) = 2$, and one line does indeed partition the plane into two regions. Now assume that the formula works for $n = k - 1$. Consider k lines drawn in the plane. Delete one line. By induction, the plane is divided into $\tfrac{1}{2}((k-1)^2 + (k-1) + 2) = \tfrac{1}{2}(k^2 - k + 2)$ regions.

Now reinsert the (deleted) kth line. It must cross every other line exactly once, so it crosses $(k-1)$ lines, and lies in k regions. It divides each of these regions into two parts, so the k regions are replaced by $2k$ new regions; the total is

$$
\begin{aligned}
\tfrac{1}{2}(k^2 - k + 2) - k + 2k &= \tfrac{1}{2}(k^2 - k + 2 - 2k + 4k) \\
&= (k^2 + k + 2),
\end{aligned}
$$

and the result is true by induction. □

Here is an example that uses induction from the starting point 4, rather than 0 or 1.

Example AA.7: Prove that $n! \geq 2^n$ whenever $n \geq 4$.

Suppose the proposition $P(n)$ means $n! \geq 2^n$. Then $P(4)$ means $4! \geq 2^4$, or $24 \geq 16$, which is true. Now suppose k is an integer greater than or equal to 4, and $P(k)$ is true: $k! \geq 2^k$. Multiplying by $k+1$, we have $(k+1)1 \geq (2^k(k+1) \geq 2^k 2 = 2^{k+1}$, so $P(k)$ implies $P(k+1)$, and the result follows by induction. □

Example AA.8: Prove by induction that $5^n - 2^n$ is divisible by 3 whenever n is a positive integer.

Suppose $P(n)$ means 3 divides $5^n - 2^n$. Then $P(1)$ is true because $5^1 - 2^1 = 3$. Now suppose k is any positive integer, and $P(k)$ is true: say $5^k - 2^k = 3x$, where x is an integer. Then $5^{k+1} - 2^{k+1} = 5 \cdot 5^k - 2 \cdot 2^k = 3 \cdot 5^k + 2 \cdot 5^k - 2 \cdot 2^k = 3 \cdot 5^k + 2 \cdot 3x$, which is divisible by 3. So the result follows by induction. □

AA.4 Binary Relations on Sets

A (*binary*) *relation* ρ from a set S to a set T is a rule that stipulates, given any element s of S and any element t of T, whether s bears a certain relationship to t (written $s \rho t$) or not (written $s \not\rho t$). For example, if S is the set of living males and T is the set of living females, the relation ρ might be "is the son of"; if s denotes a certain man and t denotes a certain woman, we write $s \rho t$ if s is the son of t, and $s \not\rho t$ otherwise.

Alternatively, we can define a binary relation ρ from the set S to the set T as a set ρ of ordered pairs (s, t), where s belongs to S and t belongs to T, with the notation that $s \rho t$ when $(s, t) \in \rho$ and $s \not\rho t$ otherwise. This means that, formally, a binary relation from S to T can be defined as a subset of the Cartesian product $S \times T$.

For example, suppose $S = \{1, 2, 3\}$ and $T = \{1, 2, 4\}$. The sets corresponding to $s < t$ and $s^2 = t$ are L and R, respectively, where

$$L = \{(1, 2), (1, 4), (2, 4), (3, 4)\}$$

and

$$R = \{(1, 1), (2, 4)\}.$$

If ρ and σ are binary relations from S to T, we define the union $\rho \cup \sigma$ and the intersection $\rho \cap \sigma$ from S to T in the obvious way: $(s, t) \in \rho \cup \sigma$ if and only if $(s, t) \in \rho$ or $(s, t) \in \sigma$ or both, and $(s, t) \in \rho \cap \sigma$ if and only if $(s, t) \in \rho$ and $(s, t) \in \sigma$. The notation agrees exactly with the usual set-theoretic notation.

If ρ is a binary relation from R to S, and σ a binary relation from S to T, then the *composition* $\rho\sigma$ is a binary relation from R to T, where $r(\rho\sigma)t$ if and only if there exists at least one element $s \in S$ such that $r\rho s$ and $s\sigma t$.

If ρ is a binary relation from R to S, and ν a binary relation from P to T, then the composition $\rho\nu$ is not defined unless $P = S$. But when it is defined, composition of binary relations is associative.

Every binary relation has an *inverse relation*: if ρ is a binary relation from S to T, so that $\rho \subseteq S \times T$, then ρ^{-1} is a binary relation from T to S, or $\rho^{-1} \subseteq T \times S$, defined by

$$\rho^{-1} = \{(t, s) : (s, t) \in \rho\}.$$

In other words, $t \rho^{-1} s$ if and only if $s \rho t$.

The inverse of a composition may be expressed as a composition of inverses:

$$(\rho\nu)^{-1} = \nu^{-1}\rho^{-1}.$$

Suppose ρ is a binary relation from a set A to A itself or, in other words, $\rho \subseteq A \times A$. Then we say that ρ is a binary relation *on* A. One obvious relation on a set is the *identity* relation: ι_A is the identity relation on A if $a \iota_A b$ is true precisely when $b = a$, that is, $\iota_A = \{(a, a) : a \in A\}$.

Suppose ρ is a relation on A.

(i) ρ is called *reflexive* if and only if $a \rho a$ for every $a \in A$. ρ is *irreflexive* if and only if $a \rho a$ is *never* true for any element a of A.

(ii) ρ is called *symmetric* if and only if $a \rho b$ implies $b \rho a$ or, in other words, if $(a, b) \in \rho$ implies $(b, a) \in \rho$. Since, by the definition of inverse relation, we know that $a \rho b$ if and only if $b \rho^{-1} a$, we could say that ρ is symmetric if and only if $\rho = \rho^{-1}$. ρ is called *antisymmetric* if and only if $a \rho b$ and $b \rho a$ together imply $a = b$ or, in other words, $\rho \cap \rho^{-1} \subseteq \iota_A$. (Clearly, a reflexive relation will trivially be antisymmetric.) If $a \rho b$ and $b \rho a$ can never both be true, then ρ is called *asymmetric*, or sometimes *skew*. (If $a \rho b$ and $b \rho a$ are never both true, then $(a \rho b \wedge b \rho a) \to a \rho a$ is vacuously true, so all asymmetric relations are antisymmetric, but this gives us no information.)

(iii) ρ is called *transitive* if and only if $a \rho b$ and $b \rho c$ together imply $a \rho c$, and *atransitive* if and only if, whenever $a \rho b$ and $b \rho c$, $a \not\rho c$, or in other words if and only if $\rho \cap \rho\rho = \emptyset$. Clearly, ρ is transitive if and only if $\rho\rho = \rho$.

As an example, the relation of *adjacency* on the set of integers, defined by $i \rho j$ if and only if $|i - j| = 1$ for i and j positive integers, is irreflexive (since $|i - i| = 0$), symmetric (since $|i - j| = |j - i|$) and not transitive (since $1 \rho 2$ and $2 \rho 3$ but $1 \not\rho 3$). It is in fact atransitive, for $i \rho j$ and $j \rho k$ together imply $i = j \pm 1$ and $k = j \pm 1$, so $i = k$ or $i = k \pm 2$, but i cannot equal $c \pm 1$. So $a \not\rho c$.

A relation α on a set A is called an *equivalence relation* on A if and only if it is reflexive, symmetric, and transitive.

The obvious equivalence relation is equality, on any set. In sets other than number sets, equal objects are often called "equal in all respects." More generally, an equivalence relation can be considered as a statement that two objects are "equal in some (specified) respects." One example, on the integers, is the relation α, where $a \alpha b$ is true if and only if $a = \pm b$—a and b have the same absolute value. Another is congruence, on the set of all plane triangles.

Equivalence relations give us an alternative way to discuss multiple elements. You can view a multiset as an object that is based on a set of distinct elements on which an equivalence relation has been defined, so that the elements of a multiset are the equivalence classes of the underlying set.

The word '*congruence* is also used as follows: the relation "congruence modulo n" for a positive integer n is defined on the set $\mathbb{Z}$ of integers by specifying that a is congruent to b modulo n, written "$a \equiv b (\mathrm{mod}\, n)$," if and only if $n \mid (a - b)$. In other words, two integers are congruent modulo n if and only if they belong to the same residue class modulo n. It is easy to show that congruence modulo n is an equivalence relation: since $n \mid (a - a)$, the relation is reflexive; if $n \mid (a - b)$ then $n \mid (b - a)$, so the relation is symmetric; if $n \mid (a - b)$ and $n \mid (b - c)$ then we have $a - b = kn$, and $b - c = ln$, for some $k, l \in \mathbb{Z}$, hence $a - c = (k + l)n$ and $n \mid (a - c)$, which shows that the relation is transitive. So congruence modulo n is an equivalence relation on the integers.

Appendix 1 Exercises A

1. Suppose $A = \{a, b, c, d, e\}$, $B = \{a, c, e, g, i\}$, $C = \{c, f, i, e, o\}$. Write down the elements of

   (i) $A \cup B$.   (ii) $A \cap C$.

   (iii) $A \backslash B$.   (iv) $A \cup (B \backslash C)$.

2. Consider the sets

$$
\begin{aligned}
S_1 &= \{2, 5\}, \\
S_2 &= \{1, 2, 4\}, \\
S_3 &= \{1, 2, 4, 5, 10\}, \\
S_4 &= \{x \in \mathbb{Z}^+ : x \text{ is a divisor of } 20\}, \\
S_5 &= \{x \in \mathbb{Z}^+ : x \text{ is a power of 2 and a divisor of } 20\}.
\end{aligned}
$$

   For which i and j, if any, is $S_i \subseteq S_j$? For which i and j, if any, is $S_i = S_j$?

3. If $S \cup T = U$ and $S \cap T = \emptyset$, prove that $T = \overline{S}$.

4. Consider the data: *All mathematicians are loners; all surgeons are rich; no loners are rich.* Which of the following conclusions can be drawn?

   (i) No mathematicians are rich.

   (ii) All surgeons are loners.

   (iii) No one can be both a mathematician and a surgeon.

5. Show that

$$\overline{R} + \overline{S} = \overline{R + S}$$

for any sets R and S.

6. Show that the following three statements are equivalent: $S \subseteq T, S \cup T = T, S \cap T = S$.

7. Prove the given proposition by induction.

    (i) The sum of the first n odd positive integers is n^2.

    (ii) $\displaystyle\sum_{r=1}^{n} r^2 = \frac{1}{6}n(n+1)(2n+1)$.

8. Prove that the sum of the cubes of any three consecutive integers is a multiple of 9.

9. The numbers $x_1, x_2, \ldots$ are defined as follows. $x_1 = 1$, $x_2 = 1$, and if $n \geq 2$ then $x_{n+1} = x_n + 2x_{n-1}$. Prove that x_n is divisible by 3 if and only if n is divisible by 3.

10. Suppose ρ and σ are relations defined on $S = \{1, 2, 3, 4, 5\}$ by the sets
$$\rho = \{(2,2),(3,2),(3,4),(4,1),(4,3),(5,5)\};$$
$$\sigma = \{(2,3),(1,2),(3,1),(3,2),(4,4),(4,5),(5,4),(5,5)\}.$$

    (i) Find the set form of the composition $\rho\sigma$.

    (ii) What is the set form of the relation $\sigma\rho$?

    (iii) What are the set forms of the relations $\rho\rho$ and $\sigma\sigma$?

11. Suppose A is the set $\mathbb{Z} \times \mathbb{Z}$. Prove that the relation ρ on A, defined by

$$(x, y) \, \rho \, (z, t) \text{ if and only if } x + t = y + z,$$

is an equivalence relation.

Appendix 1 Exercises B

1. Suppose

$$A = \{2, 3, 5, 6, 8, 9\} \quad B = \{1, 2, 3, 4, 5\} \quad C = \{5, 6, 7, 8, 9\}.$$

Write down the elements of

(i) $A \cap B$.

(ii) $A \cup C$.

(iii) $A \backslash (B \cap C)$.

(iv) $(A \cup B) \backslash C$.

2. Suppose

$$A = \{a, b, c, d, e, f\} \quad B = \{a, b, g, h\} \quad C = \{a, c, e, f, h\}.$$

Write down the elements of

(i) $A \cap B$.

(ii) $A \cup C$.

(iii) $A \backslash (B \cap C)$.

(iv) $(A \cup B) \backslash C$.

3. Are the given statements true or false?

(i) $2 \in \{2, 3, 4, 5\}$.   (ii) $3 \notin \{2, 3, 4, 5\}$.   (iii) $3 \in \{2, 4, 5, 6\}$.

(iv) $\{1, 2\} = \{2, 1\}$.   (v) $\{1, 3\} \in \{1, 2, 3\}$. (vi) $\{1, 3\} = \{1, 2, 3\}$.

4. In each case, are the sets S and T disjoint? If not, what is their intersection?

(i) S is the set of perfect squares $1, 4, 9, \ldots$; T is the set of cubes $1, 8, 27, \ldots$ of positive integers.

(ii) S is the set of perfect squares $1, 4, 9, \ldots$; T is the set Π of primes.

5. R, S and T are any sets, and U is the universal set.

(i) Prove: if $R \subseteq S$ and $R \subseteq T$, then $R \subseteq (S \cap T)$.

(ii) Prove: if $R \subseteq S$, then $R \subseteq (S \cup T)$.

6. Find a simpler expression for $S \cup ((\overline{R \cup S}) \cap R)$.

7. For any sets R and S, prove $R \cap (R \cup S) = R$.

8. In each case, represent the set in a Venn diagram.

(i) $(R \backslash S) \backslash T$

(ii) $R \backslash (S \backslash T)$

(iii) $(R \cap S) \backslash (S \cap T)$

(iv) $(R \cup T) \backslash (S \cup T)$

9. Consider the data: *the city zoo keeps all species of brown bears; all large bears are dangerous; the city zoo does not keep any dangerous bears.* Which of the following conclusions can be drawn?

(i) No large bears are brown.

(ii) All species of large bears are kept in the city zoo.

(iii) Some large bears are kept in the city zoo.

(iv) No large bears are kept in the city zoo.

(v) No dangerous bears are kept in the city zoo.

10. Draw Venn diagrams for use in the following circumstances:
  (i) All members of this class are math majors;
  (ii) No members of this class are math majors.

11. Prove the two distributive laws

$$A \cap (B + C) \quad = \quad (A \cap B) + (A \cap C);$$
$$(A + B) \cap C \quad = \quad (A \cap C) + (B \cap C).$$

12. Show that if $S \cup T = \emptyset$, then $S = T = \emptyset$ and that if $S \cap T = U$, then $S = T = U$.

13. Prove that $R \backslash (S \backslash T)$ contains all members of $R \cap T$, and hence prove that

$$(R \backslash S) \backslash T = R \backslash (S \backslash T)$$

is *not* a general law (in other words, relative difference is *not* associative).
  (i) First, prove this without using Venn diagrams.
  (ii) Second, prove the result using Venn diagrams.

14. Prove that union is not distributive over relative difference: in other words, prove that the following statement is not always true:

$$(R \backslash S) \cup T = (R \cup T) \backslash (S \cup T).$$

  (i) First, prove this without using Venn diagrams. (Hint: Use the fact $(R \backslash S) \cup S = R \cup S$.)
  (ii) Now prove the result using Venn diagrams.

15. Prove the associative law for symmetric difference.

16. Prove that $A + B + C$ consists of precisely those elements that belong to an *odd* number of the sets A, B and C.

17. Prove the given proposition by induction.

  (i) $\displaystyle \sum_{k=1}^{n} k^3 = \left[\sum_{k=1}^{n} k \right]^2 = \frac{1}{4} n^2 (n+1)^2.$

  (ii) $1 + 4 + 7 + \cdots + (3n - 2) = \frac{1}{2} n (3n - 1).$

18. The numbers $a_0, a_1, a_2, \ldots$ are defined by $a_0 = 3$ and

$$a_{n+1} = 2a_n - a_n^2 \text{ when } n \geq 0.$$

Prove that $a_n = 1 - 2^{2^n}$ when $n > 0$, although this formula does not apply when $n = 0$.

19. Prove the given proposition by induction.

    (i) $2 + 6 + 12 + \cdots + n(n+1) = \sum_{k=1}^{n} k(k+1) = \frac{1}{3}n(n+1)(n+2)$.

    (ii) $\dfrac{1}{1 \cdot 3} + \dfrac{1}{3 \cdot 5} + \dfrac{1}{5 \cdot 7} + \cdots + \dfrac{1}{(2n-1)(2n+1)} = \dfrac{n}{2n+1}$.

    (iii) $1 + 3 + 3^2 + \cdots + 3^n = \frac{1}{2}(3^{n+1} - 1)$.

    (iv) $n^2 \geq 2n + 1$ whenever $n \geq 3$.

20. The numbers $a_0, a_1, a_2, \ldots$ are defined by $a_0 = \frac{1}{4}$ and

    $$a_{n+1} = 2a_n(1 - a_n) \text{ when } n > 0.$$

    Prove that
    $$a_n = \frac{1}{2}\left(1 - \frac{1}{2^{2^n}}\right).$$

21. Prove that the following divisibility results hold for all positive integers n.

    (i) 2 divides $3^n - 1$.

    (ii) 6 divides $n^3 - n$.

    (iii) 5 divides $2^{2n-1} + 3^{2n-1}$.

    (iv) 24 divides $n^4 - 6n^3 + 23n^2 - 18n$.

22. In each case a binary relation from $S = \{1, 2, 3, 4, 5, 6, 7, 8, 9\}$ to itself is defined. For each of these relations, what is the corresponding subset of $S \times S$? For each relation, what is the inverse relation?

    (i) $x \, \rho \, y$ means $y = x^2$.

    (ii) $x \, \beta \, y$ means $xy = 6$.

    (iii) $x \, \delta \, y$ means $3x < 2y$.

23. The relation ρ is defined on $\{1, 2, 3, 4, 5\}$ by

    $$\rho = \{(1,1), (2,2), (2,3), (3,3), (3,4), (4,3), (5,5)\}.$$

    Prove that ρ has none of the properties: reflexive, transitive, symmetric, antisymmetric.

24. Define the relations ρ, σ and τ on the set $\{1, 2, 3, 4\}$ by the sets
    $$\rho = \{(1,1), (2,2), (3,3), (4,4)\};$$
    $$\sigma = \{(1,1), (1,3), (2,2), (3,3), (3,1), (4,4)\};$$
    $$\tau = \{(2,3), (1,2), (1,3), (3,2)\}.$$

    Which of these relations are

(i) reflexive;   (iv) antisymmetric;
(ii) irreflexive;   (v) asymmetric;
(iii) symmetric;   (vi) transitive?

Are any of them equivalence relations?

25. A group of people, denoted by $P, Q, \ldots$, are at a lecture. Consider the following relations: Is each reflexive, symmetric, antisymmetric, asymmetric or transitive? Is it an equivalence relation?

   (i) $P \mu Q$ means P and Q come from the same city.
   (ii) $P \nu Q$ means P is the father of Q.
   (iii) $P \rho Q$ means P and Q spoke to each other at the lecture.
   (iv) $P \sigma Q$ means P and Q met for the first time at the lecture.
   (v) $P \tau Q$ means P spoke to Q at the lecture.

26. Verify: If ρ is a binary relation from R to S, and ν a binary relation from P to T, then the composition $\rho \nu$ is not defined unless $\rho(R) \subseteq P$. But when it is defined, composition of binary relations is associative.

27. Let α and β be equivalence relations on a set A.
   (i) Show that $\alpha \beta$ is an equivalence relation if and only if $\alpha \beta = \beta \alpha$.
   (ii) Show that $\alpha \cap \beta$ is an equivalence relation.

28. Prove that the inverse of a composition may be expressed as a composition of inverses:
$$(\rho \nu)^{-1} = \nu^{-1} \rho^{-1}.$$

29. Suppose p is a prime. The relation ρ on $\mathbb{Z}$ is defined by $a \rho b$ if and only if $a^2 \equiv b^2 (\mathrm{mod}\, p)$. Show that ρ is an equivalence relation.

Appendix B: Matrices and Vectors

B.1 Definitions

A *matrix* is a rectangular array whose elements—the *entries* of the matrix—are members of an algebraic structure with the operations of addition and multiplication, where multiplication is commutative. Most commonly the entries will be chosen from the real numbers, but one could restrict them to integers, or to the two numbers 0 and 1, or allow members of some integral domain or a more general structure. (We refer to a *real matrix*, a *zero-one matrix*, and so on.) The possible entries are called the *scalars* for the particular class of matrices.

The horizontal layers in a matrix are called *rows* and the vertical ones are *columns*. A matrix with m rows and n columns is called an $m \times n$ matrix; we refer to $m \times n$ as the *shape* or *size* of the matrix, and the two numbers m and n are its *dimensions*.

Matrices are usually denoted by single uppercase letters. As an example, consider the matrix

$$M = \begin{bmatrix} 1 & 0 & 1 & -2 \\ -1 & 1 & -4 & 1 \\ 1 & -3 & 0 & 2 \end{bmatrix}.$$

The first row of M is

$$\boxed{\begin{array}{cccc} 1 & 0 & 1 & -2 \end{array}}$$

and the second column is

$$\boxed{\begin{array}{c} 0 \\ 1 \\ 1 \end{array}}.$$

We refer to the entry in the ith row and jth column of a matrix as the (i, j)-*element* (or *entry*). The $(2, 3)$ element of M is -4. When no confusion is possible, we write a_{ij} to denote the (i, j)-element of a matrix A, using the lowercase letter corresponding to the (uppercase) name of the matrix, with the row and column numbers as subscripts. A common shorthand is to write $[a_{ij}]$ for A. We use the ordinary equality sign between matrices; $A = B$ means that A and B are the same shape and $a_{ij} = b_{ij}$ for every i and j.

In general if M and N are two matrices with the same shape, the *sum* of two matrices $M = [m_{ij}]$ and $N = [t_{ij}]$ is the matrix $S = [s_{ij}]$ defined by

$$s_{ij} = m_{ij} + t_{ij}.$$

There is no matrix $M + N$ if M and N are of different shapes. For example, if

$$N = \begin{bmatrix} 1 & -3 & 1 & 0 \\ -2 & 1 & -1 & 2 \\ 1 & 3 & 2 & -3 \end{bmatrix}, \quad R = \begin{bmatrix} 1 & -1 & 1 & 1 \\ -1 & 2 & -2 & 1 \end{bmatrix},$$

then

$$M + N = \begin{bmatrix} 1 & 0 & 1 & -2 \\ -1 & 1 & -4 & 1 \\ 1 & -3 & 0 & 2 \end{bmatrix} + \begin{bmatrix} 1 & -3 & 1 & 0 \\ -2 & 1 & -1 & 2 \\ 1 & 3 & 2 & -3 \end{bmatrix}$$

$$= \begin{bmatrix} 2 & -3 & 2 & -2 \\ -3 & 2 & -5 & 3 \\ 2 & 0 & 2 & -1 \end{bmatrix}$$

but $M + R$ is not defined.

It is easy to see that this addition satisfies the commutative and associative laws: if M, N and P are any matrices of the same shape, then

$$M + N = NM \quad \text{and} \quad (M + N) + P = M + (N + P).$$

Because of the associative law, we usually omit the brackets and just write $M + N + P$.

We can take the sum of a matrix with itself, and we write $2M$ for $M + M$, $3M$ for $M + M + M$, and so on. This can be extended to multipliers other than positive integers: if a is any scalar, aM will mean the matrix derived from M by multiplying every entry by a. That is,

$$aM \text{ is the matrix with } (i, j) \text{ entry } am_{ij}.$$

This matrix aM is the *scalar product* (or simply *product*) of a with M. It has the same shape as M. The scalar product obeys the laws

$$(a + b)M = aM + bM, \quad a(bM) = (ab)M, \quad a(M + N) = aM + aN$$

for any matrices M and N of the same shape and any scalars a and b. Notice that if 1_s is the multiplicative identity element of the scalars (for example, for real matrices, 1_s represents 1), then $1_s M = M$.

O_{mn} denotes a matrix of shape $m \times n$ with every entry zero. Usually we do not bother to write the subscripts m and n, but simply assume that the matrix is the correct shape for our computations. O is called a *zero matrix*, and works for matrix addition like the number zero under ordinary addition: if M is any matrix, then

$$M + O = M$$

provided O has the same shape as M.

Obviously $0M = O$ for any matrix M (where 0 is the scalar zero and O is the zero matrix). From the first law for scalar multiplication, above,

$$M + (-1)M \;=\; 1M + (-1)M \;=\; (1 + (-1))M \;=\; 0M \;=\; O,$$

so $(-1)M$ acts like a negation of M. We shall simply write $-M$ instead of $(-1)M$, and call $-M$ the *negative* of M. We can then define *subtraction of matrices* by

$$M - N \;=\; M + (-N),$$

just as you would expect. Again, subtraction only works if the matrices are the same shape. The usual conventions are followed: $-M + N$ means $(-M) + N$, and to show the negative of $M + N$ you have to write $-(M + N)$. To evaluate $M - N + P$ you first subtract N from M, then add P; because of the associative law, no special brackets are needed.

If A is an $m \times n$ matrix, then we can form an $n \times m$ matrix whose (i, j) entry equals the (j, i) entry of A. This new matrix is called the *transpose* of A, and written A^T. A matrix A is called *symmetric* if $A = A^T$.

A matrix with one of its dimensions equal to 1 is called a *vector*. An $m \times 1$ matrix is a *column vector* of length m, while a $1 \times n$ matrix is a *row vector* of length n. The individual rows and columns of a matrix are vectors, which we call the *row vectors* and *column vectors* of the matrix.

We shall write vectors with boldface lowercase letters, to distinguish them from matrices and scalars. (We treat vectors separately from matrices because, in many cases, it is not necessary to distinguish between row and column vectors.) In some books a vector is denoted by a lowercase letter with an arrow over it, $\vec{v}$, rather than a boldface letter $\boldsymbol{v}$. The ith entry of a vector is denoted by a subscript i, so the vector $\boldsymbol{v}$ has entries $v_1, v_2, \ldots$; we usually write $\boldsymbol{v} = (v_1, v_2, \ldots)$.

The two standard operations on vectors follow directly from the matrix operations. One may *multiply by a number*, and one may *add vectors*. If k is any number, and $\boldsymbol{v} = (v_1, v_2, \ldots, v_n)$, then $k\boldsymbol{v} = (kv_1, kv_2, \ldots, kv_n)$. If $\boldsymbol{u} = (u_1, u_2, \ldots, u_n)$, and $\boldsymbol{v} = (v_1, v_2, \ldots, v_n)$, then $\boldsymbol{u} + \boldsymbol{v} = ((u_1 + v_1), (u_2 + v_2), \ldots, (u_n + v_n))$. If $\boldsymbol{u}$ and $\boldsymbol{v}$ are vectors of different lengths, then $\boldsymbol{u} + \boldsymbol{v}$ is not defined. We again write $-\boldsymbol{v}$ for $(-1)\boldsymbol{v}$, so $-\boldsymbol{v} = (-v_1, -v_2, \ldots, -v_n)$, and $\boldsymbol{u} - \boldsymbol{v} = \boldsymbol{u} + (-\boldsymbol{v})$. We define a *zero vector* $\boldsymbol{0} = (0, 0, \ldots, 0)$ (in fact, a family of zero vectors, one for each possible dimension), and $\boldsymbol{v} + (-\boldsymbol{v}) = 0$.

Sometimes there is no important difference between the vector $\boldsymbol{v}$ of length n, the $1 \times n$ matrix (row vector) whose entries are the entries of $\boldsymbol{v}$, and the $n \times 1$ matrix (column vector) whose entries are the entries of $\boldsymbol{v}$. But in the next section we shall sometimes need to know whether a vector has been written as a row or a column. If this is important, we shall write $row(\boldsymbol{v})$ for the row vector form of $\boldsymbol{v}$, and $col(\boldsymbol{v})$ for the column vector form. If we simply write $\boldsymbol{v}$, you usually can tell from the context whether $row(\boldsymbol{v})$ or $col(\boldsymbol{v})$ is intended.

B.2 Vector and Matrix Products

We define the *scalar product* (also called the *dot product*) of two vectors $u = (u_1, u_2, \ldots, u_n)$ and $v = (v_1, v_2, \ldots, v_n)$ to be

$$u \cdot v = (u_1 v_1 + u_2 v_2 + \cdots + u_n v_n) = \sum_{k=1}^{n} u_k v_k.$$

Example B.1: Suppose $t = (1, 2, 4), u = (-1, 3, 0)$ and $v = (1, -2, 1)$. Calculate $t \cdot u, (t - u) \cdot v$ and $(u \cdot t)$.

$$
\begin{aligned}
t \cdot u &= 1.(-1) + 2.3 + 3.0 = (-1) + 6 + 0 = 5, \\
(t - u) \cdot v &= (2, -1, 4) \cdot (1, -2, 1) = 2 + 2 + 4 = 8, \\
u \cdot t &= (-1).1 + 3.2 + 0.3 = (-1) + 6 + 0 = 5.
\end{aligned}
$$
□

Clearly the dot product is commutative. There is no need to discuss the associative law, because dot products involving three vectors are not defined. For example, consider $t \cdot (u \cdot v)$. Since $(u \cdot v)$ is a scalar, not a vector, we cannot calculate its dot product with anything.

The *product* of two matrices is a generalization of the scalar product of vectors. Suppose the rows of the matrix A are $a_1, a_2, \ldots$, and the columns of the matrix B are $b_1, b_2, \ldots$. Then AB is the matrix with (i, j) entry $a_i \cdot b_j$. These entries will only exist if the number of columns of A equals the number of rows of A, so this is a necessary condition for the product AB to exist.

Example B.2: Suppose

$$A = \begin{bmatrix} 1 & 2 \\ 1 & -1 \end{bmatrix}, B = \begin{bmatrix} -1 & 1 & 3 \\ 2 & -2 & 0 \end{bmatrix}.$$

What are AB and BA?

$$AB = \begin{bmatrix} -1+4 & 1-4 & 3+0 \\ -1-2 & 1+2 & 3+0 \end{bmatrix} = \begin{bmatrix} 3 & -3 & 3 \\ -3 & 3 & 3 \end{bmatrix}.$$

However, BA does not exist. □

As another example consider the matrix M of the preceding section. Suppose Q is any $m \times n$ matrix. Since M is 3×4, the product MQ will be a $3 \times n$ matrix if $m = 4$, but it is not defined when $n \neq 4$. In general, we have

THEOREM B.1 (Conformability)
Suppose A is an $m \times n$ matrix and B is an $r \times s$ matrix. If $n = r$, then AB exists and is an $m \times s$ matrix. If $n \neq r$, then AB does not exist.

If the dimensions are such that the product AB exists, the matrices A and B are called *conformable*.

Matrix multiplication satisfies the associative law: provided the relevant products all exist,

$$A(BC) = (AB)C$$

and the product is usually written ABC. However, the commutative law is a different story. If A is 2×3 and B is 3×4, then AB is a 2×4 matrix but BA does not exist. It is also possible that AB and BA might both exist but might be of different shapes; for example, if A and B have shapes 2×3 and 3×2, respectively, then AB is 2×2 and BA is 3×3. And even when AB and BA both exist and are the same shape, they need not be equal. For example,

$$\begin{bmatrix} 1 & 2 \\ 1 & 3 \end{bmatrix} \begin{bmatrix} 1 & 1 \\ 2 & 0 \end{bmatrix} = \begin{bmatrix} 5 & 1 \\ 7 & 1 \end{bmatrix};$$

$$\begin{bmatrix} 1 & 1 \\ 2 & 0 \end{bmatrix} \begin{bmatrix} 1 & 2 \\ 1 & 3 \end{bmatrix} = \begin{bmatrix} 2 & 5 \\ 2 & 4 \end{bmatrix}.$$

If $AB = BA$ we say the matrices A and B *commute*.

If AA is to exist, then A must have the same number of rows as columns; such matrices are called *square*. The common dimension is called the *order* of the matrix.

If A is square, we can evaluate the product AA. We call this A *squared*, and write it as A^2, just as with powers of numbers. We define other positive integer powers similarly: $A^3 = AAA = AA^2$, and in general $A^{n+1} = AA^n$. As with numbers, indices apply only to the symbol nearest: AB^n means $A(B^n)$, not $(AB)^n$. (Similarly, AB^T means $A(B^T)$, not $(AB)^T$.)

The zero matrix behaves under multiplication the way you would expect: provided zero matrices of appropriate size are used,

$$OA = O \text{ and } AO = O.$$

This is not just one rule, but an infinite set of rules. If we write in the subscripts, then the full statement is

If A is any $r \times s$ matrix, then $O_{m,r} A = O_{m,s}$ for any positive integer m, and $AO_{s,n} = O_{r,n}$ for any positive integer n.

There are also matrices that act like the number 1: multiplicative identity elements. We define I_n to be the $n \times n$ matrix with its $(1,1), (2,2), \ldots, (n,n)$ entries 1 and all other entries 0. For example,

$$I_3 = \begin{bmatrix} 1 & 0 & 0 \\ 0 & 1 & 0 \\ 0 & 0 & 1 \end{bmatrix}.$$

If A is any $r \times s$ matrix, then $I_r A = A = A I_s$. We call I_n an *identity matrix* of order n.

Another way of combining matrices is to form the *Cartesian product* of two matrices. If A is a $p \times q$ matrix and B is an $r \times s$ matrix, their cartesian product $A \times B$ (or $A \otimes B$) is formed as follows. In A, replace entry a_{ij} by the scalar product $a_{ij}B$. That is, each cell is transformed into an $r \times s$ block of cells. Therefore $A \times B$ is a $pr \times qs$ matrix.

B.3 Inverses

If the matrices A and B satisfy $AB = BA = I$, we say that B is an *inverse* of A.

Which matrices have inverses? In the real numbers, everything but 0 has an inverse. In the integers, only 1 and -1 have (integer) inverses, but we can obtain inverses of other non-zero integers by going to the rational numbers. The situation is more complicated for matrices.

Suppose A is an $r \times c$ matrix. Then AB has r rows and BA has c columns. If the two are to be equal, then AB is an $r \times c$ matrix; and if this is to equal an identity matrix, it must be square. So $r = c$, and A is also a square matrix. So only a square matrix can have an inverse.

Moreover, there are non-zero square matrices without inverses. One obvious example is a matrix with one row or one column all zeroes. But there are less trivial examples, even if we restrict our attention to the 2×2 case.

Example B.3: Show that the matrix

$$A = \begin{bmatrix} 2 & 1 \\ 1 & 1 \end{bmatrix}$$

has an inverse, by construction.

If

$$B = \begin{bmatrix} x & y \\ z & t \end{bmatrix}$$

is an inverse, then $AB = I$, so

$$\begin{bmatrix} 2 & 1 \\ 1 & 1 \end{bmatrix}\begin{bmatrix} x & y \\ z & t \end{bmatrix} = \begin{bmatrix} 2x + z & 2y + t \\ x + z & 2y + t \end{bmatrix} = \begin{bmatrix} 1 & 0 \\ 0 & 1 \end{bmatrix}.$$

Comparing the $(1,1)$ entries of the two matrices, so $2x + z = 1$; from the $(2,1)$ entries, $x + z = 0$. So $x = 1$ and $z = -1$. Similarly, $y = -1, t = 2$. So

$$B = \begin{bmatrix} 1 & -1 \\ -1 & 2 \end{bmatrix}.$$

On the other hand,

$$C = \begin{bmatrix} 3 & 1 \\ 3 & 1 \end{bmatrix}$$

has no inverse: if

$$D = \begin{bmatrix} x & y \\ z & t \end{bmatrix}$$

were an inverse, then

$$\begin{bmatrix} 2 & 1 \\ 2 & 1 \end{bmatrix}\begin{bmatrix} x & y \\ z & t \end{bmatrix} = \begin{bmatrix} 3x+z & 3y+t \\ 3x+z & 3y+t \end{bmatrix} = \begin{bmatrix} 1 & 0 \\ 0 & 1 \end{bmatrix}.$$

Comparing $(1,1)$ entries again, $3x+z = 1$, but the $(2,1)$ entries give $3x+z = 0$. This is impossible.

A matrix that has an inverse will be called *invertible* or *non-singular*; a square matrix without an inverse is called *singular*.

We said "an inverse" above; however, if a matrix has an inverse, it is unique. We say "the inverse" (if one exists) and write A^{-1}.

THEOREM B.2 (Uniqueness of Inverse)
If matrices A, B, C satisfy $AB = BA = I$ and $AC = CA = I$, then $B = C$.

Proof. Suppose A, B and C satisfy the given equations. Then

$$C = CI = C(AB) = (CA)B = IB = B$$

so B and C are equal. □

In fact, either of the conditions $AB = I$ or $BA = I$ implies the other, and therefore implies that B is the inverse of A.

As we said, the usual notation for the inverse of A (if it exists) is A^{-1}. If we define $A^0 = I$ whenever A is square, then powers of matrices satisfy the usual index laws

$$A^m A^n = A^{m+n}, (A^m)^n = A^{mn}$$

for all non-negative integers m and n, and for negative values also provided that A^{-1} exists. If x and y are non-zero reals, then $(xy)^{-1} = x^{-1}y^{-1}$. The fact that matrices do not necessarily commute means that we have to be a little more careful, and prove the following theorem.

THEOREM B.3 (Invertibility of Product)
If A and B are invertible matrices of the same order, then AB is invertible, and

$$(AB)^{-1} = B^{-1}A^{-1}.$$

Proof. We need to show that both $(B^{-1}A^{-1})(AB)$ and $(AB)(B^{-1}A^{-1})$ equal the identity. But $(B^{-1}A^{-1})(AB) = B^{-1}(A^{-1}A)B = B^{-1}IB = B^{-1}B = I = AA^{-1} = AIA^{-1} = A(BB^{-1})A^{-1} = (AB)(B^{-1}A^{-1})$. □

There are two cancellation laws for matrix multiplication. If A is an invertible $r \times r$ matrix and B and C are $r \times s$ matrices such that $AB = AC$, then

$$AB = AC \quad \Rightarrow \quad A^{-1}(AB) = A^{-1}(AC)$$
$$\Rightarrow \quad (A^{-1}A)B = (A^{-1}A)C \Rightarrow IB = IC \Rightarrow B = C,$$

so $B = C$. Similarly, if A is an invertible $s \times s$ matrix and B and C are $r \times s$ matrices such that $BA = CA$, then $B = C$.

The requirement that A be invertible is necessary. We can find matrices A, B and C such that AB and AC are the same size, A is non-zero and $AB = AC$, but B and C are different. One very easy example is

$$\begin{bmatrix} 1 & 0 \\ 0 & 0 \end{bmatrix} \begin{bmatrix} 1 & 1 \\ 2 & 1 \end{bmatrix} = \begin{bmatrix} 1 & 0 \\ 0 & 0 \end{bmatrix} \begin{bmatrix} 1 & 1 \\ 1 & 2 \end{bmatrix}.$$

Some other examples are given in the exercises.

Moreover we can only cancel on one side of an equation; we cannot mix the two sides. Even if A is invertible it is possible that $AB = CA$ but $B \neq C$ (see the exercises for examples of this, also).

B.4 Determinants

THEOREM B.4 (Invertibility of 2 by 2 Matrices)
The matrix

$$A = \begin{bmatrix} a & b \\ c & d \end{bmatrix}$$

is singular if $ad - bc = 0$. Otherwise it is invertible, with inverse

$$B = \frac{1}{(ad - bc)} \begin{bmatrix} d & -b \\ -c & a \end{bmatrix}.$$

Proof. If $ad - bc \neq 0$, it is easy to see that $AB = I$. So we assume $ad - bc = 0$ and prove that A has no inverse.

If $a = 0$, then either b or c must be zero, and A will have a zero row or column (and consequently no inverse). So we need only consider cases where $a \neq 0$, and we can write $d = a^{-1}bc$. Similar considerations show that we can also assume c is non-zero. Now suppose

$$\begin{bmatrix} a & b \\ c & a^{-1}bc \end{bmatrix} \begin{bmatrix} x & y \\ z & t \end{bmatrix} = \begin{bmatrix} 1 & 0 \\ 0 & 1 \end{bmatrix}.$$

Comparing $(1,1)$ entries, $ax + bz = 1$, so $x = a^{-1} - a^{-1}bz$. On the other hand, the $(2,1)$ entries give $cx + a - 1bcz = 0$, or $x = -a^{-1}bz$. So $a^{-1} = 0$, an impossibility. □

The quantity $ad - bc$ is called the *determinant* of the matrix A, written $\det(A)$. Determinants are defined recursively for square matrices of any order: if A is the $n \times n$ matrix $[a_{ij}]$, where $n > 2$, write A_{ij} for the matrix derived from A by deleting row i and column j, and define

$$\det(A) = \sum_{j=1}^{n} (-1)^{i+j} \det(A_{ij}).$$

This is called the *expansion along the ith row*; it may be shown that the value does not depend on the choice of i, and there is a similar formula for expansion along columns. For further discussion of determinants, see standard texts on matrix theory and linear algebra, such as [3]; among other things, a square matrix is non-singular if and only if its determinant is non-zero.

Appendix B Exercises A

1. Carry out the following matrix computations.

(i) $3 \begin{bmatrix} 1 & 1 & -1 \\ -2 & 0 & 8 \end{bmatrix}$

(ii) $\begin{bmatrix} 2 & -1 \\ -1 & 0 \end{bmatrix} + \begin{bmatrix} 2 & -1 \\ -1 & 2 \end{bmatrix}$

(iii) $3 \begin{bmatrix} 2 & -1 \\ 4 & -2 \\ -2 & 1 \end{bmatrix} + 2 \begin{bmatrix} 2 & 3 \\ 2 & 2 \\ -1 & 0 \end{bmatrix}$

2. Find x, y and z so that $\begin{bmatrix} x - 2 & 3 & 2z \\ y & x & y \end{bmatrix} = \begin{bmatrix} y & 2z & 3 \\ 6z & y + 2 & 6z \end{bmatrix}$.

3. Carry out the following vector computations.

(i) $3(2, 4, 1)$ (ii) $4(3, 1, -1)$

(iii) $(1, 3) + (2, 4)$ (iv) $2(2, -1, 2, 3) - 3(1, 3, -2, -3)$

4. A is a 2×3 matrix; B is 2×3; C is 1×4; D is 3×2; E is 3×4; F is 4×3; G is 3×3. Say whether the indicated matrix exists. If it does exist, what is its shape?

(i) $A + B$ (ii) CE^T (iii) $D(A + B)$

(iv) F^T (v) $2FC$ (vi) $AD + DA$

(vii) F^2 (viii) $DA + 3G$ (ix) G^2

5. Carry out the following matrix computations.

(i) $\begin{bmatrix} 1 & 2 & 1 \\ 2 & -1 & 1 \end{bmatrix} \begin{bmatrix} 3 & -1 \\ 2 & -2 \\ -1 & 1 \end{bmatrix}$ (ii) $\begin{bmatrix} 3 & -2 \\ 3 & 0 \end{bmatrix} \begin{bmatrix} 3 & -2 \\ -4 & 4 \end{bmatrix}$

(iii) $\begin{bmatrix} 1 & -1 \\ 1 & -1 \\ -1 & 1 \end{bmatrix} \begin{bmatrix} 2 & 2 & 1 \\ 1 & -1 & 1 \end{bmatrix}$ (iv) $\begin{bmatrix} 2 & -1 \\ 1 & 0 \\ 2 & 3 \end{bmatrix} \begin{bmatrix} 1 & 0 \\ -1 & -1 \end{bmatrix}$

(v) $\begin{bmatrix} 2 & -1 & 4 \\ 0 & 1 & 2 \\ 1 & 1 & 0 \end{bmatrix} \begin{bmatrix} 1 \\ -2 \\ 1 \end{bmatrix}$

6. Find a matrix A such that

$$A \begin{bmatrix} 1 & 2 \\ 2 & 3 \end{bmatrix} = \begin{bmatrix} -1 & 1 \\ 4 & 1 \end{bmatrix}.$$

7. The matrix A is given. Find A^2 and A^3.

(i) $\begin{bmatrix} 1 & -1 \\ -1 & 0 \end{bmatrix}$ (ii) $\begin{bmatrix} 1 & 2 & 1 \\ 0 & -3 & 1 \\ 0 & 0 & -2 \end{bmatrix}$

8. Find the determinant of the matrix, and use it to invert the matrix or show that it is singular.

(i) $\begin{bmatrix} 5 & 6 \\ 2 & 3 \end{bmatrix}$ (ii) $\begin{bmatrix} 2 & 1 \\ 4 & 2 \end{bmatrix}$

9. In each case show that $AB = AC$. What is the (common) value?

(i) $A = \begin{bmatrix} 2 & 1 \\ 2 & 1 \end{bmatrix}$, $B = \begin{bmatrix} 0 & 1 \\ 3 & 7 \end{bmatrix}$, $C = \begin{bmatrix} 1 & 4 \\ 1 & 1 \end{bmatrix}$

(ii) $A = \begin{bmatrix} 1 & 1 & 1 \\ 1 & 0 & 1 \\ 1 & 2 & 1 \end{bmatrix}$, $B = \begin{bmatrix} 2 & 1 & 2 \\ 2 & 1 & 0 \\ 2 & 2 & 2 \end{bmatrix}$, $C = \begin{bmatrix} 4 & 2 & 1 \\ 2 & 1 & 0 \\ 0 & 1 & 3 \end{bmatrix}$

Appendix B Exercises B

1. Carry out the following matrix computations.

(i) $2\begin{bmatrix} 8 & -4 \\ 1 & 3 \\ -2 & 1 \end{bmatrix} - 5\begin{bmatrix} 1 & 1 \\ 1 & -2 \\ -2 & 0 \end{bmatrix}$.

(ii) $3\begin{bmatrix} 3 & -1 \\ 2 & 2 \\ -2 & 3 \end{bmatrix} - 2\begin{bmatrix} 2 & 1 \\ 4 & -2 \\ -1 & 0 \end{bmatrix}$

(iii) $3\begin{bmatrix} 4 & -1 \\ 2 & 2 \end{bmatrix} - 3\begin{bmatrix} 1 & -1 \\ -1 & 3 \end{bmatrix}$

(iv) $2\begin{bmatrix} 2 & -1 & -1 \\ 2 & 3 & 1 \end{bmatrix} + 3\begin{bmatrix} 1 & -1 & 2 \\ 1 & 1 & -2 \end{bmatrix}$

2. Suppose $\begin{bmatrix} x & -1 \\ -1 & 3 \end{bmatrix} = \begin{bmatrix} y+2 & -1 \\ -1 & x \end{bmatrix}$.

What are the values of x and y?

3. Carry out the following vector computations.

(i) $-(1, -1)$

(ii) $3(1, -2)$

(iii) $2(2, 1) - 2(1, -1)$

(iv) $2(1, 2, 1) - 2(1, -1, -1)$

(v) $3(-2, 2, 3) + 3(2, 1, -1)$

(vi) $2(1, -2, 2) + 3(2, 3, -1)$

(vii) $(1, 0, 3) + 2(2, 1, 3)$

(viii) $3(1, 6, 1, 0) - 3(2, 0, 4, -1)$

4. Suppose

$$A = \begin{bmatrix} 1 & -1 \\ -1 & 1 \end{bmatrix}, \quad B = \begin{bmatrix} 0 & -1 \\ 2 & 2 \end{bmatrix}, \quad C = \begin{bmatrix} 3 \\ 2 \end{bmatrix},$$

$$D = \begin{bmatrix} 1 & 3 \\ -1 & 0 \end{bmatrix}, \quad E = \begin{bmatrix} 1 & -2 \\ 1 & 0 \end{bmatrix}, \quad F = \begin{bmatrix} 3 \\ -1 \end{bmatrix}.$$

Calculate the following, or say why they do not exist.

(i) $3A + B$

(ii) $C + D - E$

(iii) $A - D - E$

(iv) $2C - 4F$

(v) $C - 3F$

(vi) $3A - 3A$

5. A is a 2×5 matrix; B is 2×5; C is 1×3; D is 5×2; E is 3×5; F is 5×3; G is 5×5. Say whether the indicated matrix exists. If it does exist, what is its shape?

 (i) $2A - B$ (ii) AD (iii) CF
 (iv) CF^T (v) DA (vi) AGF
 (vii) A^2 (viii) D^2 (ix) G^2

6. In this exercise,

$$A = \begin{bmatrix} 1 & -1 \\ -2 & 3 \end{bmatrix}, \quad B = \begin{bmatrix} 3 & 0 & -1 \\ -1 & 4 & 4 \end{bmatrix}, \quad C = \begin{bmatrix} 6 \\ 2 \end{bmatrix},$$

$$D = \begin{bmatrix} -1 & 1 & -1 \\ 1 & 3 & 3 \\ -2 & 2 & 0 \end{bmatrix}, \quad E = \begin{bmatrix} 1 & -1 \\ 1 & 0 \\ 2 & 2 \end{bmatrix}, \quad F = \begin{bmatrix} 2 \\ -1 \\ 2 \end{bmatrix},$$

$$G = \begin{bmatrix} 2 & 1 & -1 \end{bmatrix}, \quad H = \begin{bmatrix} 2 & 2 \end{bmatrix}, \quad K = \begin{bmatrix} -1 & 2 \end{bmatrix}.$$

Carry out the matrix computations, or explain why they are impossible:
 (i) BF (ii) AC (iii) CF
 (iv) BG (v) D^2 (vi) $EK + KB$

7. Suppose the square matrices A and B satisfy $AB = I$. Prove that $BA = I$.

8. In each case, find the products AB and BA. Do the two matrices commute?

 (i) $A = \begin{bmatrix} 1 & -1 \\ 1 & 1 \end{bmatrix}, \quad B = \begin{bmatrix} 2 & 1 \\ -1 & 2 \end{bmatrix}.$

 (ii) $A = \begin{bmatrix} 1 & 2 \\ 3 & 5 \end{bmatrix}, \quad B = \begin{bmatrix} 5 & 3 \\ 2 & 1 \end{bmatrix}.$

 (iii) $A = \begin{bmatrix} 3 & 1 & -2 \\ 1 & 1 & -1 \\ -1 & -1 & 2 \end{bmatrix}, \quad B = \begin{bmatrix} 1 & 0 & 1 \\ -1 & 2 & 0 \\ 0 & 1 & 2 \end{bmatrix}.$

 (iv) $A = \begin{bmatrix} 0 & 1 \\ 2 & 1 \end{bmatrix}, \quad B = \begin{bmatrix} 1 & 3 \\ -1 & 1 \end{bmatrix}.$

 (v) $A = \begin{bmatrix} 2 & 1 \\ 1 & -1 \end{bmatrix}, \quad B = \begin{bmatrix} -2 & 4 \\ 3 & -2 \end{bmatrix}.$

9. Consider the matrix $\quad A = \begin{bmatrix} 1 & 3 \\ 5 & 3 \end{bmatrix}.$

    (i) Find A^2 and A^3.

    (ii) Evaluate $A^3 - 2A - I$.

    (iii) Show that $A^2 - 4A - 12I = O$.

10. The matrix A is given. Find A^2 and A^3.

    (i) $\begin{bmatrix} 1 & 2 \\ -1 & 1 \end{bmatrix}$ (ii) $\begin{bmatrix} 1 & -1 \\ -1 & 1 \end{bmatrix}$

    (iii) $\begin{bmatrix} -1 & 2 \\ -1 & -1 \end{bmatrix}$ (iv) $\begin{bmatrix} 1 & 2 & 0 \\ 0 & 1 & 3 \\ 1 & 0 & -1 \end{bmatrix}$

    (v) $\begin{bmatrix} 2 & 0 & 3 \\ 0 & -2 & -1 \\ -1 & 1 & 2 \end{bmatrix}$ (vi) $\begin{bmatrix} 3 & 0 & -1 \\ 1 & 1 & 3 \\ -3 & 0 & -1 \end{bmatrix}$

11. Suppose A is a square matrix with one row all zeroes. Prove that A has no inverse. Prove the corresponding result when one column is all zeroes.

12. Find a matrix A such that

$$A \begin{bmatrix} 3 & 2 \\ 2 & 1 \end{bmatrix} = \begin{bmatrix} 1 & 3 \\ 2 & -1 \end{bmatrix}.$$

13. Find the determinant of the matrix, and use it to invert the matrix or show that it is singular.

    (i) $\begin{bmatrix} -2 & 2 \\ 2 & 1 \end{bmatrix}$ (ii) $\begin{bmatrix} 3 & 6 \\ 2 & 5 \end{bmatrix}$

    (iii) $\begin{bmatrix} 3 & 2 \\ -1 & 1 \end{bmatrix}$ (iv) $\begin{bmatrix} 4 & 6 \\ 5 & 7 \end{bmatrix}$

    (v) $\begin{bmatrix} 3 & 2 \\ 6 & 4 \end{bmatrix}$ (vi) $\begin{bmatrix} 4 & 3 \\ 1 & 1 \end{bmatrix}$

14. A and B are any two matrices such that AB exists. Prove that $B^T A^T$ exists, and that

$$B^T A^T = (AB)^T.$$

15. For the following matrices, show that A^{-1} exists, but $AB = CA$, even though $B \neq C$.

$$A = \begin{bmatrix} 2 & -1 \\ 1 & 1 \end{bmatrix}, \quad B = \begin{bmatrix} 4 & 0 \\ 4 & 2 \end{bmatrix}, \quad C = \begin{bmatrix} 2 & 0 \\ 2 & 4 \end{bmatrix}.$$

16. Suppose M_x denotes the 2×2 matrix

$$\begin{bmatrix} 1 & x \\ 0 & 1 \end{bmatrix},$$

where x may be any real number.

(i) Compute $M_x M_y$, and show that the matrices M_x and M_y commute for any real numbers x and y.

(ii) Find $M_x{}^2$, $M_x{}^3$ and $M_x{}^4$.

(iii) Find a formula for $M_x{}^n$, where n is any positive integer.

(iv) What is $M_x{}^{-1}$?

Appendix C: Some Combinatorial People

This appendix contains brief biographical sketches of some of the people who have contributed to the development of combinatorics. If you are interested in further information, we have cited relevant articles, in Wikipedia and elsewhere. Several good books on the history of mathematics have concentrated on biographical sketches; in particular, we recommend [6] and [16]. And nowadays, if you want further information about any mathematician, simply type their name into your favorite web browser.

Leonhard Euler [133]

Leonhard Paul Euler (April 15, 1707–September 18, 1783) was a Swiss mathematician who spent much of his life in Berlin and St. Petersburg and produced more mathematical papers than any other mathematician in history. He developed the modern notion of a function and did much pioneering work in infinite series, among many other disciplines. Among many other formulas, he is known for "Euler's identity" ($e^{i\pi} + 1 = 0$), and introduced the notation e for the base of the common logarithm and i for $\sqrt{-1}$. He also became very well known for a popularization of mathematics that consisted of a collection of letters written at the request of Frederic of Prussia to the Princess of Anhalt-Dessau, generally known as the *Letters to a German Princess*.

His contributions to combinatorics include the invention of graph theory with his solution of the Königsberg Bridges problem in 1736. The term *Latin square*, as well as *Graeco-Latin square* are due to his notation in writing pairs of orthogonal Latin squares. His related problem of the 36 military officers led to significant research in the subject, although the idea itself predated Euler, having first appeared for the 4×4 case as a problem of playing cards. He also developed the Gamma function, which is a continuous analog of the factorial function, and developed a formula relating the numbers of edges, faces, and corners of polyhedra that is a mainstay of topological graph theory.

One story of Euler is fairly well-known; while living in Russia, be became unhappy with the political situation there, and took the opportunity to move to the Academy in Berlin. Early in his stay there, it was pointed out to him that he seemed very quiet and untalkative, and was asked why this might be. According to the story, he replied, "Madam, I come from a country where, if we speak, we are hanged." After twenty-five years in Berlin, he returned to St. Petersburg in 1766. This was possible in part because of a change in

Russia's leadership (as Catherine the Great took power), and Euler felt that he was no longer welcome in Berlin.

Another fairly famous anecdote of Euler comes from his time in St. Petersburg. The French philosopher Denis Diderot was visiting the court of Catherine the Great, and allegedly making a stir with his outspoken atheism. The story goes that Euler was asked to embarrass Diderot, and a joke was arranged. Diderot was told that the famous mathematician had a mathematical proof of the existence of God. Euler, according to this story, gave his "proof" with the statement that "Sir, $(a + b^n)/z = x$, therefore God exists!" Supposedly, Diderot was so ignorant of mathematics that he did not realize that the statement was nonsense. Since Diderot was an accomplished mathematician himself, this story is widely assumed to be completely untrue; however, Burton ([16]) suggests that the incident may have occurred, in which case Diderot would have realized at once that it was merely a practical joke.

He lost all sight in his right eye in 1738 (likely due to an infection), and lost all sight in the other eye in 1766 due to a cataract. Remarkably, his blindness did not impair his mathematical productivity in any way, and indeed some of his most productive years occurred after his blindness. In part this was due to an ability to perform mental calculations, and in part it was due to an amazing memory; he was able to recite Virgil's *Aeneid* from memory, for example. His complete works are being published; there are 76 volumes, with a few more to appear, consisting of his personal letters.

Évariste Galois [129]

Évariste Galois (October 25, 1811–May 31, 1832) was a French mathematician who was the first to find necessary and sufficient conditions to solve a polynomial in radicals. He developed a considerable part of what today we call group theory, including the original use of the word *group* as an algebraic object. He also pioneered the study of finite fields.

He was distracted from mathematics by his political interests, as he was expelled from his university and arrested twice for various activities. For reasons that will probably never be fully understood, a month after being released from jail the last time, Galois became involved in a duel to the death over a woman. He died of his wounds, but (having expected to die) spent considerable time before the duel writing up results that he had polished while imprisoned. These are preserved in a letter to a colleague who had helped him publish his work previously.

Howard Garns [130]

Howard Garns (March 2, 1905–October 6, 1989) was an American architect who gained fame only after his death as the creator of Number Place, the number puzzle that became a worldwide phenomenon under the name Sudoku.

Garns developed the idea while working at the Daggett architecture firm in Indianapolis around 1960, but it was not published until 1979, in Dell Pencil Puzzles and Word Games.

Garns was born in Connersville, Indiana, and by his teens had moved to Indianapolis with his father, W. H. Garns, an architect. He attended the University of Illinois, and received a Bachelor of Science in architectural engineering in 1926. He worked for his father's firm until the Second World War, when he became a captain in the US Army Corps of Engineers. He joined the Daggett architecture firm after the war.

William Rowan Hamilton [138]

William Rowan Hamilton (August 4, 1805–September 2, 1865) was an Irish physicist and mathematician whose work included the development of *quaternions* (which may be thought of as a generalization of the complex number system) as well as the solvability of a polynomial of degree 5 by radicals.

In his youth he displayed a remarkable gift for languages, having learned a large number of them including Hebrew, Arabic, Sanskrit, Malay, and many European languages. His schooling at Trinity College in Dublin was cut short prior to graduation when Hamilton was offered the position of Professor of Astronomy at the University of Dublin. Although he was not primarily an astronomer, the expectation was that he would continue to contribute to science as a whole, which he certainly did as he contributed significantly to optics and mechanics.

His name appears in combinatorics only in relation to the *Hamilton circuit*, based on his study of cycles on a dodecahedron. This was inspired by and connected to his work on non-commutative algebra, of which quaternions were the first example.

Richard Hamming [134]

Richard Wesley Hamming (February 11, 1915–January 7, 1998) was an American mathematician. He was a professor at the University of Louisville during World War II, and joined the Manhattan Project in 1945, where he programmed one of the earliest electronic digital computers. The program was designed to discover if the detonation of an atomic bomb would ignite the atmosphere. The United States did not use the bomb until the computations showed that atmospheric disaster would not occur. (It was then tested in New Mexico and subsequently used twice against Japan.)

Hamming worked at the Bell Telephone Laboratories, where he collaborated with Claude E. Shannon, from 1946 to 1976. He then moved to the Naval Postgraduate School, where he worked as an Adjunct Professor until 1997, then became Professor Emeritus.

He was a founder and president of the Association for Computing Machinery.

Thomas Kirkman [10, 61, 81, 84]

Thomas Penyngton Kirkman (March 31, 1806–February 4, 1895) was an English clergyman and mathematician.

He was born in Bolton, near Manchester, England, and attended school there until age fourteen, when he was forced to leave school to work in his father's office. He returned to University nine years later, attending Trinity College, Dublin. He entered the Church of England and became a curate in Bury and in Lymm. In 1839 he became vicar in the Parish of Southworth in Lancashire, and stayed in that post until 1891.

Kirkman's first mathematical paper [57], published when he was age 40, answered a problem posed by Woolhouse in the *Lady's and Gentleman's Diary* in 1844 [140]. He proved the existence of balanced incomplete block designs with $k = 3$ and $\lambda = 1$. Steiner did not know of Kirkman's work, and in 1853 he published an article [110] that asked whether such systems existed; the solution was published by Reiss [86] in 1859. Kirkman subsequently wrote "...how did the Cambridge and Dublin Mathematical Journal Vol II p. 191, contrive to steal so much from a later paper in Crelle's Journal Vol LVI p. 326 on exactly the same problem in combinatorics?" [81] Despite this, these designs came to be known as "Steiner triple systems." In 1918, Cummings [25] said that Kirkman's paper "appears to have been overlooked by all writers on this subject," and even after Cummings' paper was published, Kirkman's contribution was ignored. (For example, in [37], the author says that Steiner "initially posed the problem" in [110], says that Reiss solved it, and neither Woolhouse nor Kirkman is mentioned.)

Kirkman is best known for the schoolgirl problem that we discussed in Section 14.1. He later published in many other areas of mathematics; he published extensively on the enumeration of polyhedra, then moved into group theory—finding a recursive method for compiling lists of transitive groups and giving a complete list of transitive groups of degree ≤ 10—and finally studied the theory of knots. Overall, his contributions to mathematics—not just to combinatorics—are very significant.

Joseph Kruskal [132]

Joseph Bernard Kruskal, Jr. (January 29, 1928–September 19, 2010) was an American mathematician and statistician whose work has influenced the disciplines of computer science, combinatorics, graph theory, and linguistics. He was born in New York City, the youngest of five brothers (two others of whom also went on to become noted mathematicians). He attended the University of Chicago, and earned his Ph.D. from Princeton University in

1954. In addition to the minimal spanning tree algorithm from Chapter 11, his name is attached to the Kruskal tree theorem and the Kruskal–Katona theorem.

Édouard Lucas [128]

François Édouard Anatole Lucas (April 4, 1842–October 3, 1891) was a French mathematician who produced a formula for finding the nth term of the Fibonacci sequence, and did a great deal of other work on sequences and related problems. He devised methods for testing the primality of numbers. When he was fifteen years old, Lucas began testing the primality of $2^{127} - 1$ by hand, using Lucas Sequences. Nineteen years later he finally proved that it was prime; it was then the largest known Mersenne prime.

Lucas was also interested in recreational mathematics, among other things inventing the Tower of Hanoi puzzle which we discussed in Chapter 1.

Lucas worked in the Paris observatory, served in the military and finally became a professor of mathematics in Paris.

Samuel Morse [7]

Samuel Finley Breese Morse (April 27, 1791–April 2, 1872) was an American painter of historic scenes.

In 1825, the city of New York commissioned Morse to paint a portrait of Lafayette, in Washington. While he was there, a horse messenger delivered a letter from his father that read one line, "Your wife is dead." Morse immediately left Washington for his home at New Haven, leaving the portrait of Lafayette unfinished. But when he arrived home she had already been buried. In his distress at the fact that for days he was unaware of his wife's failing health and her lonely death, he moved on from painting to pursue a means of rapid long distance communication.

In 1832, Morse met Charles Thomas Jackson of Boston, an expert on electromagnetism. Morse used these ideas to develop the concept of a single-wire telegraph. The original Morse telegraph is in the National Museum of American History at the Smithsonian Institution. He then developed the preliminary version of the code that bears his name, for use with his device. He originally planned to send only numerals. A version very like the present Morse code was developed by Alfred Vail in the early 1840s.

Morse was a leader in the anti-Catholic and anti-immigration movements of the mid-nineteenh century. In 1836, he ran unsuccessfully for mayor of New York. He worked to unite Protestants against Catholic institutions (including schools), wanted to forbid Catholics from holding public office, and promoted changing immigration laws to limit immigration from Catholic countries.

Isaac Newton [87, 135]

Isaac Newton (January 4, 1642–March 31, 1727) is possibly the best-known mathematician, physicist, philosopher, alchemist, and astronomer in history. Attending Cambridge University beginning in 1661, he developed many of the methods of calculus and applied them to the broad range of scientific disciplines in which he made his mark. In combinatorics, he is known almost solely for the generalized binomial theorem, which is the source of many generating functions as described in Chapter 6. This result, a generalization of the well-known Pascal's triangle method of raising a binomial to an integer power, gives an infinite series for an expression of the form $(a + b)^r$ for any real exponent r.

The bulk of his creative mathematical work was done during his years at Cambridge, although much of it remained unpublished until years later. In particular, Newton himself cites the years of the bubonic plague (1665 through 1667) as the time in which most of his mathematical work was accomplished. During this time, Cambridge was closed and he returned to the family farm at Woolsthorpe. He returned to Cambridge in 1667 and became Lucasian Professor of Mathematics in 1669. His work was not entirely on mathematics; at this time, he developed his theories of optics, and invented the reflecting telescope, as well as continuing to work on his mathematical model of gravity. He also continued to study alchemy and wrote religious tracts.

He took up a position with the Royal Mint in 1696, which required him to relocate to London, and in 1701 gave up his position at Cambridge in order to concentrate on his work there. He was, by some accounts, quite active, personally aiding in locating and shutting down counterfeiters. This aspect of his life is the subject of Philip Kerr's mystery novel *Dark Matter* [56].

Newton was made president of the Royal Society in 1703, and knighted by Queen Anne in 1705.

Blaise Pascal [16, 126]

Blaise Pascal (June 19, 1623–August 19, 1662) was a French mathematician with a strong theological bent. A child prodigy, when he was 11 he wrote a work on the sounds produced by vibrating objects. His father forbade him from further study of science and mathematics so as not to interfere with Pascal's study of Latin and Greek, a ban that was lifted when Pascal (at 12) was caught proving a theorem in geometry (that the angles in a triangle add to 180°) independently.

At 16, he produced but never published an essay on conic sections that was received with much praise, and shortly thereafter invented a mechanical device for addition and subtraction capable of adding a column of eight figures. This device was at first intended merely to assist his father Etienne with his work dealing with tax records, but Pascal produced many others, some few of which remain today. His hopes of making a profit from selling these devices were never realized, however.

In 1654 an acquaintance sent Pascal a letter asking certain technical questions concerning probabilities involved in games of chance. The acquaintance, the Chevalier de Méré, was friends with many eminent mathematicians (probably because he made his living by gambling) and addressed a question to Pascal concerning the proper way to divide the stakes between two players if a game is interrupted. Pascal found questions of this sort easy, and his research led him to re-discover the triangle of binomial coefficients that today bears his name. He published *Traité du Triangle Arithmétique* around 1654, although it was not available until much later. In this work, he develops many of the relations and patterns involving binomial coefficients that we know and use today. This work also contains the first published statement of the principle of mathematical induction, which Pascal used to prove a formula equivalent to $\binom{n}{r+1} = \frac{n-r}{r+1}\binom{n}{r}$.

Emanuel Sperner []

Emanuel Sperner (December 9, 1905 - January 31, 1980) was a German mathematician. He was born in Waltdorf, in what is today Poland. He obtained his doctorate in 1928 from the University of Hamburg for a dissertation that contained one of the two results for which he is known today; this is Sperner's Lemma, a result concerning the colorings of a triangulation of a simplex. While geometrical in nature, this result is used to prove powerful results in topology as well. In 1928, he also published what is now called Sperner's Theorem, a major result in the combinatorial discipline of extremal set theory. His later work served to make him a major figure in modern geometry as well.

He later worked in Beijing for a time before returning to Germany, where he took a position at the University of Königsberg until the second World War. He then worked for the German Navy for a time before resuming his academic career at Oberwolfach, where he is regarded as one of the founding members of the Oberwolfach Mathematical Institute. Other appointments followed at Bonn and then at Hamburg, where he retired in 1974. He took a number of visiting positions thereafter to institutions in the United States, South Africa, and Brazil, and spent his final years in Sulzburg-Laufen, where he died in 1980.

Gaston Tarry [80]

Gaston Tarry (September 27, 1843–June 21, 1913) was a French amateur mathematician. He was born in Villefranche-de-Rouergue, in south central France. He joined the French Financial Administration and became part of the French administration in Algeria, until his retirement in 1902. He was never employed as a mathematician, but did a great deal of mathematics, nearly all combinatorial in nature.

In 1886, Tarry published a general method for finding the number of Euler circuits in a graph or multigraph [112]. Then, in 1895, he published a systematic method so solve a maze [113], using what is essentially a depth-first search algorithm. In 1900 and 1901 [114] he solved Euler's 36 Officer Problem, showing that there is no pair of orthogonal Latin squares of order 6.

Tarry then went on to work on magic squares, producing a substantial number of results.

John Venn [93]

John Venn (August 4, 1834–April 4, 1923) was an English logician and mathematician. In 1857 he was named a Fellow of Caius College of Cambridge University; he was ordained a priest 1859 and for a year was curate at Mortlake. He then began lecturing at Caius College. He introduced what are now called Venn diagrams in two papers in 1880 [118, 119]. He was elected a Fellow of the Royal Society in 1883.

The use of diagrams in formal logic certainly originated well before his time. However, Venn comprehensively surveyed and formalized their usage, and was the first to generalize them.

Solutions to Set A Exercises

Chapter 1

1. The sets are shown in grey.

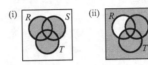

2. Select three "active" players. They proceed as if they were the only players, in a team of three. All other players pass.

3. Let's write S_n for the required sum, $S = 1 + 3 + \ldots + (2n - 3) + (2n - 1)$.

   (i) Partition the first $2n$ integers into the sets of odd and even integers:

   $$1 + 2 + 3 + \ldots + (2n - 1) + 2n$$
   $$= [1 + 3 + \ldots + (2n - 3) + (2n - 1)] + [2 + 4 + \ldots + (2n - 2) + 2n];$$

   so $S_n = [1 + 2 + 3 + \ldots + (2n - 1) + 2n] - [2 + 4 + \ldots + (2n - 2) + 2n]$.

   Now Gauss's formula tells us that

   $$1 + 2 + 3 + \ldots + (2n - 1) + 2n = 2n(2n + 1)/2 = n(2n + 1),$$

   and obviously

   $$2 + 4 + \ldots + (2n - 2) + 2n = 2[1 + 2 + \ldots + (n - 1) + n]$$
   $$= 2[n(n + 1)/2] = n(n + 1).$$

   So $S_n = n(2n + 1) - n(n + 1) = n^2$.

   (ii) Clearly $S_1 = 1 = 1^2$. Assume $S_n = n^2$ for $n = 1, 2, \ldots, r$.
   Then $S_{r+1} = S_r + (2r + 1) = r^2 + (2r + 1) = (r + 1)^2$.

4. ABC, ADE, AFG, BDF, BEG, CDG, CEF.

5. Case $n = 1$ is easy. Assume $\sum_{i=1}^{r} i(i + 1) = \frac{1}{3}r(r + 1)(r + 2)$. Then

   $$\sum_{i=1}^{r+1} i(i + 1) = \frac{1}{3}r(r + 1)(r + 2) + (r + 1)(r + 2)$$
   $$= \left[\frac{r}{3} + 1\right](r + 1)(r + 2) = \frac{1}{3}(r + 1)(r + 2)(r + 3).$$

6. We see that τ is reflexive, since $x + 2x$ is divisible by 3 for integer x. Similarly, if x, y, and z are integers with $x \tau y$ and $y \tau z$, we write $x + 2y = 3m$, $y + 2z = 3n$ and add these equations to get $x + 3y + 2z = 3(m + n)$. Thus $x + 2z = 3(m + n - y)$, so the relation is transitive. Finally, suppose $x + 2y = 3n$, and consider $3x + 3y = (x + 2y) + (y + 2x)$ to see that τ is symmetric.

7. $4 + 3 + 2 = 9$, by the addition principle.

8. $4 \times 3 \times 6 = 72$, by the multiplication principle.

9. $3 \times 3 + 6 \times 4 + 10 \times 2 = 53$ ways.

10. $8 \times 7 \times 6 = 336$ permutations of three items from a set of 8.

11. $26^2 \times 10^4 = 6,760,000$.

12. Write Y for the set of all subsets with n or fewer elements and Z for those with more. Clearly Y and Z are disjoint, and equal in size (Z consists of all the complements in X of elements of Y). Y and Z have 2^{2n+1} elements between them (see Example 1.2), so each has 2^{2n} elements.

13. (i) 2, 1, 0, 3, 4, 3, 2, 5, 2   (ii) 2,2,2,2,2,2,2   (iii) 2, 4, 3, 2, 5, 2

14. There are n edges; one vertex of degree n, and n of degree 1.

15. Minimum 11 (the complete graph K_{10} on 10 vertices has only 45 edges, so the minimum is at least 11; K_{11} has 55 edges, and if we delete any seven of them the resulting graph will have no isolated vertex, so 11 can be realized.) Maximum 96.

16. Consider a (multi)graph whose vertices are the people, with an edge for each handshake. (i) If no one shook hands five or more times, the sum of the degrees would be at most 32, so there could be at most 16 handshakes. (ii) No, because if it were true the graph or multigraph would have an odd number of odd vertices.

17. (i) Yes, for example $(1,4,2,3)$. (ii) No, because the four sets $123, 12, 13, 23$ have only three elements between them.

Chapter 2

1. Since there are 52 letters possible for the last three, the answer is $26 \times 26 \times 10^2 \times 52^3$, or $9,505,100,800$.

2. Solution: Eight choices for first digit, two for second digit and 10 for last digit gives us 160 area codes possible. If we remove the "x11" possibilities, there are eight that we disallow; 211, 311, 411, 511, 611, 711, 811, and 911. Subtracting leaves us 152 area codes.

3. Solution: There are 24 choices for the first window, 23 for the second, and so on. The result is $24!/19! = 5,100,480$.

4. This is equivalent to finding (i) the total number of 4-digit numbers, (ii) the number of 4-digit numbers with no repeated digits. The answer to (i) is 9000 (all the numbers from 1000 to 9999). For (ii), there are 9 choices (any of 1 to 9) for the first digit (x say), 9 for the second (0 is now available but x is not), 8 for the third and 7 for the fourth, so the answer is $9 \times 9 \times 8 \times 7 = 4536$.

5. Consider the strings of 6 digits, from 000000 to 999999. (000000 represents 1,000,000; otherwise $abcdef$ represents abc, def with leading zeroes deleted.) There are six places to put 3; for each choice there are five places to put 5; for each of these there are four places to put 7; there are seven choices for each other number. The total is $6 \times 5 \times 4 \times 7^3 = 41160$.

6. Since the sites are mutually exclusive (the student will join only one site), the addition principle applies; we add the numbers for each site. For a given site, befriending one acquaintance does not affect whether or not the student befriends another, so that the multiplication principle applies. Thus, the first site provides the student with 2^3 or 8 possibilities; the second, $2^5 = 32$; the third, 64; and the fourth, 16. We add these to find 120 possibilities.

7. There are $C(4,2) = 6$ ways to choose the two sites. If the numbers of acquaintances were different, we would have to add up the possibilities for each of the six choices. Because the numbers are the same, each chosen site gives us 2^4 or 16 possible ways to befriend or not befriend the acquaintances on that site. We therefore have $6·16·16 - 1536$ possible outcomes.

8. This must be broken into cases, unfortunately. There is one way to choose the two sites with four acquaintances; each site admits 16 possibilities, so we have 256 possibilities for this case. There are four ways to choose one site with four acquaintances and one with five; there are 16 possibilities for the former and 32 for the latter, for 2048 possibilities. Then there is one way to choose both of the five-acquaintance sites, and each of those sites admits 32 possibilities, giving 1024 possible outcomes in this case. Since the three cases are mutually exclusive, we add the results to obtain 3328 possibilities altogether.

9. If she neglects the underage friend, she has three friends to choose from and seven restaurants, for 21 possibilities. If she chooses the underage friend, she has one friend to choose from and four restaurants. Since she is taking only one friend, the two are mutually exclusive, so she has 25 possibilities. If she can afford to take two friends, then she has $C(3,2) = 3$ ways if she leaves the underage friend behind, and seven restaurants. If she takes the underage friend, she has $C(3,1) = 3$ ways to choose the second friend, and four restaurants. We get $21 + 12 = 33$ possibilities.

10. Solution: There are $\binom{6}{3}$ ways to select three tutors from a set of 6; this gives us 20 ways to assign three tutors to the 2:00 PM shift.

11. $6{\cdot}5 = 30$, $7{\cdot}6 = 42$, $8{\cdot}7 = 56$.

12. 1, 7, 21, 35, 35, 21, 7, 1.

13. $1{\cdot}x^{10} + 5{\cdot}x^7 + 10{\cdot}x^4 + 10{\cdot}x + 5{\cdot}x^{-2} + 1{\cdot}x^{-5}$.

14. We can rewrite Corollary 2.3.3 as
$$1 + \sum_{k=1}^{n} (-1)^k \binom{n}{k} = 0.$$
So $\sum_{k=1}^{n}(-1)^k \binom{r}{k} = -1$. Divide each term by -1.

15. $C(6,3)$ ways to get to the west corner of block $b2$ times $C(5,2)$ ways to go from the east end of $b2$ to the northeast corner gives us 200 ways. Since there are $C(12,5) = 792$ total possible paths, we subtract to get 592 paths that do not pass through block $b2$.

16. There are 16 letters, of which 3 are S, 2 are L, 5 are E, 2 are P, and there are one each of I, N, G, and T. This gives us $16!/(3!{\cdot}2!{\cdot}5!{\cdot}2!{\cdot}1!{\cdot}1!{\cdot}1!{\cdot}1!)$ or $7,264,857,600$.

17. We can write 3 as a sum in 10 ways: $1+1+1$, $1+2+0$, $1+0+2$, $0+1+2$, $2+1+0$, $2+0+1$, $0+2+1$, $3+0+0$, $0+3+0$, and $0+0+3$. Computing $3!/(1!1!1!) = 6$, $3!/(2!1!0!) = 3$, and $3!/(3!0!0!) = 1$, we obtain

$6{\cdot}3x{\cdot}5y{\cdot}7z$
$+ 3\left((3x)^2{\cdot}5y + (3x)^2{\cdot}7z + (5y)^2{\cdot}3x + (5y)^2{\cdot}7z + (7z)^2{\cdot}3x + (7z)^2{\cdot}5y\right)$
$+ 1\left((3x)^3 + (5y)^3 + (7z)^3\right).$

18. We visualize 10 cards and 2 dividers to separate the allotment of the teenagers; this is 12 symbols of which we wish to choose 2 to be dividers, or $C(12,2) = 66$. If each teenager is to receive at least one, we distribute one card to each and have seven cards and two dividers; this gives us $C(9,2) = 36$ ways to distribute the cards.

19. Apply Theorem 2.8; there are $n = 3$ covers to be purchased, and $k = 8$ distinct items (colors) from which to choose them. This gives us $C(3 + 8 - 1, 8 - 1) = 120$ ways.

20. 365142; 351642; 641532

Chapter 3

1. 0.26.

2. (i) 1/4 (ii) 3/4.

3. 1/6.

4. (i) $(2,0),(2,1),(2,2),(1,0),(1,1),(1,2),(0,0),(0,1),(0,2)$.
(ii) $A = (2,0),(2,1),(1,0)$, $B = (2,0),(1,1),(0,2)$,
$C = (2,0),(2,2),(1,1),(0,0),(0,2)$.
(iii) The probabilities of the different results are $P(2,0) = P(2,2) = P(0,0) = P(0,2) = 1/16$, $P(2,1) = P(1,0) = P(1,2) = P(0,1) = 1/8$ and $P(1,1) = 1/4$. So $P(A) = 5/16, P(B) = 3/8, P(C) = 1/2$.

5. (i) (i) 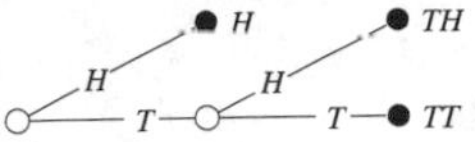

(ii) $\{H, TH, TT\}$ (iii) 1/4.

6. $\binom{4}{1} \times 0.5^1 \times 0.5^3 = 4 \times 1/2^4 = 1/4$.

7. Call the dice die 1 and die 2. There are six possible rolls on each die, or 36 possibilities, and 6 are doubles. So the probability is $6/36 = 1/6$.

8. 1/10.

9. $2,3,4,5,6,7,8,9,10$; $P(2) = P(10) - 1/24$; $P(3) = P(8) = 1/12$; $P(3) = P(10) = 1/8$; $P(5) = P(6) = P(4) = 1/6$.

10. (i) Yes (ii) No; add to 1.4, not 1 (iii) Yes (iv) No, contains a negative (v) No, contains a "probability" greater than 1 (and adds to more than 1).

11. (i) 1/2 (ii) 1/13 (iii) 1/26 (iv) 7/13.

12. (i) Solid dots are outcomes: 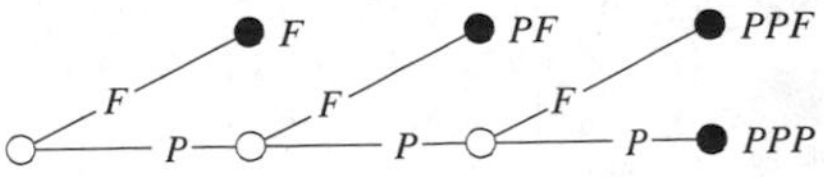

(ii) 4 (iii) $\{F, PF\}$.

Chapter 4

1. 7. (This is a pretty trivial exercise, but notice: if we associate pigeon-holes with the six possible rolls, then seven rolls must result in at least one pigeonhole with two or more occupants.)

2. Write $A_1 = a_1$, $A - 2 = a_1 + a_2$, ..., $A_n = a_1 + a_2 + \ldots + a_n$. If n divides a_i, put $s = 1, t = i$; we are done. If n never divides a_i for any i, write S_k for the set of all a_i congruent to k modulo n. There are n sums and they belong to $n - 1$ sets (since S_0 is empty). So one set contains at least two elements.

3. By pigeonhole principle, $k = 5$; for if each car carried only four or fewer people, there would be at most 84 patrons in the club. We cannot conclude $k = 6$ since if all the cars carried at most five people, we might have had as many as 105 patrons.

4. Let S_i be the number of returns completed when i days have passed. Clearly $S_1 \geq 1$ and $S_{50} \leq 80$. Also, the numbers $\{S_1 + 19, \ldots S_{50} + 19\}$ are between 20 and 99. So the set of 100 integers $\{S_i : 1 \leq i \leq 50\} \cup \{S_i + 19 : 1 \leq i \leq 50\}$ is a set of 100 integers between 1 and 99, and so at least two of them must be equal. It follows that there are h, k such that $S_h = S_k + 19$, as required.

5. Let S_i be the total number of journal entries turned in by the end of the ith class day. Clearly $S_1 \geq 1$ and $S_{60} \leq 90$. Each of the 120 numbers $S_1, \ldots S_{60}, S_1 + 29, \ldots S_{60} + 29$ lies between 1 and $90 + 29 = 119$, so there must be some h and k with $S_h = S_k + 29$, as required.

6. Suppose $\sum_{t \in T} t = k$. Then $k \equiv \ell (\text{mod } n)$ for some $\ell, 0 \leq \ell \leq n - 1$. If $\ell \neq 0$, choose $x = \ell$; if $\ell = 0$, choose $x = n$.

7. "Ends in 01" is equivalent to "equals 1 mod 100." Write $\overline{p}$ to represent p mod 100. There are only finitely many different integers mod 100 (100 of them, to be precise!), so there must be repetitions among the powers of $\overline{p}$. Say $\overline{p}^i = \overline{p}^j$ is such a repetition, where $i > j$. Since p is not divisible by 2 or 5, $\overline{p}$ has an inverse mod 100. Then $\overline{p}^{i-j} = \overline{p}^i \overline{p}^{-j} \equiv 1$ mod 100, and $i - j$ is positive.

8. Given n, define $S_i = \{y : \frac{i}{n} \leq y < \frac{i+1}{n}\}$. The sets $S_0, S_1, \ldots, S_{n-1}$ partition the real numbers from 0 to 1 (including 0 but not 1), so the remainder on dividing kx by n lies in one of them, for any k. Write x_k for this remainder. If one of $x_1, x_2, \ldots$ lies in S_0, we are finished. If not, consider $x_1, x_2, \ldots, x_n$. There are n numbers, and each lies in one of the sets $S_1, S_2, \ldots, S_{n-1}$. So two of them—say x_a and x_b—lie in the same set, S_i say. Say $b > a$. Then $x_b - x_a$ is the difference of two

positive numbers, each greater than or equal to $\frac{i}{n}$ and less than $\frac{i+1}{n}$. So $|x_b - x_a| < \frac{i+1}{n} - \frac{i}{n} < \frac{1}{n}$; but $|x_{b-a}| = |x_b - x_a|$, and we are finished.

9. The two are equal (simply exchange the labels A_1 and A_2 in the definition, in Theorem 4.12).

10. Suppose the edges of a complete graph are colored in red and blue so that there is no red triangle or blue K_5. Select a vertex x. If there are five red edges at that vertex, no two of the other endpoints of those edges can be joined by a red edge, or else we would have a red triangle. But the edges joining those five other endpoints cannot all be blue, or we would have a blue K_5. So there are at most four red edges at x.

Now suppose there were nine blue edges at x. Consider the K_9 formed by the other nine endpoints and the edges joining them. By Example 3.9, this must contain either a red triangle or a blue K_4. The former is not allowed; the latter together with x would form a blue K_5 in the original graph.

So there are at most four red and eight blue edges touching x. Therefore the complete graph has at most 13 vertices ($4 + 8 + 1$—don't forget to count x).

11. R and B form a symmetric sum-free partition.

12. Clearly the sets partition $(\mathbb{Z}_4 \times \mathbb{Z}_4)^*$. To check symmetry, write $-R$ for the list of negatives of elements of R: $R = \{02, 10, 30, 11, 33\}$, so (retaining the order) $-R = \{02, 30, 10, 33, 11\} = R$, and similarly for B and G, so we have symmetry. To check whether R is sum-free, observe that the addition table

+	02	10	30	11	33
02	00	12	32	13	31
10	12	20	00	21	03
30	32	00	20	01	23
11	13	21	01	22	00
33	31	03	23	00	22

contains no element of R among the sums; similarly for B and G.

13. Theorem 3.8 says $R(4, 5) \leq 35$, so the answer is "No."

Chapter 5

1. $199 - 66 - 18 + 6 = 121$.

2. $5000 - 1250 - 1000 - 833 + 250 + 416 + 166 - 83 = 2666.$

3. (i) $35 + 35 - 15 = 55$; (ii) $35 + 35 - 2 \times 15 = 40.$

4. (i) 0; (ii) $31 + 36 - 47 = 20$; (iii) $31.$

5. (i) 189; (ii) 367; (iii) $1308.$

6. (i) 3; (ii) $50.$

7. $11! - 9! - 9! - 7! + 7! + 5! + 5! - 3! = 39,191,274.$

8. $91 - 28 - 21 - 21 + 0 + 0 + 0 - 0 = 21.$

9. $455 - 84 - 35 - 220 + 0 + 20 + 4 - 0 = 140.$

10. $165 - 56 - 10 - 35 - 35 + 0 + 4 + 4 + 0 + 0 + 1 - 0 - 0 - 1 - 0 + 0 = 37.$

11. There are 77 squares less than $6,000$ and 18 cubes less than $6,000$; there are four sixth powers less than $6,000$. We get $6,000 - 77 - 18 + 4 = 5,909.$

12. $4 \times 20 - 6 \times 9 + 4 \times 2 - 0 = 34.$

13. $4 \times 24 - 5 \times 12 + 7 + 1 = 44.$

14. $D_6 = 6 \times 44 + 1 = 265$; $D_7 = 7 \times 265 - 1 = 1,854$; $D_8 = 8 \times 1,854 + 1 = 14,833.$

15. (i) $D_6 = 265$. (ii) $\binom{6}{1} \times D_5 = 264$. (iii) $6! - 265 = 455$. (iv) 0 because when five women get their own coats back the last woman has no other coat but her own to get back.

16. Choose the integers to be fixed in $\binom{7}{3}$ ways; then choose the derangements of the others in D_4 ways. We get $35 \times 9 = 315.$

Chapter 6

1. (i) $(x^3 + 4x^2 + x)/(1 - x)^4$. (ii) $(4x^2 + 3x - 1)/(1 - x)^3$. (iii) $x/(1 - x)^4$. (iv) $1/(1 - 3x) - 3/(2 - 4x)$.

2. (i) We expand $(1 + x + x^2 + x^3)^3$ and the coefficient of x^n is the result. This is $1 + 3x + 6x^2 + 10x^3 + 12x^4 + 12x^5 + 10x^6 + 6x^7 + 3x^8 + x^9$.

(ii) Similarly, $(x + x^2 + x^3)^3 = x^3 + 3x^4 + 6x^5 + 7x^6 + 6x^7 + 3x^8 + x^9$.

(iii) For at least two banana-walnut muffins, we expand $(x^2 + x^3)(1 + x + x^2 + x^3)^2$, to get: $x^2 + 3x^3 + 5x^4 + 7x^5 + 7x^6 + 5x^7 + 3x^8 + x^9$.

(iv) For an unlimited supply, $(1 + x + x^2 + \ldots)^3 = (1 - x)^{-3}$. We may use Theorem 2.7 to see that we can do this in $C(n+3-1, 3-1) = (n^2 + 3n + 2)/2$ ways. To find the recurrence, expand the denominator; this is $1 - 3x + 3x^2 - x^3$. This gives the recurrence $a_n = 3a_{n-1} - 3a_{n-2} + a_{n-3}$.

3. (i) $a_n = 4^n$. (ii) $a_n = (n-2)2^n$. (iii) $a_n = (-3)^n$. (iv) $a_n = a_n = 2^n$. (v) $a_n = 4^n - (-3)^n$. (vi) $a_n = n2^{n-1}$.

4. 1, 1, 2, 3, 5, 8, 13, 21, 34, 55, 89, 144.

5. We get $(1 + x^3 + x^6 + \cdots + x^{3k} + \ldots)(1 + x + x^2 + \cdots + x^k + \ldots) = [(1 - x^3)(1-x)]^{-1}$. Our recurrence relation is thus $a_n = a_{n-1} + a_{n-3} - a_{n-4}$. A closed form is $\lceil (n+1)/3 \rceil$.

6. We get $\sum_{n \geq 0}(2^n - n)x^n = 1/(1 - 2x) + x/(1 - x)^2$. This simplifies to $f_a(x) = (1 - x - x^2)/(1 - 4x + 5x^2 - 2x^3)$. This gives us the recurrence $a_n = 4a_{n-1} - 5a_{n-2} + 2a_{n-3}$.

7. (i) $f_a(x) = x^2 e^x/2$. (ii) $f_b(x) = e^{3x} - e^{2x}/2$. (iii) $f_c(x) = 2e^x$. (iv) $f_d(x) = e^{2x} + 2e^x$.

8. This is the sequence $\{1, 0, 1, 0, \ldots\}$ or $a_n = (1 + (-1)^n)/2$.

9. (i) $f_a(x) = e^{4x}$. (ii) $f_b(x) = (e^x - 1)^4$. (iii) $f_c(x) = (x + x^2/2! + x^3/3!)^4$.

Chapter 7

1. A proper way of lining up customers ensures that we never have more people offering us a twenty dollar bill than we have had people who pay with ten dollar bills. This corresponds exactly to a sequence of left and right parentheses where the number of right parentheses never exceeds the number of left parentheses that have occurred. This is just C_n.

2. There are six ways to place three chords at one corner; six ways to draw chords in the shape of the (possibly reversed) letter N; and two ways to put chords in the form of an equilateral triangle.

3. Each dot is touched by exactly one arc; write a "(" below a dot whose arc connects it to a dot to its right, and ")" below a dot whose arc connects it to a dot on its left. Then a set of arcs corresponds exactly to a proper sequence of parentheses.

4. $C_5 = C_0C_4 + C_1C_3 + C_2^2 + C_3C_1 + C_4C_0 = 14 + 5 + 4 + 5 + 14 = 42$.

5. $C_9 = \binom{18}{9}/10$. This is $18!/[10(9!)^2] = (18 \times 17 \times 16 \times 15 \times 14 \times 13 \times 12 \times 11)/(2 \times 3 \times 4 \times 5 \times 6 \times 7 \times 8 \times 9)$. Careful cancellation makes this problem not too difficult; $(17 \times 16 \times 14 \times 13 \times 12 \times 11)/(8 \times 7 \times 6 \times 4) = (17 \times 13 \times 2 \times 11) = 4,862$.

6.
$$\begin{bmatrix} 1 & 0 & 0 & 0 \\ 1 & 1 & 0 & 0 \\ 1 & 3 & 1 & 0 \\ 1 & 7 & 6 & 1 \end{bmatrix}^{-1} = \begin{bmatrix} 1 & 0 & 0 & 0 \\ -1 & 1 & 0 & 0 \\ 2 & -3 & 1 & 0 \\ -6 & 11 & -6 & 1 \end{bmatrix}.$$

7. $\left\{ {7 \atop 2} \right\} = 63$; $\left\{ {7 \atop 3} \right\} = 301$; $\left\{ {7 \atop 4} \right\} = 350$.

8. $\left\{ {4 \atop 2} \right\} = 7$.

9. $\left\{ {6 \atop 3} \right\} = 90$; $\left\{ {6 \atop 3} \right\} \times 3! = 540$.

10. $\binom{6}{3}\left\{ {3 \atop 3} \right\} + \binom{6}{2}\left\{ {4 \atop 3} \right\} + \binom{6}{1}\left\{ {5 \atop 3} \right\} + \binom{6}{0}\left\{ {6 \atop 3} \right\} = 350$; $350 \times 3! = 2100$.

11. $\left\{ {5 \atop 3} \right\} \times 3! = 150$.

12. $\left\{ {5 \atop 2} \right\} = 15$.

13. $\binom{8}{4}\left\{ {6 \atop 4} \right\} \times 4! = 109,200$.

14. $B_3 = 5$; there are five ways, namely $1 \times 154 = 2 \times 77 = 7 \times 22 = 11 \times 14 = 2 \times 7 \times 11$.

15. $B_8 = 4140$.

16. $\left[{7 \atop 4} \right] = -735$, $\left[{7 \atop 5} \right] = 175$, and $\left[{7 \atop 6} \right] = -21$.

17. $\left| \left[{6 \atop 2} \right] \right| = 274$; if one table is distinguished, we have $2 \times 274 = 548$.

Chapter 8

1.

	e	ρ	ρ^2	ρ^3	ρ^4
e	e	ρ	ρ^2	ρ^3	ρ^4
ρ	ρ	ρ^2	ρ^3	ρ^4	e
ρ^2	ρ^2	ρ^3	ρ^4	e	ρ
ρ^3	ρ^3	ρ^4	e	ρ	ρ^2
ρ^4	ρ^4	e	ρ	ρ^2	ρ^3

2. (i) $\begin{pmatrix} 1 & 2 & 3 & 4 & 5 \\ 5 & 3 & 4 & 2 & 1 \end{pmatrix}$

(ii) $\begin{pmatrix} 1 & 2 & 3 & 4 & 5 \\ 2 & 4 & 3 & 5 & 1 \end{pmatrix}$

3. (i) $(13)(254)$; (ii) (12534).

4. For simplicity, we number the corners 1, 2, 3, 4 starting at the upper left and moving clockwise. We may form a reflection across the horizontal axis, corresponding to the permutation $(14)(23)$; a reflection across the vertical axis, corresponding to $(12)(34)$, and a rotation by $180°$, corresponding to $(13)(24)$. Together with the identity e, this forms a group.

5. Denote $C_2 = \{e, \alpha\}$ and $C_3 = \{e, \rho, \rho^2\}$. The group elements are

$$\{(e,e), (e,\rho), (e,\rho^2), (\alpha,e), (\alpha,\rho), (\alpha,\rho^2)\}$$

and its multiplication table follows.

	(e,e)	(e,ρ)	(e,ρ^2)	(α,e)	(α,ρ)	(α,ρ^2)
(e,e)	(e,e)	(e,ρ)	(e,ρ^2)	(α,e)	(α,ρ)	(α,ρ^2)
(e,ρ)	(e,ρ)	(e,ρ^2)	(e,e)	(α,ρ)	(α,ρ^2)	(α,e)
(e,ρ^2)	(e,ρ^2)	(e,e)	(e,ρ)	(α,ρ^2)	(α,e)	(α,ρ)
(α,e)	(α,e)	(α,ρ)	(α,ρ^2)	(e,e)	(e,ρ)	(e,ρ^2)
(α,ρ)	(α,ρ)	(α,ρ^2)	(α,e)	(e,ρ)	(e,ρ^2)	(e,e)
(α,ρ^2)	(α,ρ^2)	(α,e)	(α,ρ)	(e,ρ^2)	(e,e)	(e,ρ)

6. The cycle index is $C(z_1, z_2, z_3) = (z_1^3 + 3z_1 z_2 + 2z_3)/6$. Replacing with r and b gives us simply $r^3 + r^2 b + rb^2 + b^3$. Each color distribution gives us only one equivalence class of colorings.

7. If the bands are labeled 1, 2, 3, 4 from left to right, the group consists of two elements, e and $(14)(23)$. The cycle index is thus $C(z_1, z_2) = (z_1^4 + z_2^2)/2$. Expanding in r and b gives us $(2r^4 + 4r^3 b + 8r^2 b^2 + 4rb^3 + 2b^4)/2$. Dropping the r^4 and b^4 terms leaves $2r^3 b + 4r^2 b^2 + 2rb^3$, so there are eight ways to color the pole if each color is used at least once.

8. Numbering the corners 1 through 5 clockwise, we find $\rho = (12345)$, $\rho^2 = (13524)$, $\rho^3 = (14253)$, and $\rho^4 = (15432)$. This gives us $C(z_1, z_5) = (z_1^5 + 4z_5)/5$. Substituting k gives us $C(k) = (k^5 + 4k)/5$. To two-color, $\hat{C}(r,b) = ((r+b)^5 + 4(r^5 + b^5))/5$.

9. Numbering the corners 1 through 7 clockwise, we find $\rho = (1234567)$, and all six powers of ρ are cycles of length seven. The various reflections all look like $\alpha = (1)(27)(36)(45)$. This will give us $C(z_1, z_2, z_7) = (z_1^7 + 6z_7 + 7z_1 z_2^3)/14$. Substituting k gives us $C(k) = (k^7 + 7k^4 + 6k)/14$. To two-color, $\hat{C}(r,b) = ((r+b)^7 + 6(r^7 + b^7) + 7(r+b)(r^2 + b^2)^3)/14$.

Chapter 9

1. See the diagram on the left in the figure below.

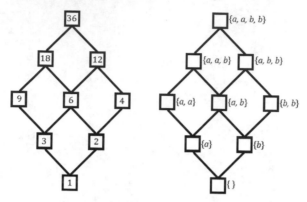

2. See the diagram on the right in the figure above.

3. Let $c = a \wedge b$; then $d = b \wedge a$. Since c is the greatest lower bound of a and b, we must have $c \preceq d$; but d is the greatest lower bound of b and a, so $d \preceq c$. It follows that $c = d$, so $a \wedge b = b \wedge a$. The proof for $a \vee b$ is similar.

4. We show $a \wedge a = a$. Notice that $a \preceq a$, so that a is indeed a lower bound. Also, if b is any other lower bound, then by definition $b \prec a$, so that a is the *greatest* lower bound. The proof for $a \vee a$ is similar.

5. If $a \wedge b = a$, then $a \preceq b$ follows immediately; if $a \preceq b$ then (since $a \preceq a$) clearly a is a lower bound; if c is any lower bound with $a \preceq c$ then $c \preceq a$ (because it is a lower bound); thus $a = c$. Thus a is the greatest lower bound.

6. Associate with vertex a the set $\{a, c, d, f\}$; b with $\{b, d, e, g\}$; c with $\{c, f\}$; d with $\{d, f, g\}$; e with $\{e, g\}$; f with $\{f\}$; and g with $\{g\}$.

7. Suppose M is an antichain consisting of only minimal elements, and let m be a minimal element not contained in M. We need to show that m is not comparable to any element of M. But this is immediate; since m is minimal, there is no x with $x \prec m$, and if $m \prec x$ then x is not minimal. Thus $M \cup \{m\}$ is a larger antichain containing M.

8. The minimal elements are g, h, j, and k.

9. The width is 5.

Chapter 10

1. Consider a graph with v vertices. All degrees are less than v. If all are different, the degree sequence is $(v-1, v-2, \ldots, 1, 0)$. The vertex of degree $v-1$ is adjacent to every other vertex, but it cannot be adjacent to the vertex of degree 0—a contradiction.
    The multigraph example is

2. (i) $(4,2,2,2,1,1) \Rightarrow (1,1,1,0,1) = (1,1,1,1,0) \Rightarrow (0,1,1,0)$ Yes.
    (ii) $(5,4,3,2,2,2) \Rightarrow (3,2,1,1,1) \Rightarrow (1,0,0,1)$ Yes.
    (iii) $(3,3,3,3,2,2) \Rightarrow (2,2,2,2,2) \Rightarrow (1,1,2,2) = (2,2,1,1) \Rightarrow (1,0,1)$
    Yes.
    (iv) No (sum is odd).
    (v) $(6,5,2,2,1,1,1,0) \Rightarrow (4,1,1,0,0,0,0)$ No.
    (vi) $(6,5,4,3,3,3,1,1,0) \Rightarrow (4,3,2,2,2,0,1,0) \Rightarrow (2,1,1,1,0,1,0) \Rightarrow$
    $(0,0,1,0,1,0)$ Yes.
    Graphs:

    (i) (ii) (iii) (vi)

3. $xabcf, xabef, xadef, xdabcf, xdabef, xdef, xadbecf, xdebcf.$

4. $a, 1 \qquad b, 2 \qquad c, 3 \qquad d, 1 \qquad e, 2 \qquad f, 3.$

5. (i) 1 to a, c, w, x, 2 to d, v, y, 3 to e, z. \qquad (ii) 1 to a, c, k, 2 to all others.

6. Go along the path $x, a, \ldots$ until the first time you meet a vertex (other than x) that appears in both paths. Say that vertex is c. Then follow the second path back from c to y. The result is a closed walk containing x. The walk will be a cycle unless it contains some repetition, and this will only happen if $a = b$. In the latter case, the walk generated is not a cycle, it goes xax, a repeated edge.

7. $\chi(G) = max(\chi(G_1), \chi(G_2)).$

8. (i) 4, \quad (ii) 4, \quad (iii) 3, \quad (iv) 3, \quad (v) 3, \quad (vi) 3.

9. (i) Color G with two colors and apply a new third color to all new vertices. (ii) No: for example, suppose G is a tree.

10. Beginning at vertex x, select some vertex v_1 adjacent to x and mark the edge xv_1 "used." If possible, choose v_1 to be of odd degree; then we are done. If not, v_1 has even degree, so there must be at least one edge incident to v_1 that is not yet used. Now choose a neighbor of v_1 other than x and call it v_2. If v_2 has odd degree, we are done. Continue this

process. Each time we select a new vertex we use one edge, so if the vertex has even degree there must be another edge that has not been used, and we can go on. But the process must end, because there are only finitely many edges. So we must eventually reach an odd vertex.

Chapter 11

1. (i) No; 1   (ii) Yes   (iii) No; 3   (iv) Yes   (v) No; 3   (vi) 3.
   Solutions to (ii) and (iv) are found by the algorithm. Here are sample solutions for the eulerizations.

   (i)   (iii)   (v)   (vi)

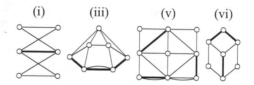

2. None is Eulerian. Numbers are 4, 1, 2, 5, 2, 2.

3. Part I shows the maze with a graph imposed on it, part II shows the solution:

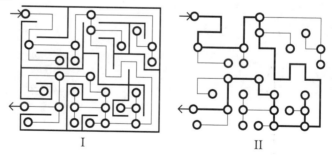

   I   II

4. These are not the only solutions:

   (i)   (ii)   (iii)   (iv)

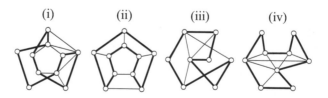

5. If a is to be on the path, edge ab is necessary. For c, bc is needed. Similarly g requires bg. So the "path" has a threeway at b.

6.

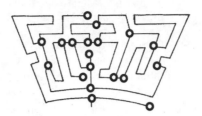

7. One well-known example is:

8. (i) Sorted edges: (31245). Nearest neighbors: (12345), (21345), (31245), (41235), (51234). (ii) Say the cycle is (*abcde*). The length of this cycle is

$$(a+b) + (b+c) + (c+d) + (d+) + (e+a) = 2(a+b+c+d+e).$$

But $\{a, b, c, d, e\} = \{1, 2, 3, 4, 5\}$ (in some order), so the sum is always 30.

9. (i) sorted: *cebda* (103); *a*: *acebd* (103); *c*: *cadbe* (103).
(ii) sorted: *dabec* (289); *a*: *abecd* (289); *c*: *cbead* (290).
(iii) sorted: *bacde* (122); *a*: *acdbe* (121); *c*: *cdabe* (124).
(iv) sorted: *dbcea* (125); *a*: *aecbd* (125); *c*: *cbdea* (121).

10. (i) Sorted edges: *xvywuz*, 26; others: *uvyw*, fails; *vywuzx*, 26; *wuvyzx*, 26; *xvywuz*, 26; *ywuvxz*, 32; *zxvywu*, 26.
(ii) Sorted edges: *wxyz*, 28; others: *wxy*, fails; *xwzy*, 28; *ywx*, fails; *zwxy*, 28.

11. In order to reach z, edges *wz* and *zx* must be included, so *wx* cannot occur. Sorted edges starts with *tu*, *uw*, *wx*. Every nearest neighbor includes *wx*. To find a Hamilton cycle, first assume the path *vywzx*; to complete, go *xutsv*.

12. The trees are

13. Say there are v vertices. There are $v - 1$ edges, so the degrees add to $2v - 2$. Each degree is at least 1. Assume one vertex has degree n. Say there are x of degree 1 and y of degree 2 or greater. Then $x + y = v - 1$; the sum of degrees is at least $n + x + 2y = n + x + 2(v - 1 - x) = 2v - 2 + n - x$. So $n - x \leq 0$, $x \geq n$.

14. (i) For the multiple path, one has n choices of edge for each edge of the underlying path, giving n^{v-1} paths in all.

(ii) For the multiple cycle, each spanning tree is a path; there are v choices for the pair of adjacent vertices that will not be adjacent in the spanning tree, and for each choice there are again n^{v-1} paths.

15. The following solutions were found using Prim's algorithm. Even so, they are not unique: For example, in (i), edge ad could have been chosen instead of de.

(i)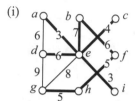

(ii)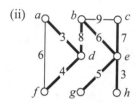

(iii)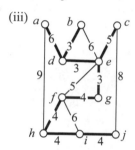

(iv)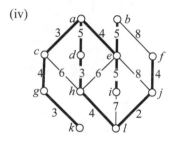

Chapter 12

1. (i) 2. (ii) 8. (iii) 3. (iv) 5.

2. Suppose $A = a_1a_2\ldots a_n$, $B = b_1b_2\ldots b_n$ and $C = c_1c_2\ldots c_n$ are three codewords. We need to prove that $d(A, C) \leq d(A, B) + d(B, C)$.

We can partition $\{1, 2, \ldots, n\}$ into four sets: $P = \{i : a_i = b_i = c_i\}$; $Q = \{i : a_i = b_i \neq c_i\}$; $R = \{i : a_i \neq b_i = c_i\}$; $S = \{i : a_i \neq b_i \neq c_i\}$. Notice that—since there are only two possible values for entries—members of S all satisfy $a_i = c_i$. Write S_{AB} for the set of places where A and B differ, and so on. Then $S_{AB} = R \cup S$, $S_{BC} = Q \cup S$, $S_{AC} = Q \cup R$. Now $|S| \geq 0$, so

$$d(A, C) = |S_{AC}| = |Q| + |R|$$
$$\leq |Q| + |R| + 2|S|$$
$$= (|R| + |S|) + (|Q| + |S|)$$
$$= |S_{AB}| + |S_{BC}| = d(A, B) + d(B, C).$$

3. Test the received word to see its distance from A, B, C, D, E. If the distance is 1 or 0, that is definitely the intended codeword. Otherwise, consider further. (i) A (distance 1). (ii) B (distance 1). (iii) C (distance is 2, but every other codeword is distance greater than 2 from the candidate.) (iv) The word is distance 2 from D and also distance 2 from E, so we cannot decode it.

4. Minimum distance is 4. So the code can detect 3 errors and always correct 1.

5. (i) 32. (ii) 1; 0.

6. For four sequences, use $\{00000, 11100, 00111, 11011\}$.
If we have a set of five there must be three that begin with the same symbol: say $0abcd$, $0efgh$, $0ijkl$. Without loss of generality we can assume that the last of these is 00000 (if, for example, $i = 1$ then we can change the second entry in all five codewords, 0 becoming 1 and 1 becoming 0, so as to achieve a set of five where $Z = 00...$, and similarly for the other positions). Then at least three of $abcd$ equal 1, as do at least three of $efgh$. Then $abcd$ and $efgh$ must have at least two common entries, and $0abcd$ and $0efgh$ are distance at most 2.

7. (i) 1010001. (ii) 0011100. (iii) 1001101. (iv) 0010111.

8. (i) 1001. (ii) 0010. (iii) 1001. (iv) 1101. (v) 1011. (vi) 1100.

9.
$$\begin{bmatrix} 1 & 0 & 0 & 1 & 1 & 0 \\ 0 & 1 & 0 & 1 & 0 & 1 \\ 0 & 0 & 1 & 0 & 1 & 1 \end{bmatrix} \; ; 1; 0.$$

(For example, you cannot decide whether 100000 was meant to be 000000 (decodes as 000) or 100100 (decodes as 100).)

10. (i) 000100011000101100111101011100 (ii) 100100000100011001010000.

11. (i) ACAACCAAAAGAAATAA (ii) AAACAATAAGACACAAAC.

Chapter 13

1.

0	1	2	3	4	5
1	2	3	4	5	0
2	3	4	5	0	1
3	4	5	0	1	2
4	5	0	1	2	3
5	0	1	2	3	4

2. Assume the symbols are 1, 2, 3, 4, 5. Then the $(2,1)$ and $(2,4)$ entries are equal (there are many other problems).

3. (i) Suppose a Latin square L has first row $(a_1, a_2, \ldots, a_n)$. Carry out the permutation $1 \mapsto a_1$, $2 \mapsto a_2$, $\ldots$, $n \mapsto a_n$ on the columns of L. Permute the rows in the corresponding fashion.

(ii) Each reduced square gives rise to $n!$ different squares by column permutation. Once a column permutation is chosen, there are $(n-1)!$ different row permutations available (leaving row 1 as the first row). The $n!(n-1)!$ resulting squares are all different (any two will differ somewhere in the first row or column).

(iii) $r_3 = 1$, $r_4 = 4$.

(iv) The reduced squares are:

1	2	3	4
2	1	4	3
3	4	1	2
4	3	2	1

1	2	3	4
2	1	4	3
3	4	2	1
4	3	1	2

1	2	3	4
2	3	4	1
3	4	1	2
4	1	2	3

1	2	3	4
2	4	1	3
3	1	4	2
4	3	2	1

4. Suppose A is a Latin square of side n based on $\{a_1, a_2, \ldots, a_n\}$, and B is a Latin square of side r based on $\{b_1, b_2, \ldots, b_r\}$. Rows $(s-1)r + 1, (s-1)r + 2, \ldots, sr$ of $A \times B$ form the array
$$[(a_{s,1}, B), (a_{s,2}, B), \ldots, (a_{s,n}, B)].$$

Now the elements in each row of $(a_{s,t}, B)$ are precisely the ordered pairs $(a_{s,t}, b_j)$, $1 \leq j \leq r$, once each. As t ranges from 1 to n, $a_{s,t}$ covers each of $a_1, a_2, \ldots, a_n$ precisely once, so the $(a_{s,t}, b_j)$ go through all the (a_i, b_j) once each, and each row of $A \times B$ contains each pair precisely once. This is the row property of a Latin square. The column property is similar. If A and B are transversal squares, the transversal property for $A \times B$ follows from the same argument.

5. (i) Where will the entry $(1,1)$ occur in (A, A^T)? If it is in position ij, it also appears in position ji. To avoid two occurrences, we need $i = j$, a diagonal position. So 1 appears on the diagonal somewhere. The same argument applies to every symbol. So each of $1, 2, \ldots, n$ is on the diagonal. There is just enough room for each symbol to appear exactly once.

(ii) Say A is a self-orthogonal Latin square of side 3 whose diagonal is $(1, 2, 3)$. The only possible $(1, 2)$ entry is 3, and it is also the only possible $(2, 1)$ entry. Therefore $(3, 3)$ appears too many times in (A, A^T).

(iii)

1	3	4	2
4	2	1	3
2	4	3	1
3	1	2	4

1	3	2	5	4
4	2	5	1	3
5	4	3	2	1
3	5	1	4	2
2	1	4	3	5

6. A seating plan is obviously equivalent to a pair of orthogonal Latin squares of order 5; one based on A, B, C, D, E (to specify the university), the other on $1, 2, 3, 4, 5$ (for example, 3 means "discipline d_3").

(i) Use any two orthogonal Latin squares of side 5.

(ii) One example is

A	C	B	D	E
E	A	C	B	D
D	E	A	C	B
B	D	E	A	C
C	B	D	E	A

1	3	5	2	4
5	2	4	1	3
4	1	3	5	2
3	5	2	4	1
2	4	1	3	5

(The second square could be anything orthogonal to the first.)

(iii) This is impossible. Without loss of generality, assume the second square has all 1s on its diagonal. Then the $(1, 2)$ entry must be 2, as this is the only cell next to $(1, 1)$; similarly the $(5, 4)$ cell is 2. Either the $(2, 1)$ or the $(2, 3)$ cell must contain 2; it must be the $(2, 1)$, because otherwise there will be no 2 in column 1, and similarly the $(4, 5)$ cell contains 2. Where does the fifth 2 go? The only row and column without a 2 are row and column 3, and the $(3, 3)$ cell is occupied.

(iv) Yes and yes:

A	C	B	D
B	D	A	C
D	B	C	A
C	A	D	B

1	2	3	4
2	1	4	3
3	4	1	2
4	3	2	1

7. Without loss of generality, suppose the subsquare occupies rows and columns 1 to s. As $s < n$, there is some symbol (x say) that is greater than s; it will not appear in the subsquare. It will appear in each of rows $1, 2, \ldots, s$, in s different columns, each numbered greater than s. At the very smallest possible, these will be in columns $s+1, s+2, \ldots, 2s$ (and maybe greater numbers will arise). So $n \geq 2s$.

8. Each column contains $n - 1$ different elements; complete the column to length n by inserting the unique missing element. In the final array, every symbol appears once in each column, so each occurs n times in total. In the original Latin rectangle, every row was a permutation, so every symbol appeared once per row, or $n - 1$ times in total. So every symbol must have occurred precisely once in the nth row.

9. We find an SDR for the sets $\{23\}, \{34\}, \{45\}, \{15\}, \{12\}$. Try 2 for the first set. The remaining members must represent $\{34\}, \{45\}, \{15\}, \{1\}$. So the fifth is 1. Now we need an SDR for $\{34\}, \{45\}, \{5\}$. Representative of set $\{5\}$ is of course 5. In this way we get $(2, 3, 4, 5, 1)$. Now the last row is determined: the square is

1	2	3	4	5
5	1	2	3	4
4	5	1	2	3
2	3	4	5	1
3	4	5	1	2

(*Note*: We did this solution in some detail because the same thinking will apply to harder problems.)

10. (i) One example:

1	5	3	4	2	6
3	2	1	5	6	4
2	3	6	1	4	5
6	1	4	2	5	3
4	6	5	3	1	2
5	4	2	6	3	1

(ii) Impossible because 5 only appears once.

11. (i) $x = 6$ because there is only one 6 in the array. One solution is

1	2	3	4	6	5
5	6	1	2	3	4
3	4	5	1	2	6
4	1	2	6	5	3
6	3	4	5	1	2
2	5	6	3	4	1

(ii) Impossible. There is only one 5 in the array, but x cannot equal 5 or there would be two 5s in column 4.

12. One example is

5	2	3	6	4	1
3	5	2	4	1	6
1	4	6	5	2	3
2	3	5	1	6	4
6	1	4	2	3	5
4	6	1	3	5	2

13. (i) Assume the diagonal is $(1, 2, 3, 4, 5)$. The $(1, 2)$ entry could be 3, 4 or 5; the case of 5 will be isomorphic to the case of 4 (apply permutation (45) to rows, columns and symbols).

In the former case, the $(1, 3)$ entry cannot be 2 (you would have a 3×3 subsquare; see Exercise 12A.7) so it might as well be 4 (the permutation (45) could be used if the entry were 5). Then you get

1	3	4	5	2
3	2	1		
4	1	3		
5			4	
2				5

and there is no possible $(2, 4)$ entry.

In the latter case, the $(1, 3)$ entry can be 2 or 5 (the permutation (45) could be used if the entry were 5). If it is 2 there is no completion (the first row must be $(1, 4, 2, 5, 3)$ and the second row cannot be completed; if it is 5, the unique solution is

1	4	5	3	2
4	2	1	5	3
5	1	3	2	4
3	5	2	4	1
2	3	4	1	5

(ii) We would need one of the squares to have 3 in the $(1, 2)$ position.

14. (i) Say there are t copies of 1 above the diagonal. Then there will also be t copies of 1 below the diagonal, so there are $n - 2t$ copies of 1 on the diagonal. This number must be non-negative; since n is odd, it cannot be zero. So it is positive. Similarly for all other entries; so there is at least one of each symbol on the diagonal. Since there are only n diagonal positions, each integer from 1 to n must appear there once.

(ii) Try a 2×2 square.

Chapter 14

1. Three blocks are 12, 23 and 13. The other two must be either two copies of the same block (e.g., $\{12, 23, 13, 12, 12\}$) or different blocks (e.g., $\{12, 23, 13, 12, 13\}$). Any example can be derived from one of these two by permuting the names of varieties.

2. Write S for the set comprising the first v positive integers, and write S_s for the set formed by deleting i from S. Then blocks $S_1, S_2, \ldots, S_v$ form the required design.

3. Clearly v, b and k are correct. To see that $r = 5$, check each element: for example, element 1 is in blocks 1, 2, 3, 4, 5, element 2 is in blocks 1, 2, 6, 7, 8, and so on. Similarly check to see that $\lambda = 2$. The constancy of k and r show that this is a block design, and the constancy of λ shows that the design is balanced.

4. The easiest example is to take the seven blocks of a $(7, 7, 3, 3, 1)$-design, twice each. For example,

$$\begin{array}{ccccccc} 123 & 145 & 167 & 246 & 157 & 347 & 356 \\ 123 & 145 & 167 & 246 & 157 & 347 & 356 \end{array}.$$

    (*Note*: This example is not simple. See Problem 14.5.)

5. Clearly $\lambda \leq r$. If λr then every block must contain every element, and the design is complete.

6. $(v, \frac{1}{2}v(v-1), v-1, 2, 1)$.

7.

v	b	r	k	λ	
15	35	7	3	1	
21	28		3		$r(k-1) = 8; \lambda(v-1) = 20\lambda > 8$
11	11	5	5	2	
14	7	4			$v > b$
12		6	3		$11\lambda = 12;$ impossible
49	56	8	7	1	

8. In each case v is even but $k - \lambda$ is not a perfect square.

9. $(v, tb, tr, k, t\lambda)$.

10. Write $C_i = \{j : i \in B_j\}$. Then the dual has blocks

$$C_0 = \{0,1,2,3,4\} \qquad C_3 = \{1,3,6,8,9\}$$
$$C_1 = \{0,1,5,6,7\} \qquad C_4 = \{2,4,6,7,8\}$$
$$C_2 = \{0,2,5,8,9\} \qquad C_5 = \{3,4,5,7,9\}.$$

Since $\{0,1\}$ lies in two blocks and $\{0,2\}$ lies in one block, the design is not balanced.

11. (i) $(12,22,11,6,5)$, $(11,22,10,5,4)$;
   (ii) $(8,14,7,4,3)$, $(7,14,6,3,2)$;
   (iii) $(33,44,12,9,3)$, $(12,44,11,3,2)$;
   (iv) $(4,12,9,3,6)$, $(9,12,8,6,5)$;
   (v) $(52,68,17,13,4)$, $(17,68,16,4,3)$.

12. One solution: $\{1,3,4,5,9\}$.

13. One solution: $\{1,2,5\}$, $\{1,6,8\}$.

Chapter 15

1. Square A; square the result; square again, and repeat until you get A^{64} (six multiplications). Then $A^{64} \times A^{16} \times A^2$ gives A^{82} in eight multiplications.

2. (i) Let $A = \begin{bmatrix} 0 & 1 \\ 3 & 2 \end{bmatrix}$. Its eigenvalues are 3 and -1. We find the corresponding eigenvectors and make them the columns of $P = \begin{bmatrix} 1 & 1 \\ 3 & -1 \end{bmatrix}$.

The matrix D becomes $\begin{bmatrix} 3 & 0 \\ 0 & -1 \end{bmatrix}$. We find P^{-1} and $PD^nP^{-1}\begin{bmatrix} 2 \\ 2 \end{bmatrix}$ to get $a_n = 3^n + (-1)^n$.

(ii) We use the same A, D, and P as for the previous exercise. Now $PD^nP^{-1}\begin{bmatrix} 1 \\ 7 \end{bmatrix}$ gives us $a_n = 3(3^n) - (-1)^n$.

(iii) Let $A = \begin{bmatrix} 0 & 1 \\ 8 & 2 \end{bmatrix}$. We get $P = \begin{bmatrix} 1 & 1 \\ 4 & -2 \end{bmatrix}$, $D = \begin{bmatrix} 4 & 0 \\ 0 & -2 \end{bmatrix}$, and $P^{-1} = \begin{bmatrix} \frac{1}{3} & \frac{1}{6} \\ \frac{2}{3} & \frac{-1}{6} \end{bmatrix}$. This gives us $b_n = 4^n - 2(-2)^n$.

(iv) We use the same A, D, and P as for the previous part. We get $b_n = 4^n + (-2)^n$.

3.

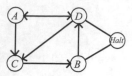

	A	B	C	D	Ha
A	0	0	1	1	0
B	0	0	0	1	1
C	0	1	0	1	0
D	1	0	0	0	1
Ha	0	0	0	0	0

	A	B	C	D	Ha
A	2	1	1	2	2
B	1	1	0	1	1
C	1	1	1	2	1
D	1	0	1	2	2
Ha	0	0	0	0	0

4.

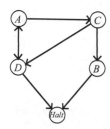

	A	B	C	D	Ha
A	0	0	1	1	0
B	0	0	0	1	1
C	0	1	0	1	0
D	1	0	0	0	1
Ha	0	0	0	0	1

	A	B	C	D	Ha
A	2	1	1	2	5
B	1	1	0	1	3
C	1	1	1	2	4
D	1	0	1	2	4
Ha	0	0	0	0	1

The matrix $\hat{M}^4$ has a $(1,5)$ entry equal to the number of ways in which the machine can go from A to *Halt* in four or fewer steps. The product $(I_5 - M^5)(I_5 - M)^{-1}$ will not yield the correct answer because $\sum_{i=0}^{\infty} M^i$ does not converge. This is because we may make paths of arbitrary length by shifting from A to D and back before ending at *Halt*.

5. (a)

$$
\begin{bmatrix}
 & 1 & 2 & 3 & 4 & 5 & 6 \\
\hline
1 & 0 & 1 & 0 & 0 & 0 & 0 \\
2 & 1 & 0 & 1 & 1 & 0 & 0 \\
3 & 0 & 1 & 0 & 0 & 1 & 0 \\
4 & 0 & 1 & 0 & 0 & 1 & 0 \\
5 & 0 & 0 & 1 & 1 & 0 & 1 \\
6 & 0 & 0 & 0 & 0 & 1 & 0
\end{bmatrix}.
$$

$(I - A)^{-1}$ does not exist; but $\sum_{i=1}^{6} A^i$ has a $(1,6)$ entry of 14.

(b)

	1	2	3	4	5	6
1	0	1	0	0	0	0
2	1	0	1	0	0	0
3	0	1	0	1	0	0
4	0	0	1	0	1	0
5	0	0	0	1	0	1
6	0	0	0	0	1	0

$(I-A)^{-1}$ exists with a $(1,6)$ entry of 1; and $\sum_{i=1}^{6} A^i$ has a $(1,6)$ entry of 1. Although $(I-A)^{-1}$ gives the correct answer, this is an accident; other entries in $(I-A)^{-1}$ are negative or fractional. Either is clearly impossible.

(c)

	1	2	3	4	5	6
1	0	1	1	0	0	0
2	1	0	0	0	0	0
3	1	0	0	0	0	0
4	0	0	1	0	1	0
5	0	0	0	1	0	1
6	0	0	0	0	1	0

$(I - A)^{-1}$ exists with a $(1,6)$ entry of 1; but $\sum_{i=1}^{6} A^i$ has a $(1,6)$ entry of 6.

(d)

	1	2	3	4	5	6
1	0	1	0	0	0	0
2	1	0	1	0	0	0
3	0	1	0	1	0	0
4	0	0	1	0	1	0
5	0	0	0	1	0	1
6	0	0	0	0	1	0

$(I - A)^{-1}$ exists with a $(1,6)$ entry of 1; and $\sum_{i=1}^{6} A^i$ has a $(1,6)$ entry of 1. Again, $(I-A)^{-1}$ accidentally gives the correct answer.

6. (a)

	1	2	3	4	5	6
1	0	1	1	0	0	0
2	1	0	1	1	1	0
3	1	1	0	1	1	0
4	0	1	1	0	1	1
5	0	1	1	1	0	1
6	0	0	0	1	1	0

$(I - A)^{-1}$ does not exist; but $\sum_{i=1}^{6} A^i$ has a $(1,6)$ entry of 184.

(b)

	1	2	3	4	5	6
1	0	1	0	0	0	0
2	1	0	1	1	1	0
3	0	1	0	1	0	0
4	0	1	1	0	1	1
5	0	0	0	1	0	1
6	0	0	0	1	1	0

$(I - A)^{-1}$ exists with a $(1,6)$ entry of 1; but $\sum_{i=1}^{6} A^i$ has a $(1,6)$ entry of 29.

(c)

$$\begin{bmatrix} & 1 & 2 & 3 & 4 & 5 & 6 \\ \hline 1 & 0 & 1 & 1 & 0 & 0 & 0 \\ 2 & 1 & 0 & 1 & 0 & 0 & 0 \\ 3 & 1 & 1 & 0 & 1 & 0 & 0 \\ 4 & 0 & 0 & 1 & 0 & 1 & 1 \\ 5 & 0 & 0 & 0 & 1 & 0 & 1 \\ 6 & 0 & 0 & 0 & 1 & 1 & 0 \end{bmatrix}.$$

$(I-A)^{-1}$ exists with a $(1,6)$ entry of $1/4$; but $\sum_{i=1}^{6} A^i$ has a $(1,6)$ entry of 29.

(d)

$$\begin{bmatrix} & 1 & 2 & 3 & 4 & 5 & 6 \\ \hline 1 & 0 & 1 & 1 & 0 & 0 & 0 \\ 2 & 1 & 0 & 1 & 1 & 0 & 0 \\ 3 & 1 & 1 & 0 & 1 & 1 & 0 \\ 4 & 0 & 1 & 1 & 0 & 1 & 1 \\ 5 & 0 & 0 & 1 & 1 & 0 & 1 \\ 6 & 0 & 0 & 0 & 1 & 1 & 0 \end{bmatrix}.$$

$(I-A)^{-1}$ exists with a $(1,6)$ entry of $11/9$; but $\sum_{i=1}^{6} A^i$ has a $(1,6)$ entry of 116.

7. (a)

$$\begin{bmatrix} & 1 & 2 & 3 & 4 & 5 & 6 \\ \hline 1 & 0 & 1 & 1 & 0 & 0 & 0 \\ 2 & 0 & 0 & 1 & 1 & 1 & 0 \\ 3 & 0 & 0 & 0 & 1 & 1 & 0 \\ 4 & 0 & 0 & 0 & 0 & 1 & 1 \\ 5 & 0 & 0 & 0 & 0 & 0 & 1 \\ 6 & 0 & 0 & 0 & 0 & 0 & 0 \end{bmatrix}.$$

$(I - A)^{-1}$ has a $(1,6)$ entry of 9, as does $\sum_{i=1}^{6} A^i$. In fact, A^n is all zero for $n \geq 6$.

(b)

$$\begin{bmatrix} & 1 & 2 & 3 & 4 & 5 & 6 \\ \hline 1 & 0 & 1 & 0 & 0 & 0 & 0 \\ 2 & 0 & 0 & 1 & 1 & 0 & 0 \\ 3 & 0 & 0 & 0 & 1 & 0 & 0 \\ 4 & 0 & 0 & 0 & 0 & 1 & 1 \\ 5 & 0 & 0 & 0 & 0 & 0 & 1 \\ 6 & 0 & 0 & 0 & 0 & 0 & 0 \end{bmatrix}.$$

$(I - A)^{-1}$ has a $(1,6)$ entry of 4 as does $\sum_{i=1}^{6} A^i$.

(c)

$$\begin{bmatrix} & 1 & 2 & 3 & 4 & 5 & 6 \\ \hline 1 & 0 & 1 & 1 & 0 & 0 & 0 \\ 2 & 0 & 0 & 1 & 0 & 0 & 0 \\ 3 & 0 & 0 & 0 & 1 & 0 & 0 \\ 4 & 0 & 0 & 0 & 0 & 1 & 1 \\ 5 & 0 & 0 & 0 & 0 & 0 & 1 \\ 6 & 0 & 0 & 0 & 0 & 0 & 0 \end{bmatrix}.$$

$(I - A)^{-1}$ has a $(1,6)$ entry of 4 as does $\sum_{i=1}^{6} A^i$.

(d)

$$\begin{bmatrix} & 1 & 2 & 3 & 4 & 5 & 6 \\ \hline 1 & 0 & 1 & 0 & 0 & 0 & 0 \\ 2 & 0 & 0 & 1 & 0 & 0 & 0 \\ 3 & 0 & 0 & 0 & 1 & 0 & 0 \\ 4 & 0 & 0 & 0 & 0 & 1 & 0 \\ 5 & 0 & 0 & 0 & 0 & 0 & 1 \\ 6 & 0 & 0 & 0 & 0 & 0 & 0 \end{bmatrix}.$$

$(I - A)^{-1}$ has a $(1,6)$ entry of 4 as does $\sum_{i=1}^{8} A^i$.

8. In each case, the matrix A is just like the matrix for the last exercise, except with a 1 for the $(6,6)$ entry. The matrix $(I - A)^{-1}$ does not exist in each case, because the geometric series will not converge.

a) $\sum_{i=1}^{6} A^i$ has a $(1,6)$ entry of 30, which is far too large. But A^6 has a $(1,6)$ entry of 9, which is the correct number of ways to move from square 1 to square 6.

b) $\sum_{i=1}^{6} A^i$ has a $(1,6)$ entry of 12, which is far too large. But A^6 has a $(1,6)$ entry of 4, which is the correct number of ways to move from square 1 to square 6.

c) $\sum_{i=1}^{6} A^i$ has a $(1,6)$ entry of 12, which is far too large. But A^6 has a $(1,6)$ entry of 4, which is the correct number of ways to move from square 1 to square 6.

d) $\sum_{i=1}^{6} A^i$ has a $(1,6)$ entry of 26, which is far too large. But A^6 has a $(1,6)$ entry of 8, which is the correct number of ways to move from square 1 to square 6.

9.
$$A = \begin{bmatrix} 1 & 0 & 1 & 0 & 1 & 0 \\ 0 & 1 & 0 & 1 & 0 & 1 \\ 0 & 0 & 1 & 0 & 0 & 1 \end{bmatrix}.$$

It is simplest to expand about the last row; in any event, we get four 1×4 matrices each containing three 1s, to give a permanent of 12.

10.
$$\mathrm{Per} \begin{bmatrix} 1 & 1 & 0 & 0 & 0 \\ 1 & 0 & 1 & 0 & 0 \\ 1 & 1 & 1 & 1 & 0 \\ 0 & 0 & 1 & 0 & 1 \\ 0 & 0 & 1 & 1 & 1 \end{bmatrix} = 5.$$

11. We use $n = 8$ and $k = 4$ to get $\mathrm{Per}(A) = \lceil 8! \, (4/8)^8 \rceil = \lceil 315/2 \rceil = 315$.

Appendix A

1. (i) $\{a, b, c, d, e, g, i\}$, (ii) $\{c, e\}$, (iii) $\{g, i\}$, (iv) $\{a, b, c, d, e, g\}$.

2. $S_1 \subseteq S_3, S_4$; $S_2 \subseteq S_3, S_4, S_5$; $S_5 \subseteq S_2, S_3, S_4$; $S_2 = S_5$.

3. $S \cup T = U \Rightarrow \overline{S} \subseteq T$; $S \cap T = \emptyset \Rightarrow T \subseteq \overline{S}$.

4. (i) and (iii).

5. $S + T$ consists of all the elements that are in S or T but not both. So $S + T = (S \cap \overline{T}) \cup (T \cap \overline{S})$ and this is a disjoint union. On the other hand, $(\overline{S + T}) = (\overline{S} \cap \overline{\overline{T}}) \cup (\overline{T} \cap \overline{\overline{S}}) = (\overline{S} \cap T) \cup (\overline{T} \cap S)$.

$S \subseteq T, S \cup T = T, S \cap T = S$.

6. (i) Assume $S \subseteq T$. If $x \in S \cup T$ then $x \in S$ or $x \in T$ (or both). But, since $S \subseteq T$, $x \in S \Rightarrow x \in T$, so in either case $x \in T$. So (assuming $S \subseteq T$), $S \cup T \subseteq T$. But $T \subseteq S \cup T$ is always true. So $S \subseteq T \Rightarrow S \cup T = T$.

(ii) $S \cup T$ consists of T together with all members of S that are not in T. If $S \cup T = T$ then there are no members of S that are not in T, so $S \cap T = S$.

(iii) Assume $S \cap T = S$. Consider any element of S. If it were not in T, then it would be a member of S that is not in $S \cap T$. Then the two would not be equal. So no such member of S exists. So $S \subseteq T$.

We have proven $S \subseteq T \Rightarrow S \cup T = T \Rightarrow S \cap T = S \Rightarrow S \subseteq T$. So the three are equivalent.

7. (i) Obviously true for $n = 1$: $1 = 1^2$. Say true for $n = r$: $1 + 3 + \ldots + (2r - 1) = r^2$. Then

$$(1 + 3 + \ldots + (2r - 1)) + (2r + 1) = r^2 + (2r + 1) = (r + 1)^2.$$

(ii) Case $n = 1$ is true. Suppose it's true for $n = k$. That is,

$$\sum_{r=1}^{k} r^2 = \tfrac{1}{6} k(k + 1)(2k + 1).$$

We need to prove case $n = k$, that is,

$$\sum_{r=1}^{k+1} r^2 = \tfrac{1}{6}(k + 1)(k + 2)(2k + 3).$$

Now, $(k + 1)(k + 2)(2k + 3) = 2k^3 + 9k^2 + 13k + 6$. But, using the assumption,

$$
\begin{aligned}
\sum_{r=1}^{k+1} r^2 &= \left(\sum_{r=1}^{k} r^2 \right) + (k + 1)^2 \\
&= \left(\tfrac{1}{6} k(k + 1)(2k + 1) \right) + (k + 1)^2 \\
&= \tfrac{1}{6} \left(2k^3 + 3k^2 + k \right) + (k + 1)^2 \\
&= \tfrac{1}{6} \left(2k^3 + 3k^2 + k \right) + k^2 + 2k + 1 \\
&= \tfrac{1}{6} \left(2k^3 + 3k^2 + k + 6k^2 + 12k + 6 \right) \\
&= \tfrac{1}{6} \left(2k^3 + 9k^2 + 13k + 6 \right).
\end{aligned}
$$

8.

$$(n-1)^3 + n^3 + (n+1)^3$$
$$= (n^3 - 3n^2 + 3n - 1) + n^3 + (n^3 + 3n^2 + 3n + 1)$$
$$= 3n^3 + 6n$$
$$= 3n(n^2 + 2).$$

We need to show that 3 divides $n(n^2 + 2)$. If 3 divides n, we are finished. If not, $n \equiv 1$ or $2 \pmod 3$, so $n^2 \equiv 1 \pmod 3$, and 3 divides $(n^2 + 2)$.

9. Rewriting slightly, $x_n = x_{n-1} + 2x_{n-2} = (x_{n-2} + 2x_{n-3}) + 2x_{n-2} = 3x_{n-2} + x_{n-3}$. So $x_n \equiv x_{n-3} \pmod 3$. Now $x_1 = x_2 = 1$, so x_n is not divisible by 3 when $n \equiv 1$ or $2 \pmod 3$. On the other hand, $x_3 = x_2 + 2x_1 = 1 + 2 = 3$, so x_n *is* divisible by 3 when $n \equiv 0 \pmod 3$.

10. (i) $\rho\sigma = \{(2,3), (3,3), (3,4), (3,5), (4,1), (4,2), (5,4), (5,5)\}$.

(ii) $\sigma\rho = \{(1,2), (2,2), (2,4), (3,2), (4,1), (4,3), (4,5), (5,1), (5,3), (5,5)\}$.

(iii) $\rho\rho = \{(2,1), (3,1), (3,2), (3,3), (4,2), (4,4), (5,5)\}$,

$\sigma\sigma = \{(1,3), (2,1), (2,2), (3,2), (3,3), (4,4), (4,5), (5,4), (5,5)\}$.

11. ρ is reflexive if and only if

$$x + y = y + x \text{ for all } x, y \in \mathbb{Z},$$

but that is just the commutative law for integers. Symmetry means that, for all x, y, z, t,

$$x + t = y + z \Rightarrow z + x = t + y,$$

which follows from the commutative law for integers and the symmetric law for equality. For transitivity, assume

$$x + t = y + z \quad z + b = t + a.$$

Adding these two equations,

$$(x + t) + (z + b) = (y + z) + (t + a).$$

Subtracting z and t from both sides,

$$x + b = y + a,$$

which is equivalent to $(x, y)\, \rho\, (a, b)$.

Appendix B

1. (i) $\begin{bmatrix} 3 & 3 & -3 \\ -6 & 0 & 24 \end{bmatrix}$. (ii) $\begin{bmatrix} 4 & -2 \\ -2 & 2 \end{bmatrix}$. (iii) $\begin{bmatrix} 10 & 3 \\ 16 & -2 \\ -8 & 3 \end{bmatrix}$.

2. $x = 11$; $y = 9$; $z = \frac{3}{2}$.

3. (i) $(6, 12, 3)$. (ii) $4(12, 4, -4)$. (iii) $(3, 7)$. (iv) $2(1, -11, 10, 15)$.

4. (i) no; (ii) 1×3; (iii) no; (iv) 4×3; (v) no; (vi) no; (vii) no; (viii) 3×3; (ix) 3×3.

5. (i) $\begin{bmatrix} 6 & -4 \\ 3 & 1 \end{bmatrix}$. (ii) $\begin{bmatrix} 17 & -14 \\ 9 & -6 \end{bmatrix}$. (iii) $\begin{bmatrix} 1 & 3 & 0 \\ 1 & 3 & 0 \\ -1 & -3 & 0 \end{bmatrix}$.

    (iv) $\begin{bmatrix} 3 & 1 \\ 1 & 0 \\ -1 & -3 \end{bmatrix}$. (v) $\begin{bmatrix} 8 \\ 0 \\ -1 \end{bmatrix}$.

6. $A = \begin{bmatrix} 5 & -3 \\ -10 & 7 \end{bmatrix}$.

7. (i) $\begin{bmatrix} 2 & -1 \\ -1 & 1 \end{bmatrix}$, $\begin{bmatrix} 3 & -2 \\ -2 & 1 \end{bmatrix}$.

    (ii) $\begin{bmatrix} 1 & -4 & 1 \\ 0 & 9 & -5 \\ 0 & 0 & 4 \end{bmatrix}$, $\begin{bmatrix} 1 & 14 & -5 \\ 0 & -27 & 19 \\ 0 & 0 & -8 \end{bmatrix}$.

8. (i) 3; $\begin{bmatrix} 1 & -2 \\ -\frac{2}{3} & \frac{5}{3} \end{bmatrix}$. (ii) 0; no inverse.

9. (i) $AB = AC = \begin{bmatrix} 3 & 9 \\ 3 & 9 \end{bmatrix}$. (ii) $AB = AC = \begin{bmatrix} 6 & 4 & 4 \\ 4 & 3 & 4 \\ 8 & 5 & 4 \end{bmatrix}$.

Hints for Problems

Chapter 1

2. What if there is an element a such that $a\rho b$ is *never* true for *any* b?

5. Think of chords.

7. Construct a graph whose vertices are the attendees (other than the hostess); there is an edge if two people shook hands. The degrees must be 0, 1, 2, 3, 4, 5, 6, 7, 8. Consider the person that shook hands eight times; the only person he/she missed must be his/her spouse. What does this tell you about the other vertices' degrees?

8. Since E only belongs to O_4, clearly if E resigns no SDR is possible as there are only four willing students and five organizations. However, if D were to resign from O_5, then O_1, O_2, O_3 would each consist of only A and C among the willing students. Then, if student E were to join either O_1 or O_3, no one resignation could prevent an SDR.

Checking this is tedious but straightforward; clearly we only have to check the possibility of a resignation of a student from the organization that he or she represents in the SDR we found for part (i).

Chapter 2

2. $1001 = 11 \times 91$; $100001 = 11 \times 9091$. Generalize this.

4. A combinatorial proof could involve selecting a committee of k individuals from a larger group of m men and n women; the right-hand side of the equation sorts the possible committees according to gender makeup.

5. You may select a committee of k people, with a chair (or facilitator). You might choose the chair first, and then choose the other $k-1$ committee members, or you might choose the committee as a whole first and then select the chair from the chosen committee members.

7. Use Corollary 2.3.3.

Chapter 3

5. Let the three dice be distinct. Then consider the number of ways to get a 9 if the low roll is a one; for instance, there are six rolls with values $\{1, 3, 6\}$. Repeat for other values of the low roll.

9. The sum of a subset must lie between 6 and 30. There are 210 subsets, and 25 sums. So the average number of subsets with a given sum is at least 8. Try to improve this.

Chapter 4

2. We are virtually using the pigeonhole principle in a reverse form; that is, if fewer than n pigeons are roosting in n pigeonholes, then at least one pigeonhole is unoccupied.

3. Look at Theorem 4.1(ii).

4. This is similar to Example 4.2.

7. A bipartite graph will contain no K_3; what is the bipartite subgraph of $K_{2R(m,n)-2}$ with the most edges?

9. Use induction.

Chapter 5

1. Look at a Venn diagram.

2. One way is to prove that D_{n+1} is odd if and only if n is even.

3. There are two cases: the previously vacant chair remains vacant or not.

5. Try induction.

Chapter 6

1. Write $f_b(x) = \sum_{k=0}^{\infty}(a_0 + a_1 + \cdots + a_k)x^k$, multiply both sides by $1 - x$, and collect like terms.

5. The two leftmost squares of the $2 \times n$ rectangle are either covered by a single 2×1 domino, in which case there are a_{n-1} ways to complete the process, or they are covered by two 1×2 dominos, in which case there are a_{n-2} ways to finish.

6. Let a_n be the number of strings of length n with this property. The leftmost bit of the string is either a 0, in which case there are a_{n-1} possibilities, or a 1, in which case the string begins with 10 and there are a_{n-2} possibilities.

Chapter 7

6. Use the explicit value for $\{{}^n_2\}$ and the fact that $x^k - y^k$ is divisible by $x - y$.

7. A distinguished element approach works here; take element $n + 1$, and choose $n - i$ others to go in its part of the partition.

Chapter 8

4. Consider for any i, j in S the permutation $\sigma_j \circ \sigma_i^{-1}$.

7. For any i, j in S, consider $\sigma_j \circ \sigma_i^{-1}$.

15. The eight permutations of D_4 are the four rotations that were found in the previous problem, together with the four reflections $(13)(45)(68)$, $(16)(27)(38)$, $(18)(25)(47)$, and $(24)(36)(57)$.

Chapter 9

1. Simply modify the proof for the existence of a maximal element.

2. The Hasse diagram has an edge between x and y provided that there is no element between them. What sort of relationships would three mutually-adjacent vertices require?

3. Consider $\mathcal{B}_4$ where the set of four elements is the set of four primes whose product is n.

4. It may help to think of the m-set as disjoint from the n-set.

6. Not every down-set is the down-set L_x generated by a single element.

7. Not all largest antichains consist only of maximal elements.

Chapter 10

3. (iv) Remember that the sum of the degrees is even.

4. Start with a cycle of length v.

5. See Exercise 10.B.2.

6. If S is a set of r vertices of G, how many edges can have at most one endpoint in S?

Chapter 11

2. (i) Consider any two adjacent vertices. Use the fact that they can have no common neighbor.
   (ii) Consider any two adjacent vertices x and y and form a new graph by identifying them (two vertices in the new graph are adjacent if and only if they were adjacent in the old graph; all vertices adjacent to x or y are adjacent to the new vertex).

3. (i) Try to include a cutpoint.
(ii) Try to include odd vertices.

6. (i) First, solve the problem for $n = 1$. To solve the general case, treat $K_{n,2n,3n}$ as n copies of $K_{1,2,3}$ together with further edges.

8. Show that any two vertices in the center of a tree are adjacent.

10. Use induction on v.

11. Use the result of the preceding problem.

Chapter 12

2. See the part of the text preceding the theorem.

5. Suppose a code has minimum distance $2n$. What can go wrong?

Chapter 13

1. What is the $(1,2)$ entry?

3. Use induction on k.

4. If you take the set all integers modulo a prime (or more generally the set of all members of a finite field) and multiply by one non-zero member, you get a permutation of the same set.

6. (iii) Use three orthogonal Latin squares of side 4.

Chapter 14

1. Any two blocks have intersection size 2. Assume the treatments are 1, 2, 3, 4, 5, 6, 7. Consider the blocks that contain 1. Without loss of generality, start with 1234. The second containing both 1 and 2 may as well be 1256 (if, for example, it is 1257, the permutation (67) is an isomorphism). Show that the other blocks containing 1 are now determined up to isomorphism (what are they?). Continue from there.

2. The proof of part (ii) is a generalization of Example 13.2.

3. The dual of a linked design must be balanced.

7. Use Exercise 14A.2 and Exercise 14A.9.

11. Divide the riders into two groups of 7. In the first seven heats, riders 1 through 7 meet each other once each.

Chapter 15

1. Use the matrix A found in part (i): show first that $A \times [a_{n-1}\, a_n\, a_{n+1}]^{\mathrm{T}} = [a_n\, a_{n+1}\, a_{n+2}]^{\mathrm{T}}$ using the recurrence.

Solutions to Problems

Chapter 1

1. Suppose ρ is reflexive and circular.

$$x\rho y \text{ and } y\rho z \Rightarrow z\rho x.$$

(i) Say $x\rho y$ is true. We know $y\rho y$ is true. Using the above equation with $z = y$,

$$x\rho y \Rightarrow x\rho y \text{ and } y\rho y \Rightarrow y\rho x.$$

So ρ is symmetric.

(ii) Apply symmetry to the equation:

$$x\rho y \text{ and } y\rho z \Rightarrow z\rho x \Rightarrow x\rho z.$$

So ρ is transitive.

3. We use the fact that $(n-1)^2 \geq 2$ when $n \geq 3$. (In fact $(n-1)^2 > 2$ for these values.) Therefore $n^2 - 2n + 1 \geq 2$ and $n^2 \geq 2n - 1 + 2 = 2n + 1$. We proceed by induction on n. The result $2^n \geq n^2$ is true for $n = 4$ (both sides $= 16$). Assume $n \geq 4$ and $2^n \geq n^2$. Then

$$2^{n+1} = 2 \times 2^n \geq 2 \times n^2 = n^2 + n^2 \geq n^2 + (2n + 1) = (n+1)^2.$$

So the result is true for all $n \geq 4$ by induction.

4. There must be at least two people. The result is true for 2 (the second person, the man, is directly behind the first person, the woman). Assume it is true for n people. Consider any line on $n + 1$ people with a woman in front and a man at the back. If the nth person is a man, then by the induction assumption the required configuration (man directly behind woman) occurs somewhere in the first n places. If not, the positions n and $n + 1$ provide the configuration.

5. The result is true (given) for $n = 3$. Assume it is true for $n = k$. Consider a polygon with vertices $A_1, A_2, \ldots, A_{n+1}$. The sum of the angles equals the sum of the angles of the polygon $A_1, A_2, \ldots, A_n$ plus those of the triangle A_1, A_n, A_{n+1}. The first equals $(n - 2)\pi$ from the induction hypothesis, the second is π, so the sum is $((n + 1) - 2)\pi$.

6. Say there are v vertices. No vertex has degree greater than $v-1$ or less than 0. So, if they are all different, there must be one vertex of each degree $0, 1, 2, \ldots, (v-1)$. The vertex of degree $(v-1)$ must be adjacent to every other vertex, but it cannot be adjacent to the one of degree 0.

7. (i) 4; (ii) 4.

8. (i) The SDR is easily found; for instance, $O_1 = A, O_2 = B, O_3 = C, O_4 = E, O_5 = D$ works.

Chapter 2

1. Write a_n for the number of positive n-digit integers with all digits distinct. If $n > 10$, there are no such integers. Otherwise, there are nine choices for the first digit (you cannot start with 0), 9 for the second, 8 for the third, $\ldots$, and $(11-a)$ for the ath place. So $a_n = 9 \times P(9, n-1)$. So the number of positive integers with all digits distinct is

$$
\begin{aligned}
& a_1 + a_2 + \ldots + a_{10} \\
= \ & 9[P(9,0) + P(9,1) + \ldots + P(9,9)] \\
= \ & 9[1 + 9 + 72 + 504 + 3,024 + 15,120 + 60,480 + 181,440 \\
& + 362,880 + 362,880] \\
= \ & 9 \times 986,410 \\
= \ & 8,877,690.
\end{aligned}
$$

There are just as many negatives, and do not forget 0. So the answer is 17,755,381.

6. There are several examples; *pappa* and *pops* are both colloquially used to refer to one's father, and there are only 10 distinct rearrangements of *pappa* compared to 12 for *pops*. Now that you have looked up the answer, try to find some words (not necessarily so similar) on your own.

8. Write O_n and E_n for the numbers of subsets of an n-set of odd and even orders, respectively. Then

$$
O_n = \sum_{k=0,k \text{ odd}}^{n} \binom{n}{k}, \quad E_n = \sum_{k=0,k \text{ even}}^{n} \binom{n}{k}.
$$

From Corollary 2.3.3, $\sum_{k=0}^{n} \binom{n}{k}(-1)^k = 0$. The terms in this sum are positive if k is even and negative if k is odd. So

$$\sum_{k=0,k \text{ odd}}^{n} \binom{n}{k} = \sum_{k=0,k \text{ even}}^{n} \binom{n}{k},$$

or $O_n = E_n$. From Corollary 2.3.2, $\sum_{k=0}^{n} \binom{n}{k} = 2^n$, that is $O_n + E_n = 2^n$. So $O_n = 2^{n-1}$ (and we have shown that $E_n = 2^{n-1}$ also).

10. (i) The 27 letters consist of 7 As, 6 Ns, 5 Bs, 4 D', 2 Ss, and one each of I, T, R. So the number of rearrangements is

$$\frac{27!}{7!6!5!4!2!} - 1 = 520,951,478,183,135,999.$$

(We subtract 1 because rearrangements were called for, so the original ordering is not included.)
(ii) Similarly we get

$$\frac{24!}{6!4!4!3!2!2!} - 1 = 62,336,074,312,511,999.$$

11. In the first case, the eight people can be seated in 7! ways. In the second case, we seat one group first (in 3! ways), and then intersperse the others between them in 4! ways, so there are $3! \cdot 4! = 144$ arrangements. In the last case, we seat the diplomat last, in one of thetwo endpoints or one of the seven positions between two of the others; this gives us $9 \cdot 144 = 1296$ seating arrangements.

12. Clearly we must choose one person from each of $n-1$ pairs, and no one from the last pair. There are $C(n, 1) = n$ ways to choose the pair that will be neglected, and there will then be 2^{n-1} ways to choose one of the two members of the $n-1$ other pairs. So we have $n \cdot 2^{n-1}$ possible committees.

13. We take the cases one at a time. Practically speaking, there are three cases; no vegetarian restaurants, no meat-only restaurants, and both. The case "both" means we have one vegetarian and one meat-eater, so we must stick to one of the five restaurants with omnivorous offerings. This means that we can choose the friends in $C(3, 1) \cdot C(2, 1) = 6$ ways, and any of the five restaurants will work, for 30 possibilities. The "no-meat-only" case means that we exclude the meat-eaters but not the vegetarians; there are $C(8, 2) = 28$ choices of two friends and eight restaurants, so there are 224 possibilities. The "no-veg-only" case means that we exclude the vegetarians but not the meat-eaters; there are $C(7, 2) = 21$ ways to pick the friends, and seven restaurants acceptable to these friends, for another 147 possibilities. Altogether, then, there are 401 possibilities.

The student will realize that this is an oversimplification of the compli-cations that occur in real life (for instance, two of the friends cannot eat highly spicy foods as are served at one of the vegetarian and two of the omnivore restaurants, and two of the friends have recently broken up so cannot go anywhere together without causing a scene, and one of the restaurants does not allow children while two of the friends cannot get a babysitter so would have to bring the baby, etc.), but we trust that this gives something of the flavor of the more complicated sorts of counting problems.

Chapter 3

1. As we saw, $P(1) = 0.2, P(2) = P(3) = P(4) = P(5) = P(6) = 0.16$. So the probability of a roll of 4 or less is $P(1) + P(2) + P(3) + P(4) = 0.2 + 0.16 + 0.16 + 0.16 = 0.68$.

2. We shall write the outcomes in the form An, where A is H or T, the result of flipping the quarter, and $n = 0$, 1 or 2, the number of heads on the pennies. The sample space is $\{H0, H1, H2, T0, T1, T2\}$. The tree diagram is

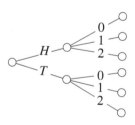

3. The diagram is

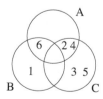

4. Suppose the game is played N times. Alice has probability $\frac{1}{4}$ of drawing the same color as the host, so she will expect to win in $\frac{1}{4}N$ games. In the remaining $\frac{3}{4}N$ games, Bruce draws. There are now two colors that have been drawn, and if Bruce duplicates either, he wins. If he

draws, the chance of him winning is $\frac{1}{2}$, so the expected number of wins is $\frac{3}{4} \times \frac{1}{2}N = \frac{3}{8}N$. So his probability is $\frac{3}{8}$. If he loses, then three colors have been drawn, so the chance of Cathy winning is $\frac{3}{4}$. There have been $\frac{1}{4}N + \frac{3}{8}N = \frac{5}{8}N$ successes, so Cathy draws in $frac58N$ cases, and she wins in $\frac{3}{8}\frac{3}{4}N = \frac{9}{32}N$ cases, so her probability is $\frac{9}{32}$. Therefore Bruce still has the best chance of winning, but now Cathy's chance is better than Alice's. The probability of no winner is $1 - \frac{1}{4} - \frac{3}{8} - \frac{9}{32} = \frac{3}{32}$.

5. The possible ways to get a total of 9 from two dice are $\{3,6\}$ and $\{4,5\}$. Suppose the lowest roll is a 1; there are six rolls with values $\{1,3,6\}$, and six with $\{1,4,5\}$. If the lowest roll is a 2, there are six rolls with values $\{2,3,6\}$, and six with $\{2,4,5\}$. If the lowest is a 3, there are three rolls with values $\{3,3,6\}$, and six with $\{3,4,5\}$. If the lowest is a 4, the only possibility for the two highest is $\{4,5\}$, and there are three rolls with values $\{4,4,5\}$. So there are $(6+6)+(6+6)+(3+6)+3 = 36$ possibilities, out of the $6^3 = 216$ rolls. So the probability is $36/216 = 1/6$.

6. (i) $\frac{5}{6}$; $\frac{25}{36}$; $\frac{125}{216}$.
   (ii) For exactly k times, there are $\binom{6}{k}$ ways to select the k times, and the probability for those k times is probability is $(\frac{1}{6})^k \times (\frac{5}{6})^{n-k}$, so the probability of exactly k doubles is $\binom{6}{k} \times (\frac{1}{6})^k \times (\frac{5}{6})^{n-k}$. So the required probability is found by summing this from $k = 2$ to 6. However, an easier way is to calculate the probability of 1 or no doubles, and subtract this from 1. That probability is

$$(\frac{5}{6})^6 + 6 \times (\frac{5}{6})^5 \times (\frac{1}{6}) = \frac{34375}{46656},$$

or about 73.7%. So the answer is $\frac{12281}{46656}$, or about 26.3%.

8. Suppose E_1 has n outcomes with values $x_1, x_2, \ldots, x_n$, and the probability of the outcome x_i is p_i. Suppose E_2 has m outcomes with values $y_1, y_2, \ldots, y_m$, and the probability of the outcome y_j is q_j. Then $v_1 = \sum_{i=1}^{n} p_i x_i$ and $v_2 = \sum_{j=1}^{m} q_j y_j$. If E_1 has outcome x_i and y_i and E_2 has outcome y_j, then the sum of the outcomes is $x_i + y_j$, and the probability of this occurrence is $p_i q_j$. The expected value of the sum of the two outcomes is

$$\sum_{i=1}^{n} \sum_{j=1}^{m} p_i q_j (x_i + y_j) = \sum_{i=1}^{n} p_i \left[\sum_{j=1}^{m} q_j x_i + \sum_{j=1}^{m} q_j y_j \right]$$
$$= \sum_{i=1}^{n} p_i [x_i + v_2] = \sum_{i=1}^{n} p_i x_i + v_2 \sum_{i=1}^{n} p_i$$
$$= v_1 + v_2.$$

9. Let $N(S)$ denote the sum of the elements of a 4-set S, and let T be the set of all 4-subsets of $\{0, 1, \ldots, 9\}$. T has $\binom{10}{4} = 210$ elements. Write T_n for the set of all members S of T with $N(S) = n$. The smallest possible value of n is 6, and the largest is 30. there is only one 4-set with sum 6, namely $\{0,1,2,3\}$, one with sum 7 ($\{0,1,2,4\}$), one with

sum 29 ($\{5, 7, 8, 9\}$), and one with sum 30 ($\{6, 7, 8, 9\}$). So the remaining 206 4-subsets have sums from 8 to 28. If the 21 sets $T_8, T_9, \ldots, T_{28}$ each contains nine or fewer elements, they would total at most 189 members between them, less than 206, which is impossible. So some T_n has 10 or more members.

Chapter 4

1. (i) Say $S = \bigcup_{i=1}^{n} A_i$. If each A_i satisfies $|A_i| \le m$ then $|S| \le nm$, a contradiction. So there is at least one i such that $|A_i| > m$.

   (ii) If $a_i \le \frac{b}{n}$ for all i then $\sum_{i=1}^{n} a_i \le b$, a contradiction.

   (iii) If $a_i > \frac{b}{n}$ for all i then $\sum_{i=1}^{n} a_i > b$, a contradiction.

   (iv) If $a_i \ge \frac{b}{n}$ for all i then $\sum_{i=1}^{n} a_i \ge b$, a contradiction.

2. The worst case is when all boxes but one contain 1 object and the remaining box contains $N - (B - 1) = N - B + 1$. So at least $N - B + 1$ must be chosen.

3. For (i) and (ii), suppose the statement is false. Then there are at most 12 people belonging to each of only 12 signs, so there are only 144 people. But we know that there are 145 people; this contradiction establishes the results. For (iii), there are $12^2 = 144$ possible pairs of signs, and 145 people, so there are at least two people who have the same pair of signs.

4. Suppose no two sets contain the same number of entries. At the very least, one will be empty, one will have order 1, one 2, ... and the total number of entries is

$$n = 0 + 1 + \ldots + (p - 1) = \frac{p(p - 1)}{2}.$$

So $n \ge \frac{p(p-1)}{2}$ is a necessary condition for all the A_i to be of different size. (If n is larger, simply add more members to the largest set.) On the other hand, if $n = \frac{p(p-1)}{2} + t$, such a set A exists—for example,

$$A_1 = \{1\}, A_2 = \{2, p + 2\}, \ldots, A_i = \{i, p + i, \ldots, (i - 1)p + i\}, \ldots,.$$

But finally

$$A_p = \{p - 1, 2p - 1, \ldots, (p + t)p - 1\}.$$

So $n \ge \frac{p(p-1)}{2}$ is necessary and sufficient.

5. Consider the n products $a_1, a_1 a_2, \ldots, a_1 a_2 \ldots a_i, \ldots, a_1 a_2 \ldots a_n$. There are two possibilities: if, say, $a_1 \ldots a_j = 1$ then we choose $s = 1, t = j$ and the problem is solved; if none of these n products equals the identity, then they take at most $(n-1)$ distinct values and hence at least two of them must be equal, so $a_1 a_2 \ldots a_j = a_1 a_2 \ldots a_{j+k}$ for some j and k, with $1 \leq j < j + k \neq n$. Put $a = j + 1, t = j + k$ and the problem is again solved.

6. Suppose both (i) and (ii) are false. Partition the intervals into sets $S_1, S_2, \ldots,$ where all members of a set are pairwise disjoint. As (ii) is false, each set contains at most n intervals. Clearly no set of intervals with a common element can contain two members of the same A_i; as (i) is false, there can be at most n sets A_i. So there are at most n^2 intervals—a contradiction.

7. Divide the $2R(m, n) - 2$ vertices into two sets X and Y of $R(m, n) - 1$ vertices each. Color the edges that connect a vertex in X to a vertex in Y in color c_1. Color the edges that connect vertices in X using only colors c_2 and c_3 in such a way that the subgraph induced by X contains no K_m in c_2 or K_n in c_3 (this coloring must exist since X has fewer than $R(m, n)$ vertices). Do the same with the vertices of Y. Because the edges of color c_1 form a bipartite graph, there is no K_3 in c_1, and by construction there is no K_m in c_2 or K_n in c_3. This shows that $R(3, m, n) \geq 2R(m, n) - 1$.

8. First, suppose n is odd, say $n = 2t + 1$. Then

$$\left\lfloor \frac{n}{2} \left(\frac{n-1}{2} \right)^2 \right\rfloor = \lfloor (t + \tfrac{1}{2}) t^2 \rfloor = \lfloor t^3 + \tfrac{1}{2} t^2 \rfloor$$

and

$$\left\lfloor \frac{n(n-1)^2}{8} \right\rfloor = \left\lfloor \frac{(2t+1) 4 t^2}{8} \right\rfloor == \lfloor t^3 + \tfrac{1}{2} t^2 \rfloor.$$

So the two are equal.

When n is even, say $n = 2t$,

$$\left\lfloor \frac{n}{2} \left(\frac{n-1}{2} \right)^2 \right\rfloor = \lfloor t(t^2 - t) \rfloor = t^3 - t^2,$$

but

$$\left\lfloor \frac{n(n-1)^2}{8} \right\rfloor = \left\lfloor \frac{2t(2t-1)^2}{8} \right\rfloor = \left\lfloor \frac{8t^3 - t^2 + 2t}{8} \right\rfloor = t^3 - t^2 + \left\lfloor \frac{t}{4} \right\rfloor,$$

with equality if and only if $\lfloor \tfrac{t}{4} \rfloor = 0$, that is $t < 4$. So the two expressions are equal unless n is even and $n \geq 8$.

Chapter 5

1. $(B \cap C) \cup (\overline{B} \cap C) = C$, and the two sets on the left are disjoint, so $|B \cap C| + |\overline{B} \cap C| = |C|$. Similarly $(A \cap B \cap C) \cup (A \cap \overline{B} \cap C) = A \cap C$ and $|A \cap B \cap C| + |A \cap \overline{B} \cap C| = |A \cap C|$. So

$$
\begin{aligned}
&|A \cup C| \\
&= |A| + |C| - |A \cap C| \\
&= |A| + (|B \cap C| + |\overline{B} \cap C|) - (|A \cap B \cap C| + |A \cap \overline{B} \cap C|).
\end{aligned}
$$

2. We prove by induction that D_n is odd if and only if n is even, or equivalently $D_{n+1} \equiv n \pmod 2$. The result is true for $n = 2$ (and also for $n = 1, D_1 = 0$, but as we explained some people say D_1 is not defined). Say it is true for $n < k$. Then D_k and D_{k-1} are of opposite parity, so $D_k - D_{k-1} \equiv 1 \pmod 2$, so $D_{k+1} = k[D_k - D_{k-1}] \equiv k \times 1 \equiv k \pmod 2$.

3. Suppose the people are numbered 1 to n and the chairs are numbered 0 to n; say person x occupies seat x initially and seat $x\prime$ after supper. Suppose the previously vacant chair remains vacant. Then $(1\prime, 2\prime, \ldots, n\prime)$ is a derangement of $(1, 2, \ldots, n)$. This can occur in D_n ways. To handle the other case, $0\prime$ is the number of the chair that is unoccupied after supper; $(0\prime, 1\prime, 2\prime, \ldots, n\prime)$ is a derangement of $(0, 1, 2, \ldots, n)$. This can occur in D_{n+1} ways.

4. Write N for $\{1, 2, \ldots, n\}$. Select any permutation $(d_1 d_2 \ldots d_n)$ of N and write $S = \{i : d_i \neq i\}$. The number of different permutations in which all symbols are in their natural position except for precisely the members of S is equal to the number of derangements of S, or $D_{|S|}$. To get the number of permutations of N with exactly k numbers out of position, sum this over all $\binom{n}{k}$ subsets S of size k, obtaining $\binom{n}{k} D_k$. Now sum over k to count all permutations.

Chapter 6

2. $\sum_{n \geq 0} a_n x^n = \sum_{n \geq 0} c_1 n (rx)^n + \sum_{n \geq 0} c_0 (rx)^n$. The second sum is clearly $c_0/(1 - rx)$; the first requires differentiation. We use

$$
\frac{d}{dx} \sum_{n \geq 0} c_1 r^n x^n = \frac{d}{dx} c_1/(1 - rx).
$$

This gives us $\sum_{n\geq 0} c_1 r^n n x^{n-1} = rc_1/(1-rx)^2$. Multiplying both sides by x gives $\sum_{n\geq 0} c_1 n r^n x^n = c_1 rx/(1-rx)^2$. Then we easily see that $\sum_{n\geq 0} c_1 n (rx)^n + \sum_{n\geq 0} c_0 (rx)^n$ is a rational function with denominator $(1-rx)^2$, as required.

7. $f_a(x) f_b(x) = (a_0 + a_1 x + a_2 x^2 + \dots) \times (b_0 + b_1 x + b_2 x^2 + \dots) = c_0 + c_1 x + c_2 x^2 + \dots$. The coefficient c_n will be the sum of coefficients of all products of terms of degree n. So, $c_0 = a_0 b_0$ because every other term has a power of x greater than 0. Distributing the product gives us terms of the form $a_i b_j x^{i+j}$, which means $c_n = a_0 b_n + a_1 b_{n-1} + \dots + a_i b_{n-i} + \dots + a_n b_0$.

9. The coefficient of x^n is

$$\sum_{i=0}^{n} \frac{a_i b_{n-i}}{i!(n-i)!} = \left(\sum_{i=0}^{n} \frac{a_i b_{n-i}}{n!}\right) \left(\frac{n!}{i!(n-i)!}\right) = \sum_{i=0}^{n} \binom{n}{i} a_i b_{n-i}.$$

10. Write $E(x) = \sum_{n=2}^{\infty} D_n \frac{x^n}{n!}$, the EGF. (We could take the lower bound as $n = 0, 1$ or 2, since $D_0 = D_1 = 0$.) Substituting,

$$E(x) = \sum_{n=2}^{\infty} n D_{n-1} \frac{x^n}{n!} + \sum_{n=2}^{\infty} (-1)^n \frac{x^n}{n!}.$$

Now

$$\sum_{n=2}^{\infty} n D_{n-1} \frac{x^n}{n!} = \sum_{n=2}^{\infty} D_{n-1} \frac{x^n}{(n-1)!} = \sum_{n=1}^{\infty} D_n \frac{x^{n+1}}{n!} = x E(x)$$

and

$$\sum_{n=2}^{\infty} (-1)^n \frac{x^n}{n!} = \sum_{n=2}^{\infty} \frac{(-x)^n}{n!} = e^{-x} - (1-x)$$

(it is the expansion of e^{-x} with the first two terms omitted). So

$$E(x) = x E(x) + e^{-x} + x - 1, \quad E(x) = \frac{e^{-x}}{1-x} - 1.$$

Chapter 7

6. Let $n = 2k+1$; then we know $\left\{\begin{smallmatrix} 2k+1 \\ 2 \end{smallmatrix}\right\} = 2^{2k} - 1 = 4^k - 1^k$. Now, the polynomial $x^k - y^k$ is divisible by $x - y$; in this case, using $x = 4$ and $y = 1$, we see that $\left\{\begin{smallmatrix} 2k+1 \\ 2 \end{smallmatrix}\right\}$ is divisible by $4 - 1 = 3$ as required.

7. Separate the $n + 1$st element. If it is alone in its partition, then there are B_n ways to partition the remaining n elements. If there is one other element in its partition, we may choose that element in $n = \binom{n}{n-1}$ ways, and there are B_{n-1} ways to partition the remaining elements. In general, we may choose i "companions" for the $n + 1$st element in $\binom{n}{n-i}$ ways, and the remaining $n - i$ elements can be partitioned in B_{n-i} ways. Summing over i gives the result.

Chapter 8

2. Say $a, b \in G$ and $x, y \in H$. Then $ab = ba$ and $xy = yx$. So
$$(a, x)(b, y) = (ab, xy) = (ba, yx) = (b, y)(a, x).$$

3. K contains the identity e; K also contains α, β, and ρ^2. Since $\alpha \circ \alpha = \beta \circ \beta = \rho^2 \circ \rho^2$, K clearly contains inverses. Now we observe that $\alpha \circ \beta = \beta \circ \alpha = \rho^2$, and $\alpha \circ \rho^2 = \rho^2 \circ \alpha = \beta$, and $\beta \circ \rho^2 = \rho^2 \circ \beta = \alpha$, we see that the set is closed under composition. This establishes that K is a subgroup of D_4.

4. If G is transitive, then the required permutations σ_i exist by definition. Given the permutations σ_i, we wish to show that for any $i, j \in S$ there exists σ_{ij} where $\sigma_{ij}(i) = j$. The permutation $\sigma_j \circ \sigma_i^{-1}$ will work.

8. We may modify Figure 8.1 by placing a 5 in the center. Now the rotation $\rho = (1234)(5)$ gives us $z_1^1 z_4^1$, and the powers of ρ are modified similarly; so $\rho^3 = (1432)(5)$ gives $z_1^1 z_4^1$, ρ^2 gives $z_2^2 z_1^1$, and e gives z_1^5. With k colors, there are thus $\frac{1}{4}(k^5 + k^3 + 2k^2)$ non-equivalent colorings. If we do not color the center spot, each permutation "loses" a z_1^1, with the result that we divide the colorings by k; the number is $\frac{1}{4}(k^4 + k^2 + 2k)$.

9. We easily see that $c_k = \frac{1}{2}(k^3 + k^2)$, since there are k^2 colorings fixed by the 180° rotation. We use the generating functions for the sequences $\{k^2\}$ and $\{k^3\}$ to find $C(x) = (2x^2 + x)(1 - x)^{-4}$. Expanding the denominator, we arrive at $1 - 4x + 6x^2 - 4x^3 + x^4$. Using Theorem 6.6, we get the recurrence relation $c_k = 4c_{k-1} - 6c_{k-2} + 4c_{k-3} - c_{k-4}$. We require the initial values $c_0 = 0$, $c_1 = 1$, $c_2 = 6$, $c_3 = 18$, and $c_4 = 40$. You should check that $c_5 = 75$, $c_6 = 126$, and perhaps $c_7 = 196$ will follow from the recurrence.

11. Each permutation consists of exactly two cycles, except of course the identity. As a result, the cycle index will consist of monomials z_1^{12}, $z_1 z_3$,

and z_2^2. We have then $C(z_1, z_2, z_3, z_4) = (z_1^4 + 8z_1z_3 + 3z_2^2)/12$ for the cycle index. Substituting k for each of the z_is gives us $(k^4 + 11k^2)/12$.

Chapter 9

2. Take the poset X with at least three different elements x, y, and z. If any two of these are incomparable, we cannot have the three mutually adjacent elements required to form a K_3, so without loss of generality assume $x \succ y$. (If not, rename x and y.) Similarly, assume $x \succ z$ (renaming if necessary). Now, either $z \prec y$ or $y \prec z$. If $z \prec y$, then we have $z \prec y \prec x$, so that x does not cover z; thus there is no line from x to z in the Hasse diagram. If $y \prec z$, then there is no line from x to y.

3. Suppose $n = p_1 p_2 p_3 p_4$, and consider the power set of the set $\{1, 2, 3, 4\}$. The isomorphism σ will map the set A of this power set to the divisor d of n precisely when d is the product of the primes whose subscripts correspond to the elements of A. Thus, $\sigma(\{2, 4\}) = p_2 p_4$, and so forth. Consider two subsets A_1 and A_2 where $\sigma(A_1) = d_1$ and $\sigma(A_2) = d_2$; if $A_1 \subset A_2$, then each prime factor of d_1 is also a factor of d_2, so that $d_1 \prec d_2$ in D_n. In the same way, if we have any two divisors d_1 and d_2 of n, where $d_1 \prec d_2$, then $\sigma^{-1}(d_1) \subset \sigma^{-1}(d_2)$ since each prime factor of d_1 is a factor of d_2.

4. Suppose that $\mathcal{B}_n$ is the poset of subsets of $S = \{1, 2, \ldots, n\}$, and $\mathcal{B}_m$ is all subsets of $T = \{n+1, n+2, \ldots, n+m\}$. Now, for $A \subset S$ and $B \subset T$ let $\sigma((A, B)) = A \cup B$. It is clearly one-to-one; a different pair (A', B') will be sent to a different union. Similarly, it is onto since any subset C of $S \cup T = \{1, 2, \ldots n+m\}$ is clearly the union of $C \cap S$ and $C \cap T$. Finally, if $A \subseteq A'$ and $B \subseteq B'$, then $A \cup B \subseteq A' \cup B'$, so σ preserves the order.

7. We can find antichains of five elements, but not of six. We see that $\{a, b, d, e, f\}$ (the antichain of all maximal elements) works; as there are only four minimal elements, the antichain of minimal elements is not a largest-sized antichain. But we can take $\{a, b, d, j, k\}$ and $\{a, b, h, j, k\}$.

Chapter 10

1. Reducing the degree sequence, we get

$$(5,5,5,3,2,2,1,1) \Rightarrow (4,4,2,1,1,1,1)$$
$$\Rightarrow (3,1,0,0,1,1) = (3,1,1,1,0,0)$$
$$\Rightarrow (0,0,0,0,0)$$

which is graphical (the empty graph). However, if the two vertices of degree 1 are adjacent, the remaining vertices form a component with degree sequence $(5,5,5,3,2,2)$, and

$$(5,5,5,3,2,2) \Rightarrow (4,4,2,1,1) \Rightarrow (3,1,0,0)$$

which is not valid (one vertex has degree 3, but there is only one other non-isolate).

2. $\lceil \frac{1}{2} v \delta(G) \rceil$

3.   (i) Here are P_4 and C_5 and their complements. The complements have been relabeled to exhibit the isomorphism.

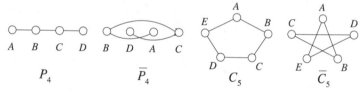

(ii) Suppose vertex x is an isolate. In the complement, x is adjacent to every other vertex, so there are no isolates.

(iii) The degree sequence consists of five numbers, between 1 and 3 inclusive, and the number of 3s equals the number of 1s. The possibilities are $(3,3,2,1,1)$, $(3,2,2,2,1)$ and $(2,2,2,2,2)$. A short exhaustive search shows there is exactly one graph with the first and third sequences, namely G (below) and C_5; this uniqueness means that they must be self-complementary. The other sequence is realized by two non-isomorphic graphs, which are complements of each other (H and $\overline{H}$, see below). So G and C_5 are the only solutions.

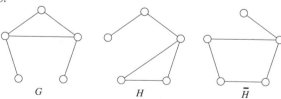

(iv) Say G has v vertices. Then $G \cup \overline{G} = K_v$, which has $v(v-1)/2$ edges. If G is self-complementary, then G and $\overline{G}$ each have $|E(G)|$ edges. As they are edge-disjoint, their union has $2|E(G)|$ edges. So $v(v-1)/2$ is even, so $v \equiv 0$ or $1 \bmod 4$.

4. Start with a cycle C_v; draw an edge from vertex i to $v/2 + i$, for $1 \le i \le v/2$.

5. No.

6. Denote the vertex-set of the graph by V. Write x_i for the vertex with degree d_i, N_i for the set of neighbors of x_i, and $V_i = \{x_1, x_2, \ldots, x_i\}$. Then $N_i \cap V_r$ is the set of neighbors of x_i that lie in V_r, and $|N_i \cap V_r|$ is the number of edges joining members of V_r to x_i. Summing over i, we get

$$\sum_{i=1}^{r} d_i = \sum_{i=1}^{v} |N_i \cap V_r| = \sum_{i=1}^{r} |N_i \cap V_r| + \sum_{i=r+1}^{v} |N_i \cap V_r|.$$

If $1 \le i \le r$, then x_i is adjacent to at most the other $r-1$ members of V_r, so $|N_i \cap V_r| \le r-1$, and

$$\sum_{i=1}^{r} |N_i \cap V_r| \le \sum_{i=1}^{r} r - 1 = r(r-1).$$

If $i > r$, then the number of neighbors of v_i that lie in V_r certainly cannot exceed either r or d_i, so $|N_i \cap V_r| \le \min(r, d_i)$, and

$$\sum_{i=r+1}^{v} |N_i \cap V_r| \le \sum_{i=r+1}^{v} \min(r, d_i).$$

Plugging these two results into the equation we get the result.

7.

8. $D(G)$ is the largest distance between any two vertices in G; say $D(x, y) = D(G)$. $R(G)$ is the smallest value of ε; let z be a vertex for which $\varepsilon(z) = R(G)$. As R is a distance between some two vertices, $R \le D$. But by definition $D(z, t) \le \varepsilon(z) = R(G)$ for every vertex t, so $D(G) = D(x, y) \le D(x, z) + D(z, y) \le 2R(G)$.

9. $n = 1$ and 2 only (otherwise, K_n contains an odd cycle).

Chapter 11

1. You would treat a loop as adding 2 to the degree of its vertex. They do not need special treatment in Euler walks—you can traverse the loop when you pass through the vertex. As roads, a loop might represent a crescent. Or a snowplow might need to clear a large circular driveway on city property.

3. Examples for the first two parts:

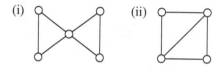

4. The cost is $n(n + 1)$.

5. Suppose the result is false. Select a graph G on $2n$ vertices, each of degree at least n, that is not Hamiltonian, but has the maximum possible number of edges among such graphs. Select two vertices, x and y, that are not adjacent. (This must be possible: the complete graph is Hamiltonian, so G is not complete.) Add edge xy to G; the resulting graph $G + xy$ must be Hamiltonian. Now select a Hamilton cycle in $G + xy$. It must contain xy, or else it would be a Hamilton cycle in G. Say it is $C = (x, x_1, x_2, \ldots, x_{2n-2}, y)$.

Suppose x_i is adjacent to x. Then x_{i-1} cannot be adjacent to y, or else $(x, x_1, x_2, \ldots, x_{i-1}, y, x_{2n-2}, x_{2n-3}, \ldots, x_i)$ would be Hamiltonian in G. So each of the n or more vertices adjacent in G to x must be preceded in C by a vertex not adjacent to y. Write S for the set of those vertices. There are at least n vertices in S, and none is adjacent to y. But there are also n vertices adjacent to y. This accounts for all $2n$ vertices.

But what about y itself? It must be a member of S; but that is impossible, because y cannot appear in the list $x_1, x_2, \ldots, x_{2n-2}$. So we have a contradiction.

6. (i) The left-hand diagram shows a Hamilton cycle in $K_{1,2,3}$. To handle $K_{n,2n,3n}$, treat it as n copies of $K_{1,2,3}$ together with further edges; the solution is shown on the right (with unnecessary edges not shown, to avoid clutter).

Alternatively, write A, B, C for the sets of size $n, 2n, 3n$, respectively. Label the vertices of A with integers $1, 2, \ldots, n$, those of B with integers $n+1, n+2, \ldots, 3n$, and those of C with integers $3n+1, 3n+2, \ldots, 6n$. One Hamilton cycle is $(1, 3n+1, 2, 3n+2, \ldots, 3n, 6n)$.

(ii) Write C for the set of $3n+1$ vertices. If the graph is to have a Hamilton cycle, no two members of C can be adjacent in the cycle, so every second vertex must be a non-member of C, so there must be at least $3n+1$ further vertices.

7. First, observe that any tree with three or more vertices has an edge, and at least one endpoint of that edge is joined to another vertex. So any tree with 3 or more vertices contains a path with three vertices.

Now suppose T is a tree on three or more vertices, and suppose there is a graph G such that $T = G^2$. Select a path abc in T. If both ab and bc are edges of G, then T will contain the 3-cycle (abc), which is impossible. So let's assume ab is not an edge of G. There must be some vertex x that is adjacent to both a and b in G. Then ax and bx are edges of T, and again T contains a cycle, in this case (abx).

8. Consider a tree T of radius R. Any two vertices of a tree are joined by a path; write $P_{x,y}$ for the path joining x to y. Say a and b are both vertices in the center of T. If they are not adjacent, there must be another vertex c that lies in $P_{a,b}$, and a vertex p such that $d(c, p) \geq R$. Then $P_{c,p}$ must pass through a (or else $d(a, p) > d(c, p) \geq R$, so $d(a, p) > \epsilon(a)$), and similarly $P_{c,p}$ must pass through b. But this is impossible if c is on P_{ab}. So all central vertices in a tree are adjacent. A tree cannot have three or more mutually adjacent vertices, so the center has one or two vertices, and if 2 then they are adjacent.

9. Suppose c is the vertex of degree 4 and x_1, x_2, x_3, x_4 are its neighbors. We can think of the tree as consisting of four subtrees T_1, T_2, T_3, T_4, with common vertex c: the vertices of T_i are all the vertices that can be reached from x_i without passing through c. The tree looks like

where the shaded sections denote the T_i. Then each T_i is a tree. T_i may just be the vertex x_i, in which case x_i is a vertex of degree 1 in T. If T_i is not a single vertex, its vertices of degree 1 (other than x_i) will all be vertices of degree 1 in T. Each T_i has at least two vertices of degree 1, unless it is the 1-vertex tree x_i; so each T_i contributes at least one vertex of degree 1 to T. So none can contribute more than one such vertex. Therefore each T_i is either a 1-vertex tree or a path. So T has one vertex of degree 4 and four of degree 1, and all others have degree 2. Therefore $T_1 \cup T_2$ and $T_3 \cup T_4$ are paths, they are edge-disjoint, and their union is T.

10. We proceed by induction. The result is trivial if $v = 1$ or 2 (there is one tree in each case: K_1 and K_2, respectively). Assume $n \ge 3$, and assume the result is true for $v < n$. Say $v = n$. Since $\sum d_i < 2n$, at least one degree is 1. Let's assume $d_n = 1$.

Write $\mathcal{S}_j$ for the set of all trees with vertices $x_1, x_2, \ldots, x_{v-1}$ where x_i has degree $d(x_i) = d_i$ when $i \ne j$ and $d(x_j) = d_j - 1$. The members of $\mathcal{S}_j$ correspond to trees in $\mathcal{T}$ with $x_j x_n$ as an edge; if $d_j = 0$ then $\mathcal{S}_j$ is empty. Since x_n must have been connected to every tree in $\mathcal{T}$, $|\mathcal{T}| = \sum_{j=1}^{n-1} |\mathcal{S}_j|$. If $d_j > 1$ then (by induction)

$$
\begin{aligned}
|\mathcal{S}_j| &= \frac{(n-3)!}{(d_1-1)!(d_2-1)!\ldots(d_j-2)!\ldots(d_{n-1}-1)!} \\
&= \frac{(n-3)!(d_j-1)}{(d_1-1)!(d_2-1)!\ldots(d_j-1)!\ldots(d_{n-1}-1)!};
\end{aligned}
$$

and if $d_j = 1$ then $|\mathcal{S}_j| = 0$ which fits the same formula. So

$$
\begin{aligned}
|\mathcal{T}| &= \sum_{j=1}^{n-1} |\mathcal{S}_j| \\
&= \sum_{j=1}^{n-1} \frac{(n-3)!(d_j-1)}{(d_1-1)!(d_2-1)!\ldots(d_{n-1}-1)!} \\
&= \frac{(n-3)!}{(d_1-1)!(d_2-1)!\ldots(d_{n-1}-1)!} \sum_{j=1}^{n-1}(d_j-1).
\end{aligned}
$$

But $d_n = 1$ (so $(d_n-1)! = 1$) and $\sum_{j=1}^{n} d_j = 2n-2$, so $\sum_{j=1}^{n-1}(d_j-1) = n-2$, and

$$
\begin{aligned}
|\mathcal{T}| &= \frac{(n-3)!}{(d_1-1)!(d_2-1)!\ldots(d_n-1)!}(n-2) \\
&= \frac{(n-2)!}{(d_1-1)!(d_2-1)!\ldots(d_n-1)!}.
\end{aligned}
$$

11. If a tree has v vertices, the sum of the degrees is $2v-2$. So Problem 11.10 told us the number of trees on v vertices with a given set of degrees. To find the total number of trees, we sum over all possible degree sets. So the sum is

$$\sum_{d_1+d_2\ldots+d_v=2v-2} \frac{(v-2)!}{(d_1-1)!(d_2-1)!\ldots(d_v-1)!}$$

and writing $k_i = d_i - 1$, this becomes

$$\sum_{k_1+k_2\ldots+k_v=2v-2} \frac{(v-2)!}{k_1!k_2!\ldots k_v!}$$

which equals v^{v-2} by the multinomial theorem (Theorem 2.6).

Chapter 12

1. The codewords correspond to the vertices of the subgraph, but there is no obvious relationship to the edges. However, the *distance* between two words equals the distance between the corresponding vertices in the underlying hypercube.

3. Minimum distance is 4. So the code can detect three errors and always correct one.

4. One method is to encode A, B, C, D, E as 0, 10, 110, 1110, 1111 respectively.

5. If a code has minimum distance $2n$ or smaller, one can select two words A and B that differ in at most $2n$ positions. Say S is the set of positions where they differ. Consider a received word that is the same as A (and B) is those positions where the two are the same, is the same as A in n of the positions where they differ, and is the same as B in the remaining positions. There is no way to know whether to decode this word to A or to B. (*Note:* If the distance is smaller than $2n$, it is tempting to say, "decode to the codeword that gives fewer errors," but there is no guarantee that this will be correct.)

Chapter 13

2. One example is

1	4	6	5	3	2
4	2	5	6	1	3
2	5	3	1	6	4
6	3	1	4	2	5
3	6	4	2	5	1
5	1	2	3	4	6

5. One solution is

1	2	3	
			4

The generalization is obvious.

6. (i)

$$
\begin{array}{ccccccccc}
1 & 1 & 1 & 2 & 2 & 2 & 3 & 3 & 3 \\
1 & 2 & 3 & 1 & 2 & 3 & 1 & 2 & 3 \\
1 & 2 & 3 & 3 & 1 & 2 & 2 & 3 & 1 \\
1 & 3 & 2 & 3 & 2 & 1 & 2 & 1 & 3
\end{array}
$$

(ii) No. You would need orthogonal Latin squares of order 6.

(iii) Use 3 MOLS of order 4 to construct a 5×16 orthogonal array, and use the columns as the codewords. Let's say two codewords are *properly separated* if they have exactly one entry in common.

Suppose the entries in position (i, j) in the three squares are a, b, c, respectively. So one codeword is (i, j, a, b, c). There will be three other columns corresponding to cells in row i, and none will have a j in position 2 (a j there would mean the corresponding entry was also in the (i, j) position, and there is only one entry per cell), or an a in position 3 (that would mean two entries a in row i of the first square). Similarly, none of them has a b in position 4, or a c in position 5. So our codeword is properly separated from the three other words with i in position 1. In the same way we see that our codeword is properly separated from the three other words with j in position 2, the three other words with a in position 3, the three other words with b in position 4, and the three other words with c in position 5. Moreover our reasoning shows that all 15 of those codewords are different. Since there are only 16 codewords, we have shown that (i, j, a, b, c) is properly separated from all the other codewords.

But (i, j, a, b, c) is completely general. So every pair of codewords is properly separated.

7. (i) The $(1, 4)$ entry must be 4 (in order to complete column 4). So the first row is 12345. So the $(5, 4)$ entry must be 3. Now consider the completion of row 5; a 2 and a 5 are needed and the 5 cannot be in column 5, so row 4 is 34512. Similarly row 3 must be 45123. The array looks like

$$
\begin{array}{ccccc}
1 & 2 & 3 & 4 & 5 \\
  & 1 &   & 3 &   \\
4 & 5 & 1 & 2 & 3 \\
3 & 4 & 5 & 1 & 2 \\
  &   &   & 4 & 5
\end{array}
$$

The 5 in row 2 must be in column 1, and to complete column 3 the $(2, 3)$ entry must be 2. So row 2 is 51234. Now the array can be completed:

$$
\begin{array}{ccccc}
1 & 2 & 3 & 4 & 5 \\
5 & 1 & 2 & 3 & 4 \\
4 & 5 & 1 & 2 & 3 \\
3 & 4 & 5 & 1 & 2 \\
2 & 3 & 4 & 5 & 1
\end{array}
$$

(ii)
$$
\begin{array}{ccccc}
1 & 2 & 3 & 4 & 5 \\
5 & 1 & 2 & 3 & 4 \\
4 & 5 & 1 & 2 & 3 \\
3 & 4 & 5 & 1 & 2 \\
2 & 3 & 4 & 5 & 1
\end{array}
\qquad
\begin{array}{ccccc}
1 & 2 & 3 & 4 & 5 \\
5 & 1 & 4 & 3 & 2 \\
3 & 5 & 1 & 2 & 4 \\
2 & 4 & 5 & 1 & 3 \\
4 & 3 & 2 & 5 & 1
\end{array}
\qquad
\begin{array}{ccccc}
1 & 2 & 3 & 4 & 5 \\
4 & 1 & 5 & 3 & 2 \\
3 & 5 & 1 & 2 & 4 \\
5 & 4 & 2 & 1 & 3 \\
2 & 3 & 4 & 5 & 1
\end{array}
$$

Chapter 14

1. (i) From the hint, there are blocks 1234 and 1256. There is one further block containing 13. Since the intersection size is 2, 4 is not included, and not both of 5 and 6 can occur, so there must be a final 7 (no other symbols are available). The block is either 1357 or 1367. We'll choose 1357 (otherwise, use the isomorphism (56)). The fourth block containing 1 can only be 1467.

There are two further blocks containing 2. Since 2 and 7 have not occurred together, 7 must be in both blocks. One must contain 3 and the other contain 4 (234 cannot occur together). We either have 2357 and 2367. But 2357 is impossible (357 has already been used) so 2367 is block, and so is 2457 (in order for 2 to occur twice with every other symbol). The final block is 3456. So we have blocks

$$1234 \quad 1256 \quad 1357 \quad 1467 \quad 2367 \quad 2457 \quad 3456.$$

Any other choice is isomorphic to this (make a different choice of the first block, use (56) or (67), or some combination).

2. Yes: one example is

$$\begin{array}{ccccccc} 123 & 145 & 167 & 246 & 157 & 347 & 356 \\ 124 & 235 & 346 & 457 & 156 & 267 & 137 \end{array}$$

3. (i) $v = 9t + 8$.

(ii) If v is even (that is, t is even) then $k - \lambda = 18 + 27t$ is a perfect square. So $2 + 3t$ is a perfect square. But all perfect squares are congruent to 0 or 1 (mod 3). So assume t is odd. There is a solution to

$$x^2 + (2 + 3t)(3y)^2 = \pm 3z^2$$

with x, y, z not all zero. Without loss of generality we can assume x, y and z have no common factor (if there were a common factor t, we could divide all three variables by t). So, modulo 3, $x^2 + 18y^2 \equiv 0$. So 3 divides x. So the largest power of 3 that divides $x^2 + (2 + 3t)(3y)^2$ must be even. But the largest power of 3 that divides $3z^2$ is odd.

4. The dual does not satisfy Fisher's inequality.

5. $v, \binom{v}{k}, \binom{v-1}{k-1}, k, \binom{v-2}{k-2}$

6. Write $B, B_1, B_2, \ldots, B_{v-1}$ for the blocks of $\mathcal{D}$. The blocks of $\mathcal{E}$ are

$$S \backslash B, S \backslash B_1, S \backslash B_2, \ldots, S \backslash B_{v-1},$$

and the blocks in the residual of $\mathcal{E}$ with regard to $S \backslash B$ are the $v - 1$ blocks

$$(S \backslash B_1) \backslash (S \backslash B), (S \backslash B_2) \backslash (S \backslash B), \ldots, (S \backslash B_{v-1}) \backslash (S \backslash B).$$

On the other hand, the blocks of the derived design of $\mathcal{D}$ with respect to B are

$$B \backslash B_1, B \backslash B_2, \ldots, B \backslash B_{v-1}.$$

Now

$$(S \backslash B_i) \backslash (S \backslash B) = B \backslash B_i$$

(check with a Venn diagram if necessary). So we have the result.

7. Write $C_0, C_1, \ldots, C_{14}$ for the blocks of $\mathcal{D}^*$.

(i) $\begin{aligned} C_0 &= \{0,1,2,3,4,5,6\} & C_1 &= \{0,2,4,7,8,10,12\} \\ C_2 &= \{0,2,4,9,11,13,14\} & C_3 &= \{0,1,6,7,8,9,14\} \\ C_4 &= \{0,1,6,10,11,12,13\} & C_5 &= \{0,3,5,7,10,11,14\} \\ C_6 &= \{0,3,5,8,9,12,13\} & C_7 &= \{2,3,6,7,8,11,13\} \\ C_8 &= \{2,3,6,9,10,12,14\} & C_9 &= \{1,2,5,7,9,11,12\} \\ C_{10} &= \{1,2,5,8,10,13,14\} & C_{11} &= \{4,5,6,7,9,10,13\} \\ C_{12} &= \{4,5,6,8,11,12,14\} & C_{13} &= \{1,3,4,7,12,13,14\} \\ C_{14} &= \{1,3,4,8,9,10,11\}; \end{aligned}$

(ii) $\mathcal{D}$ has precisely seven triplets: (012), (034), (056), (078), (0910), (01112), (0134). $\mathcal{D}^a st$ also has seven triplets: (016), (024), (035), (125), (134), (236), (456). Any isomorphism would have to carry one set of triples into the other; this is clearly impossible.

8. (i) $(n^2 + n + 1, n^2, n^2 - n)$; (ii) $(n + 1, n^2 + n, n^2, n, n^2 - n)$; (iii) an $(n + 1, n + 1, n, n, n - 1)$ design exists (see Exercise 14A.2). Take the n-multiple of it (see Exercise 14A.9).

9. (i) Say one line has k points. From the axioms, every other line contains precisely one of these, so there are $k + 1$ lines in total. If some other line had h points, there would be $h + 1$ lines in total. So $h = k$.

(ii) There are k points per line, and $k + 1$ lines; each point is on two lines, so there are $v = \frac{1}{2}k(k + 1)$ points in all. Call them $x_1, x_2, \ldots, x_v$. Say the first line is

$$x_1, x_2, x_3, \ldots, x_k;$$

the second line has one point in common with this, so (up to permutation of names of points) it is

$$x_1, x_{k+1}, x_{k+2}, \ldots, x_{2k-1}.$$

Proceeding in this way we find that the unique (up to isomorphism) solution for given k is the following set of lines:

$$
\begin{array}{ccccc}
x_1 & x_2 & x_3 & \cdots & x_k \\
x_1 & x_{k+1} & x_{k+2} & \cdots & x_{2k-1} \\
x_2 & x_{k+1} & x_{2k} & \cdots & x_{3k-2} \\
x_3 & x_{k+2} & x_{2k} & \cdots & x_{4k-3} \\
& \cdots & & \cdots & \\
x_k & x_{2k-1} & x_{3k-2} & \cdots & x_v.
\end{array}
$$

For example, with $k = 5$, the solution is

$$
\begin{array}{ccccc}
1 & 2 & 3 & 4 & 5 \\
1 & 6 & 7 & 8 & 9 \\
2 & 6 & 10 & 11 & 12 \\
3 & 7 & 10 & 13 & 14 \\
4 & 8 & 11 & 13 & 15 \\
5 & 9 & 12 & 14 & 15
\end{array}
$$

10. You would have $\lambda(2k - 1) = k(k - 1)$, so $\lambda = k(k - 1)/(2k - 1)$. Now clearly $2k > 2k - 1 > 2k - 2$, so $\frac{2}{k-1} > \frac{2k-1}{k(k-1)} > \frac{2}{k}$, and $\frac{k-1}{2} < \frac{2k-1}{k(k-1)} < \frac{k}{2}$. So λ would lie between $\frac{k-1}{2}$ and $\frac{k}{2}$, and there is no integer in that range.

11.   Say m and v are coprime. Then the integers $m, 2m, \ldots, (v-1)m$ are all different modulo v, and are congruent to $1, 2, \ldots, v-1$ in some order. Moreover, of the pairs with difference $d \pmod v$ are $\{x_1, y_1\}, \{x_2, y_2\}, \ldots, \{x_\lambda, y_\lambda\}$, then there are precisely λ pairs in mD with difference md, namely, $\{mx_1, my_1\}, \{mx_2, my_2\}, \ldots, \{mx_\lambda, my_\lambda\}$. On the other hand, if m and v are not coprime, there may be x and y in D such that mx and my are congruent mod v, or mx could be zero.

12.   (i) Call the competitors $\{1, 2, \ldots, 7, A, B, \ldots, G\}$. Using a symmetric BIBD with parameters $(7, 3, 1)$, one could try for a solution whose first seven heats are the columns of

$$
\begin{array}{ccccccc}
1 & 2 & 3 & 4 & 5 & 6 & 7 \\
2 & 3 & 4 & 5 & 6 & 7 & 1 \\
4 & 5 & 6 & 7 & 1 & 2 & 3 \\
A & B & C & D & E & F & G
\end{array}
$$

and the other seven heats might be

$$
\begin{array}{ccccccc}
A & B & C & D & E & F & G \\
B & C & D & E & F & G & A \\
D & E & F & G & A & B & C \\
* & * & * & * & * & * & *
\end{array}
$$

The fourth member of heat 8 cannot be one of $\{1, 2, 4\}$ (pairs $1A$, $2A$ and $4A$ were in heat 1), $\{2, 3, 5\}$ or $\{4, 5, 7\}$. The unique answer is 6. If the fourth members of the subsequent heats are generated by adding 1 (mod 7) we obtain the solution

$$
\begin{array}{cccccccccccccc}
1 & 2 & 3 & 4 & 5 & 6 & 7 & A & B & C & D & E & F & G \\
2 & 3 & 4 & 5 & 6 & 7 & 1 & B & C & D & E & F & G & A \\
4 & 5 & 6 & 7 & 1 & 2 & 3 & D & E & F & G & A & B & C \\
A & B & C & D & E & F & G & 6 & 7 & 1 & 2 & 3 & 4 & 5
\end{array}
$$

There may well be other solutions.

(ii) There is only one way to pair up the heats in our solution to satisfy the conditions of this part: heats 1 and 9 are paired; then 2,10; 3,11; $\ldots$; 6,14; 7,8.

Chapter 15

1. (i): $A = \begin{bmatrix} 0 & 1 & 0 \\ 0 & 0 & 1 \\ -2 & 1 & 2 \end{bmatrix}$.

    (iii): $P = \begin{bmatrix} 1 & 1 & 1 \\ 1 & -1 & 2 \\ 1 & 1 & 4 \end{bmatrix}$.

2. The induction step: Assume $A^{n-1}(i, j)$ is the number of ways to walk from i to j in $n - 1$ steps for any i and j. Then let i, j be given and determine $A^n(i, j)$. For each neighbor k of vertex j, $A^{n-1}(i, k)$ is the number of ways to walk from i to k in $n - 1$ steps. The (i, j) entry of A^n, by the definition of matrix multiplication, is the sum over all k adjacent to j of $A^{n-1}(i, k)$; this adds up all walks of length $n - 1$ from i to a neighbor of j, which correspond to walks of length n from i to j.

5. This is true if the matrix has only non-negative entries (as an adjacency matrix does). However, if we allow negative entries, the matrix $\begin{bmatrix} 1 & 1 \\ -1 & 1 \end{bmatrix}$ is a counterexample.

6. The approximation is $(2/e)^n \sqrt{2\pi n}$ which goes to 0 as $n \to \infty$.

References

[1] H. L. Abbott, Lower bounds for some Ramsey numbers, *Discrete Math.* **2** (1972), 289–293.

[2] W. S. Anglin and J. Lambek, *The Heritage of Thales*, 5th ed. (Springer-Verlag, New York, 1995).

[3] H. Anton, *Elementary Linear Algebra*, 9th ed. (Wiley, New York, 2005).

[4] K. Appel and W. Haken, (October 1977), Solution of the Four Color Map Problem, *Scientific American* **237** (1977), 108–121.

[5] M. E. Baron, A note on the historical development of logic diagrams, *Math. Gaz.* **53** (1969), 113–125.

[6] E. T. Bell, *Men of Mathematics*, 5th ed. (Simon & Schuster, New York, 1937).

[7] M. Bellis, Timeline: Biography of Samuel Morse 1791–1872, `http://inventors.about.com/od/mstartinventors/a/samuel_morse.htm`

[8] J. Bernoulli, Ars Conjectandi, Basileae, Impensis Thurnisiorum, Fratrum, 1713.

[9] K. N. Bhattacharya, A new balanced incomplete block design, *Science and Culture* **11** (1944), 508.

[10] N. L. Biggs,T P Kirkman, Mathematician, *Bull. London Math. Soc.* **13** (1981), 97–120.

[11] R. C. Bose, S. S. Shrikhande, and E. T. Parker, Further results on the construction of mutually orthogonal Latin squares and the falsity of Euler's conjecture, *Canad. J. Math.* **12** (1960), 189–203.

[12] R. L. Brooks, On colouring the nodes of a network, *Proc. Camb. Phil. Soc.* **37** (1941), 194–197.

[13] R. A. Brualdi, *Introductory Combinatorics*, 4th ed. (Pearson Prentice Hall, New Jersey, 2004).

[14] R. A. Brualdi and H. J. Ryser, *Combinatorial Matrix Theory* (Encyclopedia of Mathematics and its Applications #39) (Cambridge UP, New York, 1991).

[15] R. H. Bruck and H. J. Ryser, The non-existence of certain finite projective planes, *Canad. J. Math.* **1** (1949), 88–93.

[16] D. M. Burton, *The History of Mathematics: An Introduction*, 5th ed. (McGraw-Hill, Boston, 2003).

[17] S. Butler, M. T. Hajiaghayi, R. D. Kleinberg, and T. Leighton, Hat guessing games, *SIAM J. Discrete Math.* **22** (2008), 592–605.

[18] A. Cayley, On the colourings of maps, *Proc. Roy. Geog. Soc.* **1** (1879), 259–261.

[19] A. Cayley, A theorem on trees, *Quart. J. Math.* **23** (1889), 376–378.

[20] Y. Chang, The existence spectrum of golf designs, *J. Combin. Des.* **15** (2005), 84–89.

[21] S. Chowla and H. J. Ryser, Combinatorial problems, *Canad. J. Math.* **2** (1950), 93–99.

[22] C. J. Colbourn and J. H. Dinitz, *Handbook of Combinatorial Designs*, 2nd ed. (Chapman & Hall/CRC, Boca Raton, 2007).

[23] C. J. Colbourn and G. Nonay, A golf design of order 11, *J. Statist. Plann. Inf.* **58** (1997), 29–31.

[24] R. Cooke, *The History of Mathematics: A Brief Course*, 2nd ed. (John Wiley & Sons, Hoboken, NJ, 2005).

[25] L. D. Cummings, An undervalued Kirkman paper, *Bull. Amer. Math. Soc.* **24** (1918), 336–339.

[26] H. de Parville, Recreations mathematiques: La tour d'Hanoi et la question du Tonkin, *La Nature*, part I (1884), 285–286.

[27] E. W. Dijkstra, A note on two problems in connexion with graphs, *Numerische Math.* **1** (1959), 269–271.

[28] T. Ebert, *Applications of Recursive Operators to Randomness and Complexity*, Ph.D. Dissertation (University of California, Santa Barbara, 1998).

[29] J. Edmonds and E. L. Johnson, Matching, Euler tours, and the Chinese postman, *Math. Prog.* **5** (1973), 88–124.

[30] J. Emert and D. Umbach, Inconsistencies of 'wild card' poker, *Chance* **9** (1996), 17–22.

[31] P. Erdös and T. Gallai, Graphs with prescribed degrees of vertices [Hungarian] *Mat. Lapok.* **11** (1960), 264–274.

[32] L. Euler, Solutio Problematis ad geometriam situs pertinentis, *Comm. Acad. Sci. Imp. Petropolitanae* **8** (1736), 128–140.

[33] L. Euler, Recherches sur une nouvelle espece de quarrées magiques, *Verhand. Zeeuwsch Gen. Wet. Vlissingen* **9** (1782), 85–239.

[34] A. Evans, Embedding incomplete Latin squares, *Amer. Math. Monthly* **67** (1960), 958–961.

[35] R. A. Fisher, An examination of the different possible solutions of a problem in incomplete blocks, *Ann. Eugenics* **10** (1940), 52–75.

[36] J. Folkman, *Notes on the Ramsey Number* $N(3,3,3,3)$, Manuscript, Rand Corporation, Santa Monica, CA, 1967.

[37] W. J. Frascella, The construction of a Steiner triple system on sets of the power of the continuum without the axiom of choice, *Notre Dame J. Formal Logic* **7** (1966), 196–202.

[38] S. Gadbois, Poker with wild cards–A paradox? *Math. Mag.* **69** (1996), 283–285.

[39] M. Gardner, Euler's spoilers, *Martin Gardner's New Mathematical Diversions from Scientific American* (Allen & Unwin, London, 1969), 162–172.

[40] M. Gardner, *Time Travel and Other Mathematical Bewilderments* (Freeman, New York, 1988).

[41] M. Gardner, *Mathematical Carnival* (MAA, Washington DC, 1989).

[42] M. Gardner, *Fractal Music, Hypercards and More...* (Freeman, New York, 1992).

[43] M. Gardner, *Martin Gardner's Mathematical Games* (MAA, Washington DC, 2005).

[44] S. Giberson and T. J. Osler, Extending Theon's ladder to any square root, *College Math. J.* **35** (2004), 222.

[45] S. W. Golomb, Problem for solution, *Amer. Math. Monthly* **60** (1953), 114, 551–552.

[46] G. Gonthier, Formal proof—The four-color theorem, *Notices Amer. Math. Soc.* **55** (2008), 1382–1393.

[47] A. W. Goodman, On sets of acquaintances and strangers at any party, *Amer. Math. Monthly* **66** (1959), 778–783.

[48] R. Graham, D. Knuth and O. Patashnik, *Concrete Mathematics*, 2nd ed. (Addison-Wesley, New York, 1995).

[49] S. Hakimi, On the realizability of a set of integers as degrees of the vertices of a graph, *SIAM J. Appl. Math.* **10** (1962), 496–506.

[50] M. Hall, *Combinatorial Theory*, 2nd ed. (Wiley, New York, 1986).

[51] M. Hall and W. S. Connor, An embedding theorem for balanced incomplete block designs, *Canad. J. Math.* **6** (1954), 35–41.

[52] P. Hall, On representations of subsets, *J. London Math. Soc.* **10** (1935), 26–30.

[53] V. Havel, A remark on the existence of finite graphs [Czech], *Ĉasopis Pest. Mat.* **80** (1955), 477–480.

[54] D. M. Johnson, A. L. Dulmage, and N. S. Mendelsohn, Orthomorphisms of groups and orthogonal Latin squares I, *Canad. J. Math.* **13** (1961), 356–372.

[55] J. Karamata, Théorèmes sur la sommabilité exponentielle et d'autres sommabilités rattachant, *Mathematica* (Cluj) **9** (1935), 164–178.

[56] P. Kerr, *Dark Matter: The Private Life of Sir Isaac Newton* (Crown Press, New York, 2002).

[57] T. P. Kirkman, On a problem in combinations, *Cambridge and Dublin Math. J.* **2** (1847), 191–204.

[58] T. P. Kirkman, Note on an unanswered prize question, *Cambridge and Dublin Math. J.* **5** (1850), 255–262.

[59] T. P. Kirkman, Query VI, *Lady's and Gentleman's Diary* (1850), 48.

[60] T. P. Kirkman, Solution to Query VI, *Lady's and Gentleman's Diary* (1851), 48.

[61] W. W. Kirkman, Thomas Penyngton Kirkman, *Mem. Proc. Manchester Lit. Phil. Soc.* **9** (1895), 238–243.

[62] N. S. Knebelman, Problem for solution, *Amer. Math. Monthly* **55** (1948), 100; **56** (1949), 426.

[63] D. Knuth, *Fundamental Algorithms: The Art of Computer Programming, vol. 1*, 2nd ed. (Addison-Wesley, Reading, MA, 1973).

[64] D. Knuth, Two notes on notation, *Amer. Math. Monthly* **99** (1992), 403–422.

[65] T. Koshy, *Fibonacci and Lucas Numbers with Applications* (Wiley, New York, 2001).

[66] T. Koshy, *Catalan Numbers with Applications* (Oxford University Press, New York, NY, 2009).

[67] J. B. Kruskal Jnr., On the shortest spanning subtree and the traveling salesman problem, *Proc. Amer. Math. Soc.* **7** (1956), 48–50.

[68] P. A. Laplante, *Dictionary of Computer Science, Engineering and Technology* (CRC Press, Boca Raton, FL, 2000).

[69] L. Lovász and M. D. Plummer, *Matching Theory* (North-Holland Mathematics Studies #121) (Elsevier, New York, 1986).

[70] H. L. MacNeish, Euler squares, *Ann. Math.* **23** (1922), 221–227.

[71] R. Mandl, Orthogonal Latin squares: An application of experiment design to compiler testing, *Comm. ACM* **28** (1985), 1054–1058.

[72] H. B. Mann, The construction of orthogonal Latin squares, *Ann. Math. Statist.* **13** (1942), 418–423.

[73] M. McClure and S. Wagon, Four-coloring the US counties, *Math Horizons* **16(4)** (2009), 20–21.

[74] K. Mei-Ko, Graphic programming using odd or even points, *Chinese Math.* **1** (1962), 273–277.

[75] S. Milgram, The small world problem, *Psychology Today* **2** (1967), 60–67.

[76] L. Moser, Problem for solution, *Amer. Math. Monthly* **56** (1949) 403; **57** (1950), 117.

[77] L. Moser, Problem for solution, *Amer. Math. Monthly* **60** (1953) 262; 713–714.

[78] H. W. Norton, The 7×7 squares, *Ann. Eugenics* **9** (1939), 269–307.

[79] J. J. O'Connor and E. F. Robertson, Emanuel Sperner,
`http://www-history.mcs.st-andrews.ac.uk/history/Biographies`
`/Sperner.html`

[80] J. J. O'Connor and E. F. Robertson, Gaston Tarry,
`http://www-history.mcs.st-andrews.ac.uk/Biographies/Tarry.html`

[81] J. J. O'Connor and E. F. Robertson, Thomas Penyngton Kirkman,
`http://www-history.mcs.st-andrews.ac.uk/Biographies/Kirkman.html`

[82] C. D. Olds, *Continued Fractions* (New Mathematical Library 9) (MAA, Washington, DC, 1963).

[83] O. Ore, Note on Hamilton circuits, *Amer. Math. Monthly* **67** (1960), 55.

[84] H. Perfect, The Revd Thomas Penyngton Kirkman FRS (1806–1895): Schoolgirl parades—but much more, *Math. Spectrum* **28** (1995/96), 1–6.

[85] R. C. Prim, Shortest connection networks and some generalizations, *Bell Syst. Tech. J.* **36** (1957), 1389–1401.

[86] M. Reiss, Über eine Steinersche kombinatorische Aufgabe, *J. Reine AngMath.* **56** (1859), 326–344.

[87] V. F. Rickey, Isaac Newton: Man, myth, and mathematics, *College Math. J.* **18** (1987), 362–389.

[88] N. Robertson, D. Sander, P. Seymour, and R. Thomas, The four-colour theorem, *J. Combin. Theory* **70B** (1997), 2–44.

[89] D. F. Robinson, Constructing an annual round-robin tournament played on neutral grounds, *Math. Chronicle* **10** (1981), 73–82.

[90] S. Robinson, Why mathematicians now care about their hat color, *New York Times* (April 10, 2001), D5.

[91] S. Roman, *Introduction to Coding and Information Theory* (Springer-Verlag, New York, 1996).

[92] W. W. Rouse Ball, *Mathematical Recreations and Essays* (MacMillan, London, 1892).

[93] DS5: F. Ruskey and M. Weston, Venn Diagrams (Dynamic Survey 5), *Elec. J. Combin.* (2005).

[94] H. J. Ryser, *Combinatorial Mathematics* (Mathematical Monograph #14) (MAA, Washington DC, 1963).

[95] A. Sade, An omission in Norton's list of 7×7 squares, *Ann. Math. Statist.* **22** (1951), 306–307.

[96] S. S. Sane and W. D. Wallis, Monochromatic triangles in three colours, *J. Austral. Math. Soc.* **37B** (1988), 197–212.

[97] P. J. Schellenberg, G. H. J. van Rees, and S. A. Vanstone, Four pairwise orthogonal Latin squares of order 15, *Ars Combin.* **6** (1978), 141–150.

[98] A. Schrijver, On the history of combinatorial optimization (till 1960), in *Handbook of Discrete Optimization* (K. Aardal, G. L. Nemhauser, R. Weismantel, eds.) (Elsevier, Amsterdam, 2005), 1–68.

[99] M. P. Schutzenberger, A non-existence theorem for an infinite family of symmetrical block designs, *Ann. Eugenics* **14** (1949), 286–287.

[100] A. J. Schwenk, Acquaintance graph party problem, *Amer. Math. Monthly* **79** (1972), 1113–1117.

[101] A. Seidenberg, A simple proof of a theorem of Erdös and Szekeres, *J. Lond. Math. Soc.* **34** (1959), 352.

[102] L. E. Sigler, *Fibonacci's Liber Abaci: Leonardo Pisano's Book of Calculation* (Springer-Verlag, New York, 2002).

[103] N. M. Singhi and S. S. Shrikhande, Embedding of quasi-residual designs with $\lambda = 3$, *Utilitas Math.* **4** (1973), 35–53.

[104] N. M. Singhi and S. S. Shrikhande, Embedding of quasi-residual designs, *Geom. Ded.* **2** (1974), 509–517.

[105] B. Smetaniuk, A new construction on Latin squares I: A proof of the Evans conjecture, *Ars Combin.* **11** (1960), 155–172.

[106] M. Z. Spivey, Fibonacci identities via the determinant sum property, *College Math. J.* **37** (2006), 286.

[107] R. P. Stanley, *Enumerative Combinatorics, Volume 1* (Cambridge Studies in Advanced Mathematics #49)(Cambridge University Press, Cambridge UK, 1997).

[108] R. P. Stanley, *Enumerative Combinatorics, Volume 2* (Cambridge Studies in Advanced Mathematics #62)(Cambridge University Press, Cambridge UK, 2001).

[109] R. P. Stanley, Catalan Addendum
http://www-math.mit.edu/~rstan/ec/catadd.pdf

[110] J. Steiner, Combinatorische Aufgabe, *J. Reine Angew. Math.* **45** (1853), 181–182.

[111] J. Stewart, DijkstraApplet.html,
http://www.dgp.toronto.edu/ people/JamesStewart/270/9798s/
Laffra/DijkstraApplet.html

[112] G. Tarry, Géométrie de situation: Nombre de manieres distinctes de parcourir en une seule course toutes les all'ees d'un labyrinthe rentrant, en ne passant qu'une seule fois par chacune des all'ees, *Comptes Rend. Assoc. Fr. Avance. Sci.* **15** (1886), 49–53.

[113] G. Tarry, La problem des labyrinths, *Nouv. Ann. Math.* **14** (1895), 187–190.

[114] G. Tarry, Le problème des 36 officiers, *Comptes Rend. Assoc. Fr.* **1** (1900), 122–123; **2** (1901), 170–203.

[115] L. Teirlinck, On the use of pairwise balanced designs and closure spaces in the construction of structures of degree at least 3, *Le Matematiche* **65** (1990), 197–218.

[116] L. Teirlinck, Large sets of disjoint designs and related structures, *Contemporary Design Theory: A Collection of Surveys* (J. H. Dinitz and D. R. Stinson, editors), (Wiley, 1992), 561–592.

[117] J. Travers and S. Milgram, An experimental study of the small world problem, *Sociometry* **32** (1969), 425–443.

[118] J. Venn, On the diagrammatic and mechanical representation of propositions and reasonings, *Dublin Phil. Mag. J. Sci.* **9** (1880), 1–18.

[119] J. Venn, On the employment of geometrical diagrams for the sensible representation of logical propositions, *Proc. Camb. Phil. Soc.* **4** (1880), 47–59.

[120] W. D. Wallis, The number of monochromatic triangles in edge-colourings of a complete graph, *J. Comb. Inf. Syst. Sci.* **1** (1976), 17–20.

[121] W. D. Wallis, The problem of the hospitable golfers, *Ars Combin.* **15** (1983), 149–152.

[122] W. D. Wallis, *A Beginner's Guide to Graph Theory*, 2nd ed. (Birkhäuser, Boston, 2007).

[123] W. D. Wallis, *Introduction to Combinatorial Designs*, 2nd ed. (Chapman & Hall/CRC, Boca Raton, 2007).

[124] I. M. Wanless and B. S. Webb, The existence of Latin squares without orthogonal mates, *Des. Codes. Crypt.* **40** (2006), 131–135.

[125] E. G. Whitehead Jr., Algebraic structure of chromatic graphs associated with the Ramsey number $N(3,3,3;2)$, *Discrete Math.* **1** (1971), 113–114.

[126] Wikipedia, Blaise_Pascal,
http://en.wikipedia.org/wiki/Blaise_Pascal

[127] Wikipedia, Carl_Friedrich_Gauss,
http://en.wikipedia.org/wiki/Carl_Friedrich_Gauss

[128] Wikipedia, Édouard Lucas,
http://en.wikipedia.org/wiki/Edouard_Lucas

[129] Wikipedia, Évariste Galois,
http://en.wikipedia.org/wiki/Evariste_Galois

[130] Wikipedia, Howard Garns,
http://en.wikipedia.org/wiki/Howard_Garns

[131] Wikipedia, Jacob Bernoulli,
http://en.wikipedia.org/wiki/Jacob_Bernoulli

[132] Wikipedia, Joseph Kruskal,
http://en.wikipedia.org/wiki/Joseph_Kruskal

[133] Wikipedia, Leonhard Euler,"
http://en.wikipedia.org/wiki/Leonhard_Euler

[134] Wikipedia, Richard Hamming,
http://en.wikipedia.org/wiki/Richard_Hamming

[135] Wikipedia, Isaac Newton,
http://en.wikipedia.org/wiki/Isaac_Newton

[136] Wikipedia, Small World Experiment,
http://en.wikipedia.org/wiki/Small_world_experiment

[137] Wikipedia, Sudoku,
http://en.wikipedia.org/wiki/Sudoku

[138] Wikipedia, William Rowan Hamilton,
http://en.wikipedia.org/wiki/William_Rowan_Hamilton

[139] H. Wilf, *Generatingfunctionology*, 3rd ed. (A. K. Peters, Wellesley, 2006).

[140] W. S. B. Woolhouse, Prize question 1733, *Lady's and Gentleman's Diary* (1844), 84.

[141] B. Y. Wu and K.-M. Chao, *Spanning Trees and Optimization Problems* (Chapman & Hall/CRC, Boca Raton, 2004).

[142] F. Yates, Incomplete randomized blocks, *Ann. Eugenics* **7** (1936), 121–140.

Index